ARBORICULTURAL PRACTICES

ARBORICULTURAL PRACTICES

A Science-Based Approach

Lindsey Purcell

Purdue University

Illustrations by Jeff Harris

WAVELAND PRESS, INC.

Long Grove, Illinois

For information about this book, contact:
Waveland Press, Inc.
4180 IL Route 83, Suite 101
Long Grove, IL 60047-9580
(847) 634-0081
info@waveland.com
www.waveland.com

10-digit ISBN 1-4786-5174-1
13-digit ISBN 978-1-4786-5174-1

Printed in the United States of America

7 6 5 4 3 2

BRIEF CONTENTS

CONTENTS

FOREWORD

It has been my pleasure to know Lindsey Purcell for over 20 years as a teaching and extension colleague, and friend. He has been astutely interested in arboriculture for many years, especially with the various practical interventions we as professionals apply in the urban forest. I have witnessed his growth and development in the details of many practices, especially pruning and organizing a commonsense perspective.

Despite the importance of details which make arboriculture and urban forest tick, it's how Lindsey encourages readers to think through challenges by laying out concepts in a thought-provoking manner that will make this book excel in the industry. The text combined with the many illustrations and images together teaches the reader how to think, not what to think. This work is just as much of a strategic document as it is an operations manual, which makes it unique in the arboriculture menu of resources.

As practitioners and educators, we rely on not only research, content, and presentations, but also, our own preferred resources. There have been many credible, well-established texts and guides available for all experience levels in arboriculture. Lindsey builds on these legacy works using up-to-date research and experience-based practices. One issue discussed between us was the need for a more modern book on arboricultural practices. When he asked my thoughts on this book concept and advice on content, it was evident he knew what was necessary to create a textbook for just about any application and level. Urban forestry and arboriculture professionals, students, business owners, managers, and tree workers all can benefit and find useful information from this book. There is a distinct focus on thinking through a situation to develop an appropriate application.

One thing for sure is that it is applied arboriculture, because Lindsey has done it all during the course of over 40 years of experience. From being a climber through his roles as a Purdue University instructor and ISA chapter executive, Lindsey has brought so many perspectives to this brand-new book. What sets this textbook apart is its commitment to making tree care relevant. Lindsey consistently connects concepts back to real-world applications, helping students see the relevance of what they're learning. This approach not only makes the material more engaging but also helps students better retain what they've learned. I know this book will be successful and become an integral part of any professional arborist, urban forester, horticulturalist, and consultant's collection.

This book should be read by every urban forester and arborist, young or old.

Edward F. Gilman
Professor Emeritus
University of Florida

PREFACE

I would like to start with a question if I may? In the designed landscape, in our urban forests, what is the one living element that dominates, is typically most defended and left standing when all other green is gone? Trees. Trees are the one legacy a well-planned, sustainable design can always provide and build around. In the face of climate change and the myriad of issues that imperil our trees, urban foresters, arborists, and the green industry should strive to position themselves as a solution in the climate agenda. There is tremendous opportunity for urban forestry initiatives and arboricultural practices to align with overall community goals and sustainability initiatives. The spotlight is now on us, as stewards of one of the most important natural resources, to make a difference locally, regionally, and globally.

Well, the preparation for this book started with my experience and concluded with my relationships, combining them together to assemble the best information I could on applied arboriculture. It began with hours of literature reviews and communications with many friends, peers, and colleagues. While developing the content for this book, I soon realized it has been writing itself for many years. As a tree care professional and educator for over 40 years, I have created hundreds of publications, articles, webinars, blogs, and papers; this is the culmination of that work. My motivation is relevance, striving to be current with research and best practice and connecting it all to what is applicable to the industry. I realize there may be some subjects or concepts omitted or excluded, but I know where my education and experience can best serve others. In other words, I know how to stay in my lane! Most likely part of it was I just couldn't fit it all in one volume. It is important to humbly recognize I stand on the shoulders of giants with too many to thank and recognize for their contributions to the science and practice over the years. Researchers, speakers, professors, and colleagues have all contributed to this effort in some way and hopefully it is rendered appropriately in this text.

Finally, I have taught urban forestry and arboricultural concepts for over 21 years. Each semester began with extensive research of existing resources and publications significant to what I believed needed to be taught to future arborists. I began creating and writing my own publications and articles based on my experiences as a landscape designer, city forester, and arborist. These resources grew each year to essentially become a substantial collection bound and provided to my students. As this grew over the years, I heard many times, "You know, you should write a book!" Well, here it is. I hope that you are reading this book because you share the same passion for our profession, the science and our trees which still fascinate me, and which improve our quality of life, every day.

Reviewers:

I would like to thank those industry professionals who dedicated the time to review and contribute various chapters of this book. My respect and gratitude go out to those mentioned here:

- R. J. Laverne, PhD, Manager of Education & Training, Davey Tree Expert Company
- John Bonkowski, PhD, Plant Pathologist, Purdue University
- Kyle Daniel, Nursery and Landscape Specialist, Purdue University, West Lafayette, IN
- Zack Shier, Plant Health Care Manager, Joseph Tree, Dublin, OH
- Tchukki Andersen, Staff Arborist, Tree Care Industry Association, Manchester, NH

- Jeff Harris, ISA Certified Arborist, Arbor Rangers, Indianapolis, IN
- Jaira Harris, Company Media Editor, Arbor Rangers, Indianapolis, IN
- Andrew Mertz, Owner, Indigo Tree Consulting, Indianapolis, IN

Special thanks to Dr. Ed Gilman for his support, reviews, mentoring, and friendship during the writing process and my career.

Also, thank you, Jeff Harris, an illustrator extraordinaire, who has always been able to develop what I can't always express in writing and who recognizes the needs in my efforts.

Lastly, I want to thank my wife, Lorelie, for her everlasting love and patience during this experience. Ultimately, I dedicate all to Him from which I can always do "immeasurably more." *Ephesians 3:20-21*

CHAPTER 1

URBAN FORESTRY AND ARBORICULTURE

LEARNING OBJECTIVES

- Recognize the basic functions of arboricultural practices to create sustainable urban forest systems.
- Identify the interaction between urban forestry and arboriculture by comparing each concept.
- Discuss the environmental, economic, aesthetic, and social benefits of trees as well as the various costs that accompany them and how value can be placed on the urban forest as a resource.
- Explain how urban foresters can and should play an integral role in the urban planning process.
- List the steps necessary in planning and conducting a tree inventory and describe how it can be used as a management tool.
- Describe the components of an urban forest management plan and explain the importance of each.
- Recognize the importance of ordinances and how they help conserve the urban forest.
- Explain how tree canopy requirements and canopy goals help improve the urban environment.

The importance of trees in the urban forest should be recognized to understand the importance and need for management of individual trees. Many individual trees comprise the urban forest and each tree contributes to its success in improving our quality of life. Scientific understanding of how urban trees benefit people has increased significantly in recent years. Trees provide many benefits to communities. These can include aesthetic qualities, such as the beauty of flowers and fall colors, also functional benefits, including shade, storm water management, and filtering pollution for cleaner air and water. All these qualities combined are called ecosystem services.

According to the International Society of Arboriculture, the science of arboriculture includes the practice and study of the care of trees and other woody plants in the landscape. This primarily relates to the built environment or places in and around where people live. The professional responsibility of this task relies on the arborist. This word comes from two parts: "arbor," which simply stated is Latin for "tree," and "ist," a suffix of nouns that denotes a

Benefits of Trees

ECONOMICAL
- reduce energy use
- reduce the urban heat island effect
- improve water filtration
- provide storm water interception
- provide a source for food
- increase property values
- *and more!*

SOCIAL
- promote mental well-being & healing
- reduce neighborhood crime and violence
- promote a sense of community
- provide recreational space
- reduce exposure to ultra-violet rays
- reduce noise pollution
- *and more!*

ENVIRONMENTAL
- provide oxygen
- reduce air pollution
- absorb excess carbon dioxide
- reduce soil erosion
- promote biodiversity
- provide habitat for wildlife
- *and more!*

HOW HAVE TREES BENEFITED YOU AND YOUR COMMUNITY?

FIGURE 1.1 Benefits of trees.

person who practices or is concerned with something, or holds certain principles, doctrines, etc. Combined, these two parts form "arborist," thereby identifying a professional who possesses the technical competence, through experience and related training, to provide for or supervise the management of trees and other woody plants in residential, commercial, and public landscapes. This is the very essence of arboriculture practices.

Arboricultural practices involve an intervention with trees in a logical, ethical manner. Trees are the largest, oldest living organism on the planet and have done quite well without us interfering. However, in the urban environment, intervention in the form of proper practices is necessary to achieve the desired outcome or improved condition. Arboriculture is an essential and integral part of urban forestry which primarily is focused on creating sustainable tree communities in often harsh urban conditions.

Modern arboriculture implements best practices germinating from current research. It is about the tree system; how it grows, defends itself, and survives through its life cycle. The modern arborist carries out the principles of science and their primary duty is to maintain safe and healthy trees in an ethical and responsible manner. The practice and application of care to trees is a product of environmental stewardship. This is how the arborist contributes to urban forestry and is critical for understanding the role and importance of arboriculture in urban forestry.

Arboriculture involves the care of individual trees, while urban forestry deals with forests as systems (groups of trees) in a developed setting. The science of arboriculture focuses on proper tree planting, pruning, fertilizing, watering, and other maintenance issues. It also focuses on overall individual tree health. Urban forestry can be considered a physical science and a social science, encompassing disciplines like arboriculture. It involves landscape-level management, such as urban forest inventory, valuation, planning, policy, etc. Additionally, the human dimension perspective is critical as well. This examines the functional benefits and ecosystem services trees provide which improve our quality of life. The two fields certainly overlap, but the training and practice of both have differences based on their applications.

One of the important goals of arboriculture include the selection and establishment of tree species suitable for the site. Furthermore, to reach the goals of providing healthy, aesthetic, and safe trees, proper installation, watering, mulching, pruning, and protection are required from the arborist. Additionally, the well-informed arborist needs to understand how the science of arboriculture influences urban forestry and how urban forestry management influences the applied science of arboriculture.

FIGURE 1.2 Arborists apply current research and best practices to manage trees in the urban forest.

URBAN FORESTRY MANAGEMENT

Urban forestry is a program that invests in the future of the community. Considerations for planning and protecting trees should consider biological, management, and community needs to establish the best policies and practices.

FIGURE 1.3 Planning is critical for successful tree plantings on any scale.

Trees have proven with current research that they are more than "just a pretty face" as an aesthetic element in the landscape. Urban forestry science indicates that trees are even more valuable for their functional benefits than their form. Data from the 2018 census indicated more than 80% of Americans live in urban areas and this statistic is increasing annually. As our nation becomes more urbanized, urban foresters and arborists are working to ensure we have safe urban forest spaces to enjoy. These natural resources add more than curb appeal and economic value, they improve our quality of life.

Trees are vital to improving our environmental conditions. They are the perfect biological machine, providing valuable functional services which can save tree owners and communities millions of dollars in energy costs, stormwater management, and air quality. However, these trees must mature and survive before they can provide these important ecosystem services. Planting trees is part of the solution, but is not the answer. We can't just keep planting trees; we have to start growing and sustaining trees.

Individual trees are important but equally important is the health and planning of the overall urban forest. Our biggest challenges to the urban forest include development, climate change, and invasive pests. The U.S. Forest Service reports tree cover in urban/community areas of the United States is on the decline at a rate of about 175,000 acres per year, which corresponds to approximately 36 million trees per year due to these threats.

The consumptive culture we live in today impacts the urban forest as well. A society focused on the purchase and consumption of goods creates incredible demands on natural resources of all kinds. We are experiencing increased depletion of urban forest areas with a reduction of classified forests, natural areas, and urban woodlots. Trees in urban and suburban areas are destroyed by the increase of impervious surfaces and lawns which means keeping the forest out.

IMPACTS OF CLIMATE CHANGE

As a society, it may be seemingly impossible to do much about climate change, but the green industry does have a great influence on the effects. Trees in our urban forests can help to mitigate some of the negative impacts and social consequences of urban sprawl and make cities more resilient to these changes. These important functions are called *ecosystem services*. This is the way urban foresters measure the benefits that trees provide other than just their beauty. Ecosystems services are the many benefits that trees and plants provide to the community. They improve our quality of life by providing food, cleaner air, and water; regulating temperatures; supporting pollination; and providing recreational, health, and spiritual benefits.

For the most part, the rapid expansion of cities takes place with little consideration to land use planning strategy. The resulting human pressure has highly damaging effects on our urban trees and green spaces. The environmental impacts of climate change are intensified by urbanization such as increased pollution, increased temperatures, and larger demands on infrastructure such as stormwater systems.

Most climatologists agree closer attention to the impacts of changing climate on our urban forests is long overdue. These changes will vary across the globe and it will be a challenge to determine exactly what can be done to protect our trees and our quality of life. Also, the once simple task of selecting what will be considered a sustainable tree species during these extreme changes in weather will be equally perplexing. The current urban and suburban environments are tough for growing trees and our changing climate is going to make it even tougher.

The *greenhouse effect* is the warming that results when the atmosphere traps heat radiating from Earth toward space. Scientists attribute global warming and climate change to this greenhouse effect. In addition, human expansion has increased carbon emissions and has become one of the major contributors to our ongoing climate change issues which impacts every tree on the planet. In other words, there is no place on Earth that hasn't been affected by anthropogenic impacts. Climate change will continue to bring warmer, wetter winters and warmer, drier summers; consequently, insect disturbances are increasing. Additionally, climate change is further aggravated by existing urban stressors such as air and water pollution, soil compaction, and heat island effects. Since it is unlikely that these issues can be prevented in and around the places we live, it is essential that adaptive strategies be identified and implemented by urban forest managers before these impacts take an even larger toll on our city trees. Climate change will present both threats and opportunities, and it is not too early to begin planning for both.

The increasing levels of carbon dioxide is found everywhere on the planet. Ironically, researchers have discovered that certain tree species, especially the older, mature trees, favor these conditions. Since trees rely on carbon for energy, this elemental boost in carbon resources gives the tree a little enhancement. However, the evidence so far suggests that this will be temporary and have negative long-term consequences. The good news is that trees will receive an abundance of carbon, necessary for carbohydrate production, which is food for the plant. Several studies indicate that through a rise in the efficiency of water use, the increase of CO_2 favors the growth of trees and therefore the faster accumulation of carbon in forests. Nevertheless, the studies show that trees that grow faster have a shorter life, especially for the faster growing species.

Additionally, with the increasing levels of CO_2, we have seen major increases in pest outbreaks, higher levels of reproduction, and reduced mortality. Also, exotic insects may be able to flourish where they once were not adaptable. Carbon concentration increase in the trees also increases insect feeding activity, making insect outbreaks much worse. One of the major concerns from past years and looking forward will be drought. Water is the single most limiting factor in tree growth and sustainability. Increases in CO_2 levels increase temperatures and drought severity. Regardless of the amount of snow and rain, higher temperatures will result in increased levels of evaporation and transpiration which will reduce soil moisture and increase the likelihood of moisture deficits. According to the U.S. Drought Monitor trends from the past 30 years indicate the potential for increased drought similar to what we experienced in the past decades. It is imperative we follow and implement research which explores drought tolerant species and less water-demanding biology in trees.

FIGURE 1.4 Tree value can be calculated to include functional benefits.

CLIMATE IMPACTS SELECTION

Tree owners, managers, and decision-makers need to place more consideration into managing the impacts of these increasing environmental stressors. Tree selection is critical and "right tree, right place" is even more important. We may not be able to just pick any tree and expect it to work in the design. Tree selection requires more of a strategy, considering available maintenance inputs, water requirements, and exposure. Adaptive urban forest management will require a realistic look at our climatic future. Does policy and budget make it possible for urban forest managers to make informed decisions and implement them into a resilient, adaptive forest for our future? Also, on a private level, commercial landscapes and residential tree plantings all contribute to the overall urban forest. Responsible tree selection and realistic management inputs are vital for a flexible and resilient tree canopy which contributes locally, regionally, and globally to our health and quality of life.

Here are some means in which urban trees contribute to making cities more environmentally sustainable and liveable:

FIGURE 1.5 Trees intercept water reducing loads on city infrastructure.

- Trees can contribute to the increase of local *food and nutrition security*, providing food such as fruits and nuts for wildlife and human consumption.
- Trees play an important role in *increasing urban biodiversity*, providing plants and animals with a proper habitat, food, and protection.
- A mature tree can absorb up to 350 lbs. of CO_2 per year. As a result, trees play an important role in *climate change mitigation*. In cities with high levels of pollution, trees can improve air quality making cities healthier places to live in.

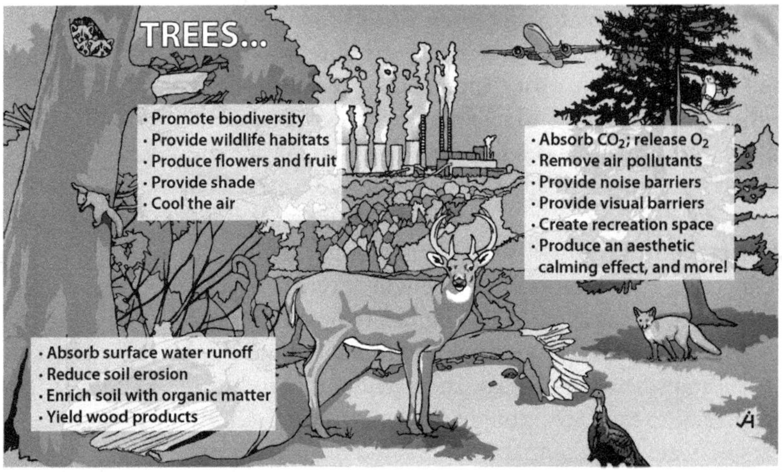

FIGURE 1.6 Tree facts.

- Strategic placement of trees in cities can help to *cool the air* between 30–40°F, thus reducing the urban "heat island" effect, helping reduce extreme heat conditions in summer weather.
- Large trees are great biological *filters for urban pollutants* and particulate pollution. They absorb pollutant gases (such as carbon monoxide, nitrogen oxides, ozone, and sulfur oxides) and filter fine particulates such as dust, dirt, and smoke out of the air by trapping them on leaves and bark.
- Research shows living in close proximity to urban green spaces and having access to them can *improve physical and mental health*, for example by decreasing high blood pressure and stress. Also, research indicates greatly improved neonatal health as well. This, in turn, contributes to the well-being of urban communities.
- Mature trees *regulate water flow* and play a key role in preventing floods and reducing the risk of sewer overflow. Stormwater management is a crucial city infrastructure issue and trees help. A mature tree, for instance, can intercept more than 5,000 gallons of water per year and without trees, every rain would contribute to floods.
- Trees also help to *reduce carbon emissions* by helping to conserve energy. For example, the correct placement of trees around buildings can reduce the need for air conditioning by 30 percent and reduce winter heating bills by 20–50%.
- Planning urban landscapes with trees can *increase property value* by up to 15% and attract tourism and business.

FIGURE 1.7 Mature trees provide significant benefits to improve our quality of life.

These are just a few examples of the functional benefits that trees provide to our everyday life. A community or neighborhood with well-planned and well-managed green infrastructure becomes more resilient, sustainable, and equitable in terms of livelihood improvement, climate change mitigation and adaptation, disaster risk reduction, and ecosystems conservation. Throughout their lifetime, trees can thus provide a benefit package worth two to three times more than the investment made in planting and caring for them. Trees aren't the answer, but they are part of the equation. Planting trees is important, but their maintenance is as equally important to help offset the effects of climate change with those important ecosystem services mentioned earlier.

Trees serve us in so many ways:

- Increase property values up to 15%.
- Improve air and water quality.
- One tree can capture up to 150,000 lbs. of carbon over its lifetime.
- Woodland soils sequester over 50% of carbon emissions.
- A medium-sized tree can intercept as much as 2,380 gallons of rainfall per year.
- Consumers are willing to pay 9–20% more for products in shopping areas with trees.
- Cars parked in lots with a tree cover of 50% emit at least 10% less evaporative pollution emissions.

URBAN FORESTRY COMPREHENSIVE PLANNING GUIDE

Urban forestry is a program that invests in the future of the community. Your final plan should consider canopy management and community needs to establish the best policies. This section will provide a survey of the basic tools and concepts to help your community with protecting and enhancing urban trees. Topics and examples will include creating and implementing a management plan, low-impact design strategies, ordinance administration, and establishing minimum canopy coverage.

Scientific understanding of how urban trees benefit people has increased significantly in recent years. One important benefit that is often overlooked is the economic values trees provide. Businesses are more competitive with consumer demand when trees help create an improved environment that is more attractive and welcoming to customers. Surveys indicate that the public prefers to patronize establishments where trees and landscaping are established. Consumers both stay longer and spend more money in business districts with green areas. Research also indicates that consumers are willing to pay more in these tree-lined shopping areas.

A process for strategic planning is needed to protect and expand the urban forest. In the following text, a guide for urban forest planning, examples of urban forest management, and resources for creating a customized plan for your community will be provided. Future generations rely on informed decision-making today to improve the canopy and influence the ecosystem services provided by the urban forest.

This begins with a plan. An Urban Forestry Management Plan (UFMP) is a roadmap that creates a shared vision for the future of the tree canopy. It's a tailored plan that guides urban forestry professionals to manage and provide for maximum long-term benefits to the community proactively and effectively. The UFMP provides recommendations based on the analysis of detailed inventories and includes additional components or documents, such as budgets, implementation schedules, policy and procedure manuals, standards and specifications, public education and monitoring plans, and existing ordinances. Annual work plans and budgets can then be developed based on the long-term UFMP.

In general, an effective UFMP will include:

- a *shared vision* for the urban forest
- *inventories and assessments* of the current status of the urban forest
- a *strategic plan* that includes goals, objectives, and actions based on identified needs
- an *implementation plan* with specific dates and assigned responsibilities of care

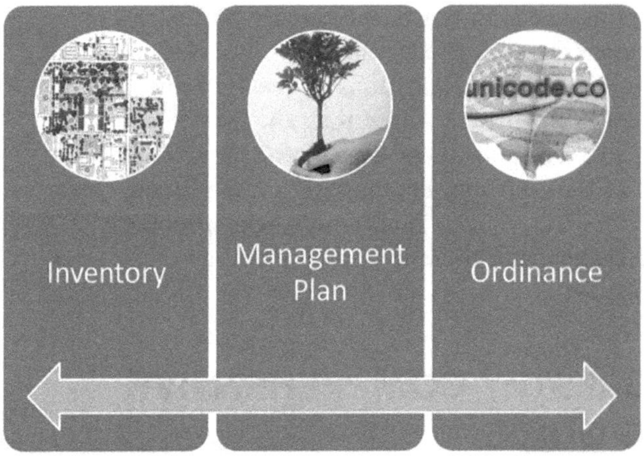

FIGURE 1.8 The basic requirements for a healthy, sustainable urban forest are an inventory, management plan, and an ordinance. These are critical attributes to the overall urban forestry management plan. It's a process of measuring existing tree canopy, creating a plan to manage those trees, and making an ordinance to protect those assets.

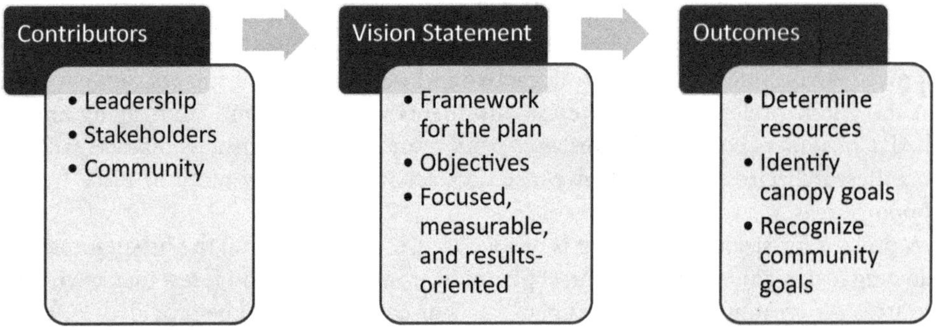

FIGURE 1.9 Step 1: Organization.

- a *monitoring and evaluation plan* with a system or matrix to check effectiveness and revise the UFMP as necessary
- a *means of enforcement* of the plan to protect the urban forest and maintain canopy goals

Establish a broad-based community working group or team. A neutral meeting facilitator is often needed to ensure that everyone is heard and that all concerns are identified. The group should meet periodically and consistently to maintain cohesiveness and effectiveness. The working group could consist of, but should not be limited to, the participants listed below:

- tree care professionals
- representatives of municipal departments such as metropolitan development, public works, and parks
- emergency management services

- public information officers (PIOs)
- local utility service providers
- local NPOs and NGOs, including neighborhood organizations
- business, health care, and other civic leadership

There will likely be trade-offs related to tree care, emergency management, fiscal issues, and other considerations. These need to be assessed by the more specialized members of the working group (e.g., tree care specialists), then reviewed and accepted by the community.

FIGURE 1.10 5 Steps for a Successful Urban Forest Management Plan.

VISION STATEMENT

To be effective, the vision statement and well-defined goals and objectives should be a community activity. Some ways of increasing community participation include discussing the plan with friends and neighbors; organizing outreach activities, such as news releases and public meetings; and developing educational programs for schools and other community groups.

When there is participation, the UFMP has the potential for greater success. It will help identify and develop alternative management options. The team will discover new information relevant to the community and urban forest and have a better understanding of challenges and opportunities. Be sure to demonstrate fairness across community demographics by representing environmental justice in the process.

The following should be accomplished during Step 1:

- Organize existing resources and determine the value to the planning process.
- Get the issues recorded so they can be used to gather support and be critiqued within the community.
- Create the vision statement and a broad list of goals and objectives.
- Identify the stakeholders, planning team, scale of the project, and partnerships.
- Recognize the financial obligations and identify funding sources.

Most communities will need information to help develop the vision, goals, and objectives into a management plan. Some key questions this information should answer include what the urban forest should provide for the community, what funding is available to help move the plan forward, and whether the community has the necessary resources available to set the plan in action.

A resource analysis is critical to determine people, funding, and inventory, including a reliable tree inventory for all streets and public spaces. The information needed for your plan can come from several sources. A systematic tree inventory is particularly useful for assessing your tree resources and establishing and measuring your goals and objectives.

Keep in mind that data collection is expensive and measure only what is needed, based on previously determined objectives. Reviewing current practices (such as tree planting, pruning, and removal) and plans (such as street tree management, emergency response plans, ordinances, etc.) can also help identify common objectives and ways to plan

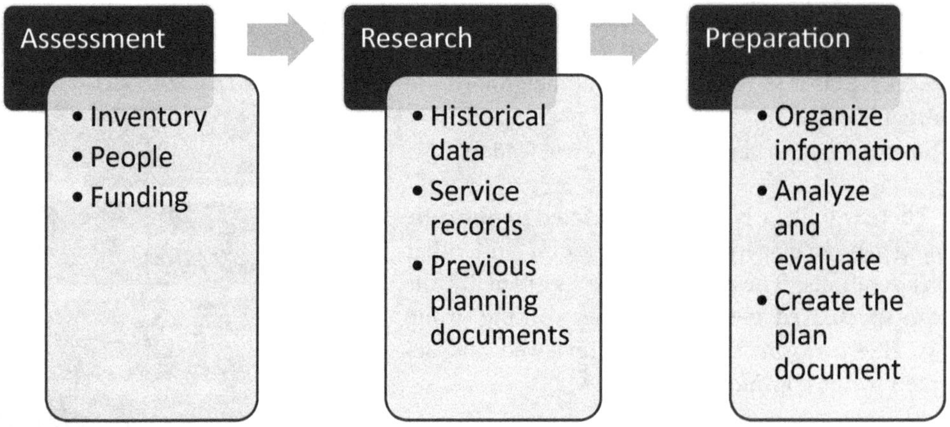

FIGURE 1.11 Step 2: Assessment and preparation.

effectively. If funding is a critical issue, the team might want to apply for an urban community forestry grant to help offset costs. Similarly, if personnel are a critical issue, the team may want to hire a consulting firm specializing in urban forestry to do the inventory and data synthesis.

The following should be accomplished during Step 2:

- Get community consensus of goals and objectives.
- Establish the scope of work and the project timeline for the plan.
- Identify potential funding opportunities to support the plan.
- Discover data from your resource analysis that supports goals and objectives in the form of maps; planting sites and their attributes (location, size, utility conflicts, soil volume, etc.); stocking levels; risk trees present; trees requiring plant health care, such as pruning; and list of potential resources (community volunteers, government and private technical assistance, grants, etc.).
- Prepare and create the plan.

Once the plan is complete, share the excitement with stakeholders, community, and staff. Transparency is important. Provide details of the scope and intent of the plan. Prioritize objectives and develop action items that support the established goals and outcomes. Be realistic with the action items, prioritizing based on needs and risk

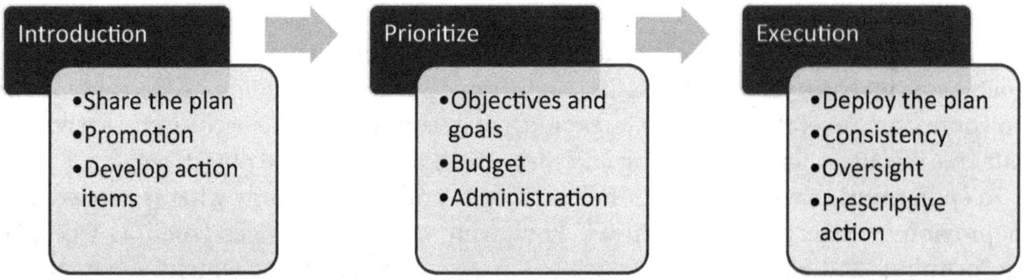

FIGURE 1.12 Step 3: Implementation.

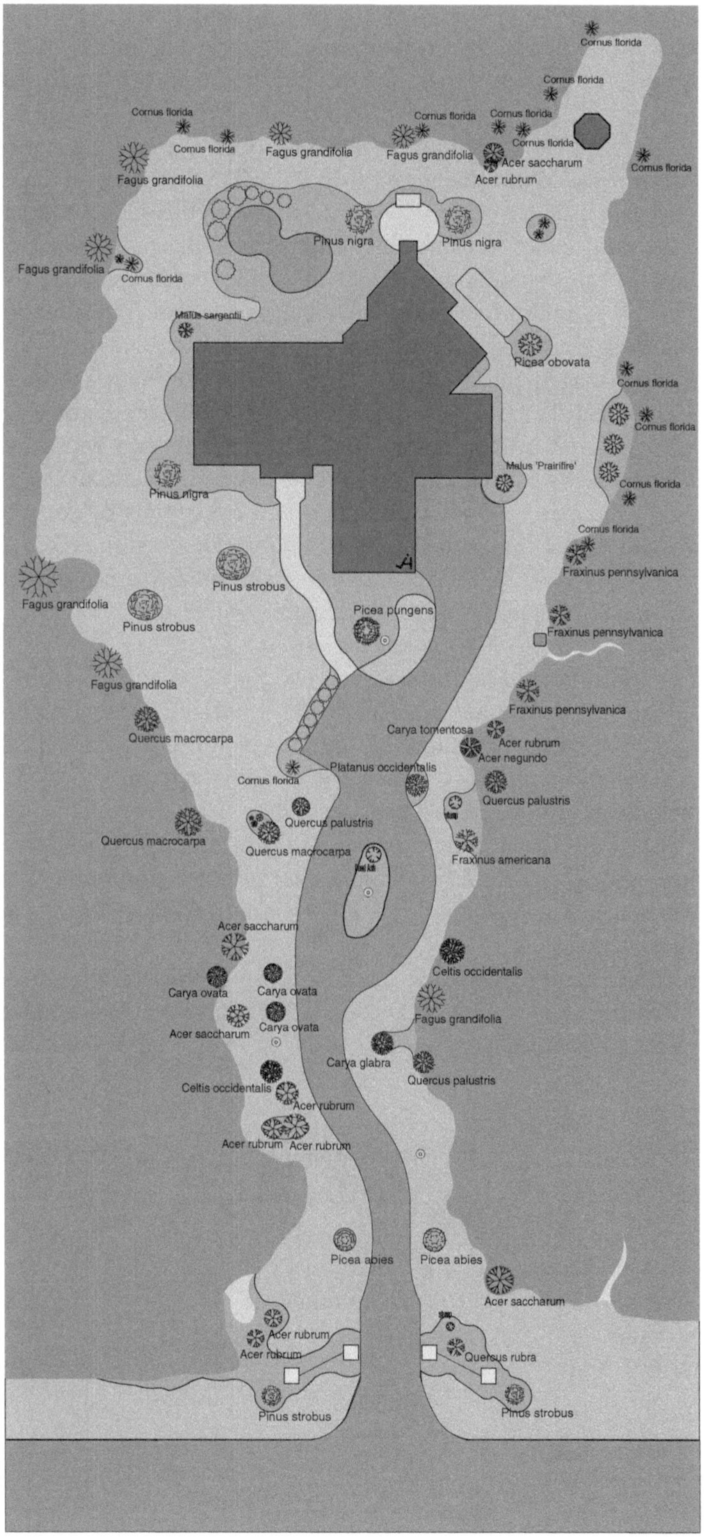

FIGURE 1.13 A tree inventory is necessary to manage the urban forest effectively.

management protocols. Some objectives can be achieved easily and quickly, but budget is an important consideration. Be certain financial support is consistent with both short-term and long-term objectives. Working with the financial principal will be critical to implementation.

Be sure to consider how the community and staff might respond to these changes. All the planning and building of consensus up to this point will help to ensure that the UFMP will run as smoothly as possible. However, you should approach this step as a learning experience and anticipate the need for contingency planning. Some objectives can be achieved within a certain timeline, but this process needs to be updated regularly because your community, environment, resources, and urban forest will change over time. Updates are important for keeping your planning team and staff aware of priorities and progress.

In most existing management scenarios, monitoring and evaluating the impact of the plan is the most neglected step. Yet, it is one of the most critical elements of any plan because it will determine if the plan's goals and objectives are realistic and actually being met. Monitor and evaluate the information as a team, learn from other team members, and modify or improve goals if necessary. The culture, politics, budgets, and constraints of every community are different and balancing the community's needs with urban forest management is challenging and complex. However, changing and adapting your objectives should serve as learning experiences and not as failures. Consistent evaluation will also provide feedback on how to improve your plan.

A plan and its vision should not have a shelf life; they should be dynamic, flexible management instruments. If the ecological, economic, or social assumptions that directed the initial plan change or become questionable, then the plan needs to be adjusted to meet the new realities. In the aftermath of a weather event, the impact on the urban forest is about the last thing on anyone's mind. Trees will be secondary to ensuring public safety, mitigating risk to people and property, cleaning debris, and restoring public services and utilities. In fair weather, however, the urban forest should be a primary community concern. Careful planning for the allocation of resources to the urban forest will provide a community with a healthy, strong, resistant forest.

Tree ordinances are among the tools used by communities striving to attain a healthy, vigorous, and well-managed community forest. By themselves, however, tree ordinances cannot ensure that the trees will be improved or even maintained. Tree ordinances

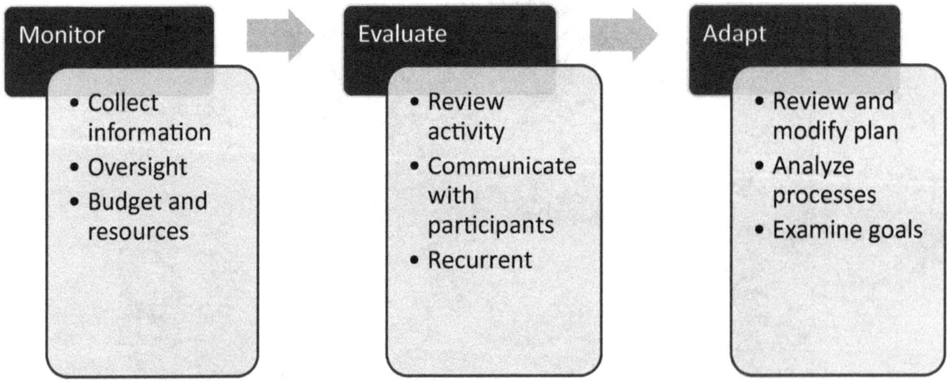

FIGURE 1.14 Step 4: Monitor and evaluate.

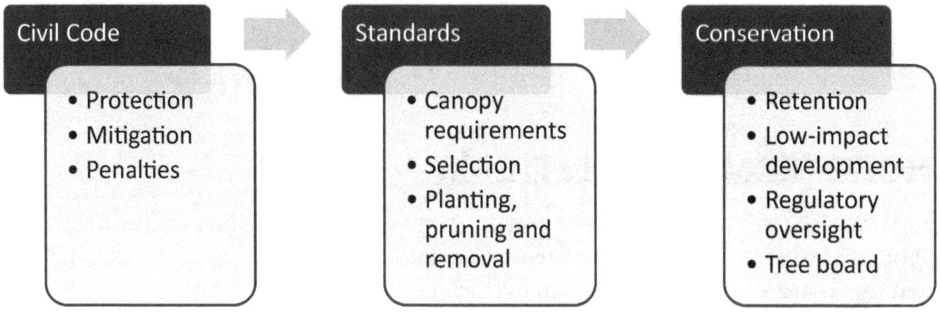

FIGURE 1.15 Step 5: Enforcement.

provide authorization and standards for present and future management activities. If these activities are not integrated into an overall management strategy, problems might arise.

TYPES OF TREE ORDINANCES

Tree ordinances fall into a couple of basic categories:

- *Street tree ordinances* primarily cover the planting and removal of trees within public rights-of-way. Recommended species, locations, and permits are a part of the ordinance. They often contain provisions governing maintenance or removal of private trees that pose a hazard to the public. Also included in this category are ordinances with tree planting requirements, such as those requiring tree planting in parking lots.
- *Tree protection ordinances* are primarily directed at providing protection for native trees or trees with historical significance. They usually require that a permit be obtained before protected trees can be removed, encroached upon, or, in some cases, pruned. They can also include specific protocols related to construction activities near trees.

Ordinances can be one or both of the above examples. It is entirely dependent on the UFMP goals and objectives.

A clear vision, community participation, monitoring and the ability to adapt your plan to adversity will ensure the sustainability of the urban forest and its services. Conservation and planning are not discrete events, but long-term processes. Success will require all individuals of a community to cooperate.

SIMPLIFIED URBAN FOREST MANAGEMENT PLAN

These are suggestions for creating a management plan for your community. The statements listed below would require details and features that reflect the needs of the community and results from the inventory.

Vision: To sustain a healthy, safe and appealing public street and park tree population in the community.

Goal: To effectively manage the urban and community forest in an effective manner through sound fiscal, personnel, and operational management; utilizing internal and contracted services; and building a team of effective proponents for the trees in the community.

SHORT-TERM ACTION ITEMS

- Remove risk trees on rights-of-way and green spaces that could threaten the public.
- Properly prune trees to improve safety and sight lines for key areas such as traffic signals and signs, street lights, pedestrian and vehicular traffic, and buildings.
- Perform timely, systematic tree inspections, looking for defects and initiating the proper mitigation strategy.
- Install new street trees to maintain designated canopy goals. This includes replacing those damaged by weather events or in decline from subsequent maintenance operations due to necessary improvements.
- Improve species diversity with planning and selection to better protect the urban forest from pest issues.
- Establish routine, systematic pruning cycles for all trees along the community rights-of-way, parks, and public green spaces.
- Identify potential partners for urban and community forestry programs and initiatives
- in the community.
- Inform the public of ongoing efforts and long-term strategies to improve engagement and inclusion.

LONG-TERM ACTION ITEMS

- Develop a street tree ordinance for the community.
- Increase public education and involvement in the planning, care, and maintenance of the community trees and urban forest.
- Develop a community tree board to help provide guidance and recommendations to the community for care and maintenance of the community forest.
- Create a fiscal budget that matches the needs of urban and community forestry operations.
- Develop working partnerships with local and regional utilities, agencies, and non-profit organizations to improve effectiveness and efficiency of urban and community forestry operations.
- Plan responsibly to increase the number of trees planted on public lands and along the
- streets.
- Develop and maintain a current inventory of all street and park trees.
- Develop a comprehensive set of specifications for plant health care applying to internal and contracted service professionals.
- Develop and implement training to internal personnel on all aspects of urban and
- community tree care.
- Maintain communication and collaboration with all community departments and state highway and transportation agencies on developing standards and criteria for care of trees growing on state roadways.

Trees and other living plants are valuable. They beautify our surroundings, purify our air, manufacture precious oxygen, act as sound barriers, and help us save energy through their cooling shade in summer and their wind reduction in winter. Many people don't realize, however, that plants have a dollar value of their own that can be measured by skilled plant appraisers. If your trees are damaged or destroyed, you may be able to recapture your losses.

Street trees and urban woodlands provide a number of environmental and social benefits, including contributing to climate change adaptation and mitigation and providing urban green space or improving health and well-being in the neighborhood.

Urban forestry managers must be capable of quantifying or measuring the benefits or ecosystem services that trees provide to a community. Benefit-based tree valuation provides alternative estimates of the fair and reasonable value of trees while illustrating the relative contribution of different benefit types. This is an important tool for enhancing urban forestry programs by providing monetary value on natural resources as a capital asset. Today, we are able to put a dollar value on many of the ecosystem services that trees provide, such as stormwater management, carbon storage, cooling effects, and many others.

TREE ORDINANCE AND CANOPY COVERAGE POLICY

A *tree ordinance* is a tool to help protect and manage a community's trees. It can be designed to regulate various aspects of tree planting, removal, and maintenance on public and private property within a municipality. By protecting trees, a well-planned, written, and implemented ordinance can enhance a community's beauty, reduce air pollution, lower air conditioning costs, and increase biodiversity.

To evaluate the necessity and feasibility of an ordinance, a community should create a working group to assess the municipality's needs and wants, resources, and existing ordinances. In the beginning, the group should develop rules governing information sharing, decision-making, and conflict resolution.

An ideal group is composed of people who mirror the demographics of the community. It should include experts on relevant topics such as forestry and public works, as well as people from other fields. A sample group could include a realtor, developer, garden club member, arborist, planner, environmental group representative, landscape architect, forest landowner, public works official, business owner, lawyer, and interested citizen.

The International Society of Arboriculture categorizes arboriculture-focused tree ordinances, which relate to maintaining trees for aesthetic and environmental benefits, into two main categories:

- *Street and public tree ordinances* regulate the planting, removal and maintenance of trees in parks or along public rights-of-way, including private trees that could endanger the public. These ordinances can include tree planting specifications (e.g., requiring tree planting in parking lots) and tree care standards (e.g., standards for pruning and removal).
- *Tree and woodlot protection ordinances* protect specific tree species, trees of a certain circumference or height or trees with historical significance on public and private property. These ordinances usually stipulate that permits are required to remove, encroach upon or prune such trees. They also provide for the replacement of removed trees.

FIGURE 1.16 Canopy coverage policies are important for improving tree benefits.

As part of the ordinance, conservation strategies may be imposed to protect and maintain desired canopy levels and goals. Researchers estimate that tree canopy cover in urban and metropolitan areas across the U.S. averages only 27% and 33%, respectively. Additionally, the trees that are present are subject to a wide variety of stressors, which significantly shortens their lifespan. As such, it is important for urban communities to take steps to protect and enhance their urban forests through urban tree canopy (UTC) goal setting processes.

In order to set UTC goals, communities must first have an idea of how much current canopy is present. The process for conducting UTC assessments and goal setting generally includes the following steps:

1. **Measure current UTC**
   - Use remote sensing or inventories to measure existing urban tree canopy.
   - Identify the different types of forest in the community, including public (street trees, riparian corridors, parks, etc.) and private (residential, commercial, industrial areas, etc.).
2. **Estimate potential UTC**
   - Use remote sensing imagery and geographic information systems analyses to identify locations with potential for reforestation to improve UTC.
   - Identify priority locations where UTC increases will support identified community needs (e.g., water quality, air quality, stormwater management).
3. **Adopt a UTC Goal**
   - Determine a goal based on the results of the assessments and specify a timeframe.
   - Formally adopt the goal to ensure that the goal comes to fruition (e.g., institutionalize UTC goals in local ordinance, regulations, and comprehensive planning efforts).
   - Create a metric to measure inputs and manage the goal.

Once the assessment and goal-setting process is complete, the next logical step is to develop a plan. In general, a UTC plan identifies the UTC goal and timeline; describes the relationship of canopy goals to local ordinances, regulations, and the community's comprehensive plan; and outlines the specific strategies for achieving UTC goals, including identifying a timeline and responsible party. Each community must develop an approach to achieve UTC goals that considers their internal capacity and resources, political climate, and stakeholder needs. The range of strategies to achieve UTC goals include:

- Permanently protecting priority forest tracts through acquisition, conservation easements, or another method.
- Preventing forest loss during development by adopting or amending site development regulations (e.g., forest conservation regulations, open space design, low-impact design, and zoning).
- Maintaining existing forest canopy by adopting regulations that restrict tree removal.
- Increasing tree planting during development by adopting or revising site development regulations such as landscaping and parking lot shading.
- Reforesting public lands, beginning with priority sites.
- Encouraging reforestation of private land by developing education, stewardship, and incentive programs.

An example might be:

To balance environmental goals and planned density goals, the community has shifted its approach of tree retention from regulating individual trees to the conservation of the overall urban forest canopy. Recognizing the functional importance of a mixed-age, mixed-species urban tree canopy, regulations are adopted to treat urban residential sites without tree canopy the same as urban residential sites with tree canopy.

TREE CANOPY AND SIGNIFICANT TREES

Tree canopy shall include all evergreen and deciduous trees six feet in height or greater, excluding invasive species or noxious weeds as designated in the tree ordinance. Significant tree means a tree with a caliper of at least 10 inches. Dogwoods and other small, understory trees are significant trees if they have a caliper of at least six inches. For multiple-stem trees, such as serviceberry, the caliper of the individual stems shall be added together to determine if a tree meets the minimum caliper for a significant tree.

Exemptions to tree canopy requirements:

- Removal of any hazardous, dead, or diseased trees, and as necessary to remedy an immediate threat to person, property, or activity as determined by an arborist.
- Construction of a single-family dwelling, duplex, accessory, or non-accessory storage structure on an individual lot created prior to implementation date.
- Construction or maintenance of public or private road network elements and public or private utilities including utility easements not related to development.
- Construction or maintenance of public parks and trails when located within an urban residential zone.

TREE CANOPY COVERAGE REQUIREMENTS

A minimum healthy tree canopy cover shall be preserved or established on each lot subject to these requirements. Tree canopy cover shall be calculated based on the trees' expected mature canopy in the urban environment.

Tree cover calculation is a process of measuring how much existing tree canopy must be preserved, planted, or a combination of planting and preservation for a land development project to comply with local policy.

- The following areas shall not be included in the site area for purposes of calculating minimum required canopy coverage: water bodies and public street rights-of-way owned by a government agency.

TABLE 1.1 Mature Canopy Cover

Size Category of Tree	Expected Mature Canopy Cover (Square Feet)
Large	1,600
Medium	900
Small	400
Very small	150

Note. Table 1.1 lists the mature tree canopy to be credited for a newly planted or immature tree of the indicated size category. The size category shall be as indicated on the Approved Tree Species List.

TABLE 1.2 Required Tree Canopy

Zoning District	Minimum Tree Canopy Coverage
Residential	30% for a subdivision; 20% per residential lot
Central Business District	15%
Commercial	20%
Light Industrial	15%
Planned Redevelopment Area (PRA)	Use canopy requirement for the zoning designation assigned by the administration based on approved use.
Planned Unit Development	40% for the development; 20% per residential lot
Planned Business Development	30% for the development; 15% per lot

Note. Table 1.2 lists the minimum required tree canopy coverage by zoning district.

- All healthy trees on the development site may be included in the tree canopy coverage for purposes of meeting the minimum requirement, including but not limited to trees planted or retained to meet landscaping and buffer yard and forest conservation.
- Evergreen trees and tree species identified as very small by the Tree Species List shall generally not be used to meet canopy requirements. They may be used if site characteristics render deciduous or larger trees impractical.
- The canopy of a preserved tree or group of trees may be calculated by measuring the actual canopy; using recent surveys, aerial photographs, satellite images, or other means; or estimated by multiplying the diameter of the crown at its widest point by the diameter of the crown perpendicular to the first measurement.
- An existing tree determined by the zoning administrator to be in poor condition shall not be credited toward required canopy coverage. The zoning administrator may rely upon the advice of an ISA-certified arborist retained by the community.

CANOPY COVER CALCULATIONS FOR PRESERVED TREES

The canopy of preserved trees shall be calculated as indicated in Table 1.3.

The planning and zoning commission shall consider the following guidelines when reviewing site plans and subdivision plans requiring tree canopy:

- To the extent possible, required canopy shall be provided through protection of existing trees utilizing the guidelines provided by the tree board.
- Where applicable, trees shall be planted and protected where they will be most beneficial in the enhancement of water quality and preservation of environmentally sensitive areas.

Tree value is a human construct—How much value would trees really have without humans? In general, people care about the trees where they live. Whether in backyards, parks, or lawn strips, they understand the value of trees and the enormous benefits they provide for

TABLE 1.3 Canopy Credit for Healthy Preserved Trees

Tree Location	Canopy Cover Credit
Trunk entirely on development site and at least 6 inches in diameter	1.5 times the current, measured canopy cover or 1.5 times the canopy cover area assigned in Table 1.1
Trunk entirely on development site and smaller than 6 inches DBH	The current, measured canopy cover or the canopy cover area assigned in Table 1.1 for the tree species
Adjacent trees with overlapping crowns, trunks entirely on development site	1.5 times the current, measured canopy that they collectively project onto the development site
Trunk directly on property line	The current, measured canopy cover that falls onto the development site or half the canopy cover area assigned in Table 1.1
Street trees, trunk on adjacent street right-of-way on the same side of the street as the development site	The current, measured canopy cover that falls onto the development site or half the canopy cover area assigned in Table 1.1
Trunk on adjacent lot or parcel	No credit for canopy cover, even if branches overhang the development site

us every day. Air quality enhancement and water protection are especially important to those living in high-density population areas. Trees can help improve the quality of life as one of the best biological machines on the planet. Other benefits include energy conservation, wildlife habitat, better aesthetics, and a sense of community—these are all values that trees can provide.

Often, we are somewhat disheartened when we realize not all people view trees with the same enthusiasm or support. Some see trees as a "nuisance," sources of leaves and needles or shade to "spoil" a perfect lawn. Some see trees as "interfering" with their property, as obstacles for developing a new home or other building, or as hindrances to widening a road.

The dilemma is that while all trees live on "someone's" property (a municipality, utility company easement, or individual), the benefits provided do not recognize property boundaries. In other words, your tree, your neighbors' tree, your neighborhood park's tree, or your urban forest are all affecting everyone's life in many ways. So, when someone proposes cutting them down, people can get upset and feel compelled to take action.

That is where city foresters and nonprofit organizations who focus on managing public trees can help. They are there to help with questions and issues regarding imperiled trees. They encourage anyone who seeks to protect trees in their community to support those organizations and learn more about programs that improve community trees.

THE COSTS: MAINTAIN OR NOT TO MAINTAIN

There are costs for maintaining trees. This concept has been demonstrated by research and is compelling enough to improve urban forestry management. These include planting, pruning, plant health care, staking, guying, and others. It is an investment in a critical capital asset in our cities and towns. It is an investment in the value of homes and businesses. It is often challenging for governments, decision-makers, and even property owners to identify the value of spending money on trees to keep them healthy and stable—thriving, not just surviving.

Researchers and practitioners in urban forestry and arboriculture can easily place values on the benefits of trees with reasonable ease, using current software and direct valuation methods such as tree appraisals. The value of these benefits is vital when trying to justify allocating funds in any budget. However, there is an even greater cost of *not* maintaining trees that is not as easily understood or quantifiable. Tree maintenance costs are clearer than are the costs of not maintaining trees in the urban forest. Tree maintenance expenditures such as planting, pruning, removal, pest management, and others are easily documented and understood. However, the costs of not maintaining trees does not only equate to deferring maintenance to a later date; rather, managers must often make the decision to not care for trees, which leaves a secondary cost much later in a tree's lifespan that was not anticipated and often more costly. Often, the decision to manage a tree or trees is based completely on acceptable levels of risk. This is determined outside the control of the arborist or urban forest manager and decided by higher-level decision-makers. Those trees outside of the tipping point for risk are then left for the next generation of urban foresters and city residents.

Realistically, the decision is based on priorities and funding. Several examples from the research demonstrate the cost of not maintaining trees results in a loss of net benefits from urban trees. The ecosystem services that improve our quality of life are diminished as maintenance is deferred or ignored. Economic cost-to-benefit ratio or net benefit analyses allow for rational decision-making and allow tree managers to determine if a tree care treatment makes economic sense. The value of benefits over time can be optimized by comparing maintenance costs incurred during that period to determine what level of maintenance provides the highest total net benefits.

Understanding both benefits (ecosystem services) and costs (ecosystem disservices) is needed when managing and planning for an urban forest of any size or location. Seeking input to determine the perceived costs and benefits of trees will prevent future challenges and allow other issues to be handled proactively and with an improvement in public understanding and appreciation of the trees in our cities and towns. Preserving low-maintenance, publicly acceptable, larger trees in good condition provides the clearest benefits to communities. Additionally, widespread investment of resources to manage and maintain the entire urban forest will benefit everyone.

ARBORESTRY: URBAN SILVICULTURE AND ARBORICULTURAL CONCEPTS FOR MANAGING URBAN WOODLOTS

Cities and towns are engaging in large-scale greening projects motivated by the wide range of documented ecological, economic, and social benefits of urban forests. Urban forested natural areas often referred to as woodlots are a critical component of the urban forest but are often disregarded leading to a lack of formal management contexts.

Typically, street and park trees are managed on an individual basis by arborists based on client, tree owner, or manager needs. Professional, certified arborists have well-defined arboricultural practices that focus on managing risk and sustainability in the built environment. However, it has become evident there are no prescriptive standards or recommendations for urban natural areas or woodlots. It becomes clear that these contiguous stands of trees, although typically creating small geographical footprints of under an acre or up to 10 acres or more, require different management strategies than working with singular, open grown trees.

It seems prudent to apply an urban version of traditional ecological management and silviculture to these woodlots, and the extent to which such approaches may need to be modified to accommodate the complex socioecological and ecological dynamics of urban systems. There are basic management differences between natural forests and urban forest woodlots, with opportunities to better manage these forest types with traditional silvicultural practices. For example, prescriptions and treatments can be developed and implemented to manage site conditions and dynamics to achieve the desired woodlot goals and composition.

FIGURE 1.17 Trees are managed for benefits and community needs.

Many small woodlots are parts of larger tracts of land. Some very small woodlots are undeveloped pieces of subdivisions or land that isn't suitable for other uses. These tracts of land act as buffers or buffer yards between housing subdivisions, managed forests, or public properties. These forested pieces are critical parts of the overall urban forest and often a transition zone between one land use and another.

FIGURE 1.18 Periodically harvesting trees can help preserve the quality of urban woodlots.

Woodlots are dynamic ecosystems. Any transition zone where urban and rural activities overlap is changing rapidly due to human activity that greatly stresses the ecosystem. Because of this, many owners who are content with their woodlots today may not be in the future because they will face changing conditions over time.

Good management and forest stewardship will help protect and improve this dynamic environment. Adding new trees, shrubs, and perennials, managing insects and diseases, harvesting products, and controlling invasive species can help preserve the integrity of urban woodlots. There are many similarities in the practices of arboriculture and silviculture. The only real difference is client needs and expectations.

As the number of human inhabitants in an area increase, however, studies show that the likelihood of a woodlot remaining sustainable decreases. Many woodlot owners want to be good stewards and protect and enhance the woodlots they own. They must plan and manage their woodlots carefully to reach their goals.

The maintenance and management of natural areas or woodlots can also be called *stewardship*. It is the desire to leave the land better than you found it. The more you know about the forested area and goals for the woodlot, the better the likelihood those goals can be reached. Good stewards of the land manage their property to maintain or improve water

FIGURE 1.19 Invasive plant control creates better woodlot health.

quality, aesthetics, wildlife habitat, timber, recreation, soils, wetlands, and other unique features, rare plants, natural communities, and forest protection.

This is at the very core of *silviculture*. By definition, silviculture is the art and science of controlling the establishment, growth, composition, health, and quality of forests and woodlands to meet the diverse needs and values of landowners and social constructs. It's about management of a system of trees, like urban forestry, but in more naturalized areas as opposed to the built environment. In contrast, arboriculture places focus on individual trees, relative to risk, vitality, and interventions.

How woodlots are managed depends on the intents and purposes of the tract. Most common management objectives are to improve wildlife habitat, aesthetics, and recreation. In some applications, woodlots may be a source of revenue for lumber and firewood. Keep in mind that you can successfully manage your woodlot to reach more than one objective at the same time. Management plans helps create a vision and determines a schedule of silvicultural and arboricultural interventions to achieve the vision. Many woodland managers and owners benefit from qualified foresters and consulting arborists who can translate objectives into actions.

The strategy of incorporating traditional forest management, arboricultural, and silvicultural practice into urban and peri-urban locations include the following basic components: (1) forest assessment, (2) management objectives and goal setting, (3) vegetation management, and (4) monitoring and evaluation. Application of this strategy and basic framework will improve the success of adopting or adapting traditional forest management methods and applying them to urban woodlots.

Below are the steps to writing and using a management plan.

1. Understand the basic principles and responsibilities of ownership and management. Review and recognize local regulations and the potential liabilities affecting resources and the community at large.
2. Determine short-term and long-term objectives for the woodlot. Start by answering some basic questions.
   - Why do you own the woodlot or urban natural area?
   - What characteristics do you like about it? Any particular features?
   - What is noticeable that may seem unfavorable or negative with the site?
   - What do you want this woodlot to provide for those who may experience it (i.e., wildlife habitat, lumber, or recreation)?
   - What intended use or need is seen for this site 5, 10, and 20 years from now?

   Management planning can begin with greater detail with outcomes-based actions relevant to these questions. Also, identify actions needed to achieve these objectives. If the objective is to manage the property for functional benefits to improve air and water quality or stormwater management, determine which trees should be managed through best practices and any afforestation initiatives to better meet those goals.
3. Assess the physical and biological characteristics of the woodlot. Clearly identify the boundaries of the property, locating permanent landmarks for orientation. These may include boulders, roadways or paths, fence posts, or other structures that will not be

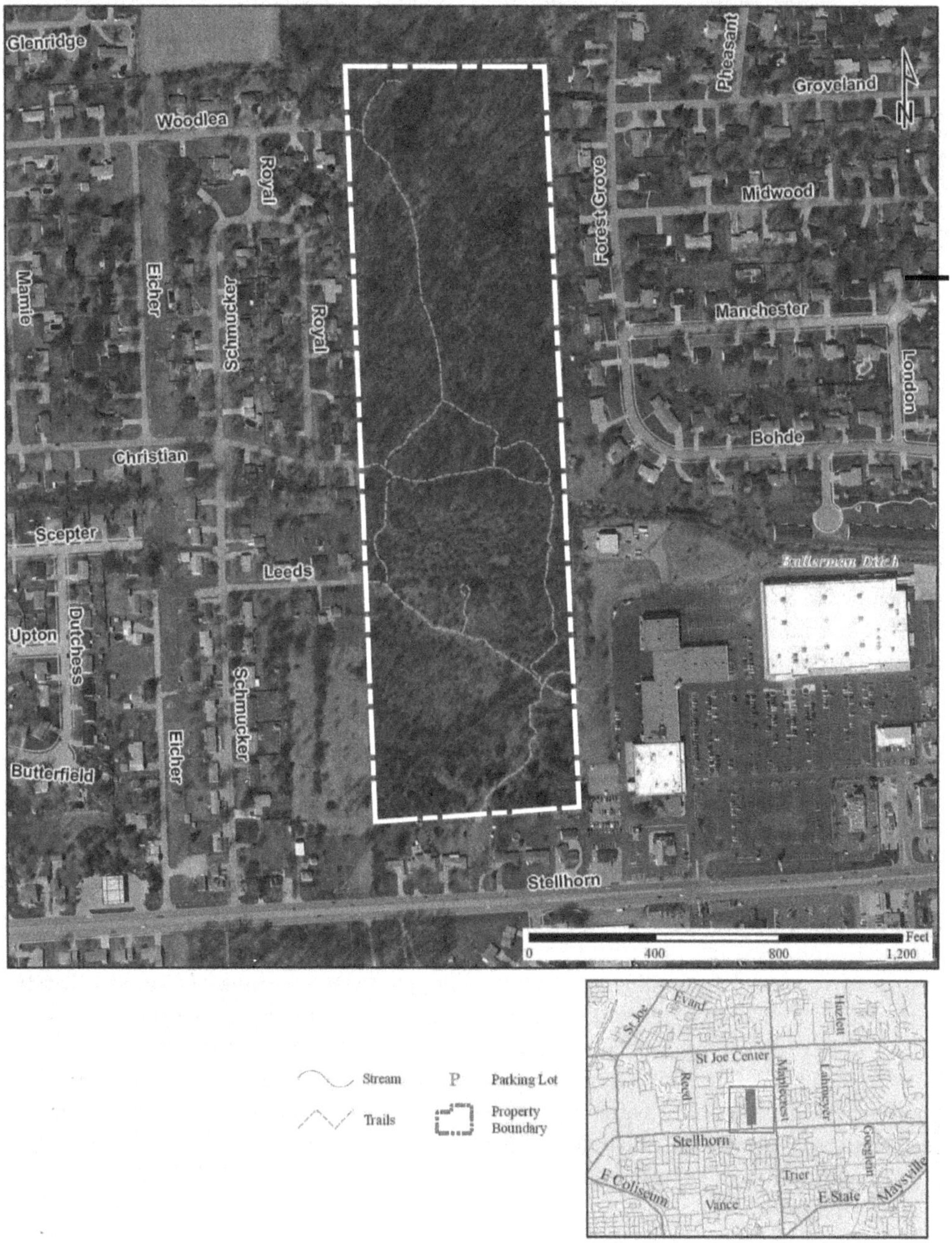

FIGURE 1.20 Urban woodlots are a critical component of the urban forest providing many benefits.

moved or removed. This helps clarify ownership for adjacent property owners who may unlawfully encroach on the site. Next, map the tree cover in the woodlot with applicable inventory management software.

4. Develop a management plan with a timetable for meeting established objectives. Consider all components of the woodlot in conjunction with established objectives to ensure goals are met within the prescribed timetable and budget.
5. Follow the management plan to achieve objectives through active monitoring. A plan is only as good as the actions that support it. Follow through on all required management activities, monitor the results, and maintain records of all activity involved with the woodlot.

THE STEWARDSHIP PLAN

In addition to a management plan, a forest stewardship plan is prepared by a professional forester in collaboration with a tree care professional such as a consulting arborist. The appropriate plan matches the woodlot capabilities with manager/owner objectives in a customized working document. A stewardship plan addresses wildlife habitat, water resources, recreation, forest protection, soils, timber, wetlands, aesthetic values, cultural features, and endangered species. Use the management plan as a guide for matching outcomes with good stewardship.

In addition to giving management direction, a plan is necessary for current assessment of land use. A tree inventory is an important section of a detailed stewardship plan. As part of the plan, it is a good recommendation to obtain details on the soils found on the site. Soil and site conditions, along with local climate history, helps to determine the species of trees and plants that will perform best over time. Soil information can also inform decisions on management of wildlife habitat, recreational development, feasibility for construction, water management, and planting.

Other core principles for managing woodlots include maintaining or improving tree health, tree risk assessment, and management protocols. Trees growing in woodlots face significant competition from surrounding trees. As trees grow larger and their crowns expand, some trees will be crowded out and die or be shaded by the canopies of the fastest growing, most competitive trees which creates decline. Professional foresters and consulting arborists can help landowners select which trees to remove and which trees will need more space to expand their crowns. Trees to be removed may be harvested for wood products, killed by girdling the trunk and left standing for wildlife habitat, or felled and allowed to gradually return to the soil. This choice is determined by management objectives for the property, potential risks posed by dead trees, and operational limitations. One of the basic needs of woodlot trees is adequate space to expand and grow. This involves choosing to eliminate some of the lesser quality trees to provide improved growing conditions for select trees. Overall, the combination of woodlot management and stewardship plans will produce a better woodlot for the community and a sustainable urban forest component which meet social and environmental needs.

TREE RISK MANAGEMENT

It is important to remember that property owners and managers have a duty to inspect and maintain their trees to an acceptable level of risk for those on the site. Woodlot and forest proprietors can be held liable or found negligent if a tree fails and results in harm or damage. As a woodlot manager, a tree risk management plan should be documented and in force.

Tree risk management begins with these basic steps:

- **Conduct a tree assessment:** Periodically inspect trees for conditions of concern leading to an increased likelihood of failure. Regular inspection compels the owner or manager to evaluate the amount of risk they are willing to assume as well. Through a tree risk assessment, it can be determined if a tree (or trees) presents an unacceptable risk level. Periodic inspections and documentation demonstrate active management and could reduce liability should a failure occur.
- **Hire a qualified arborist:** Trees should be inspected and assessed routinely by a qualified arborist, such as an ISA Certified Arborist. Additional credentials such as being Tree Risk Assessment Qualified demonstrates an even greater standard of care in the process. At a minimum, trees should be inspected every 5 years or according to the owner's risk tolerance. Another good time to inspect trees is after a storm event or when pest outbreaks are indicated by local resources, such as emerald ash borer or Asian long-horned beetle.
- **Document and maintain records:** Every inspection should be recorded and kept on file for future reference. These records are important for several reasons. Past assessments can show how a tree's health and structure has changed over time. Also, written assessments are beneficial in liability claims and potential litigation.
- **Develop a site policy and mitigation plan:** Every property manager should create and implement a site policy. This includes developing a risk management program, an inspection protocol, and a plan to mitigate the condition to reduce risk levels. Routine care and scheduled maintenance for trees is essential for health and safety.
- **Mitigate tree risk:** Tree owners should make every effort to reduce risk with recommended mitigation strategies. These include target manipulation, pruning, and other tree-based options. Consider all choices before deciding on removal. If possible, move or remove the target (people, property, or activities that could be injured, damaged, or disrupted by a tree). Perhaps reroute a trail or move a designated seating area. Modify site conditions to improve the environment around the tree. Risk mitigation requires logical consideration of all options with a focus on protecting the target and, at the same time, preserving the tree.
- **Schedule tree work:** As part of any evaluation, an arborist should note tree maintenance needs and assist with the planning and timetable. The consulting arborist can determine the timing and priority of work needed based on the defect or conditions. The work should be prioritized first for safety, then for tree health. Consider potential risk, activities around the tree, level of acceptable risk, and client needs.

Urban woodlots are an important environmental component of any community's natural resources. They provide benefits such as wildlife habitat, improved human health, visual interest, and numerous recreational opportunities. Woodlots can be focal points for development, adding interest and value to commercial and residential properties. Regardless of the level of management, every woodlot property owner has an underlying responsibility to inspect and maintain the property for the safety of others. This is the most basic of management practices and always a consideration. The best success for management includes the marriage of arboriculture and silviculture practices. The union of these two practices creates the term arborestry.

Arboriculture involves the care of individual trees, while urban forestry deals with forests as systems (groups of trees) in a developed setting.

Arborestry could be considered a combination of arboricultural practices and forest management practices used in traditional forestry. Arborists specialize in individual tree care, in urban or suburban settings, which leads to the urban forestry label. However, many urban forestry specialists manage tree systems using forestry and silviculture practices in the built environment. Traditional foresters use tree measurements, landholder goals, and knowledge of forest management to develop plans to care for a stand of trees or woodlots over time. All overlap very neatly with many common qualities. The only difference is the environment or ecosystem where the professional operates within.

CHAPTER 2

TREE BIOLOGY AND PHYSIOLOGY

LEARNING OBJECTIVES

- Describe the basic composition of a tree's vascular system and explain how water, carbohydrates, and other hormones are transported within this system.
- Recognize how the processes of photosynthesis, transpiration, and respiration function within the tree.
- Recognize the interaction between genetic potential and environmental influences and how it affects tree response.
- Explain the *compartmentalization of decay in trees* (CODIT) and what influences the process.
- Recognize factors that impact tree physiology resulting in decline, dieback, and death of urban trees.
- Explain the dormancy process and the factors that influence timing and duration.

The ability to achieve maximum benefits from our urban forests and individual trees requires efficient management, as well as an understanding of how they grow and survive. As arborists, horticulturalists, or any tree manager, we may not understand their processes fully, but we do need to recognize how their growth is influenced by their heredity, environment, practices, and man.

Trees are woody plants with a single main trunk or multiple stems. They have amazing strategies to grow, recover, and regenerate to make the places we live better. As discussed in the urban forestry chapter, trees provide ecosystem services which improve our quality of life. Sustainable growth in the built environment of the urban forest requires careful, planned intervention by arborists. Trees live longer and provide increasing benefits if we proactively care for them.

It is nearly impossible to manage the health of any living system without first understanding how it functions. Trees are systems with interrelated components, elements, and processes. They are a pathway of connectivity which includes roots, trunk, stem, and branches. There is a vascular link of transport vessels including xylem and phloem which create pathways for water, food, and other phytochemicals. At the same time, they can't be fully understood apart from the larger system without an understanding of each of these system level relationships.

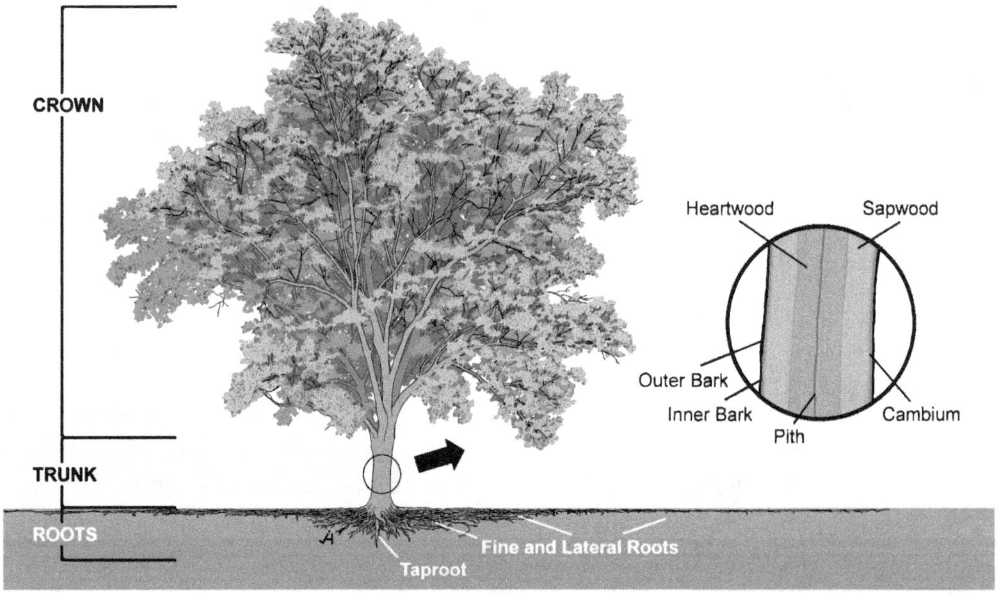

FIGURE 2.1 All parts of the tree are linked through a connective pathway.

Understanding how trees support themselves and how they adapt and acclimate has very practical implications that go to the very heart of arboricultural practices. An arborist must be capable of observing and interpreting these system processes to manage them properly. Due to the "unnatural" environment of the urban forest, trees require adaptive care which can only be achieved by knowing how a tree lives and survives. A basic survey of plant processes and functions will be discussed in this chapter as they relate to arboricultural practices common to most woody plants. These include the production and allocation of necessary resources, such as carbon and water. A look into the response of trees to interventions such as pruning will be examined in more detail as well as what influences the decline and death of trees.

MERISTEMS AND BUDS

A tree may be defined as **a** woody plant possessing a woody trunk that perennially renews its growth. Most plants classified as trees have a single self-supporting trunk containing woody tissues, and in most species the trunk produces secondary limbs, called branches. The late Dr. Alex Shigo of the U.S. Forest Service defined trees as woody, long-lived, compartmentalizing perennials. Regardless of the definition, all living organisms, including woody trees, have cells, tissues, and organs. Cells are basic building blocks of plant structure. New cells arise from the division of existing cells. This process occurs in specialized zones called *meristems.* Tree growth occurs in two ways. Primary growth is growth from the root and shoot tips resulting in increases in height and length. Secondary growth is growth which increases the thickness of stems and branches. In other words, there are two basic types of meristems: (1) primary meristems, which produce the cells that result in the elongation of shoots and roots, and (2) secondary meristems (also known as lateral meristems), which produce cells resulting in increases in diameter or thickness.

Before moving further, more detail is required on bud formation, in particular how they relate to sprouting. Buds on trees are classified by location on the tree component, status, morphology,

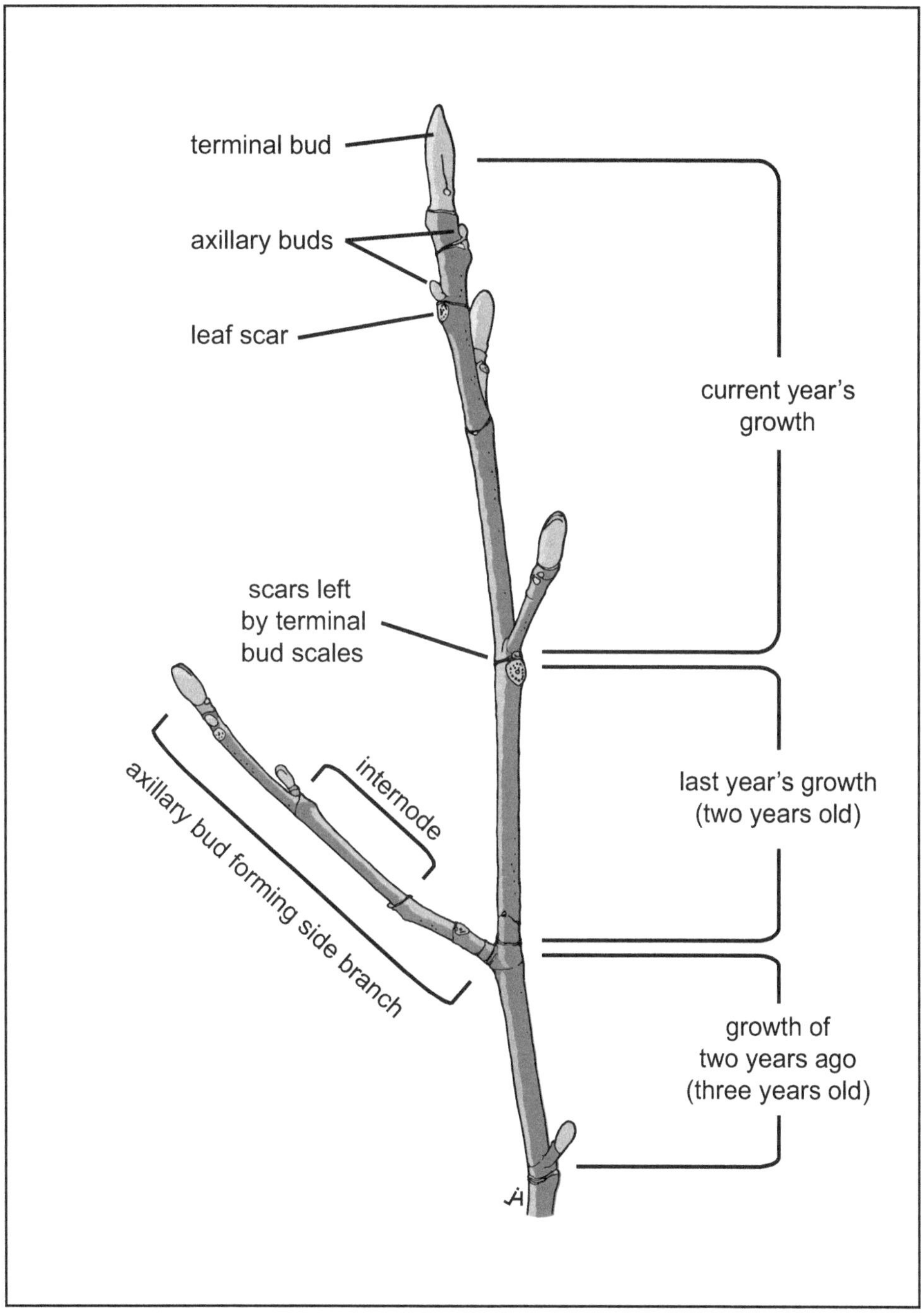

FIGURE 2.2 Components on a branch.

and function. They are located either at the end of a stem or branch (terminal bud) or at the sides of the branch or stem (lateral or axillary bud). Axillary buds are formed at the base of leaf axils, where leaves emerge, and internodal locations on tree branches. Buds that form on mature portions of a plant and not at the stem tips or in the leaf axils are called adventitious buds.

Root sprouts ("root suckers") arise from adventitious buds. Stump sprouts originate from root collars and the lower stem; epicormic branches of angiosperms and sprouts of gymnosperms following a disturbance such as fire, injury, and pruning arise from dormant rather than from adventitious buds. They can also be called epicormic sprouts, Latin for "upon the stem" (*epi* for "upon," *cormic* for "stem"). When a terminal leader or branch is damaged or removed, the lateral branch suppressing hormone auxin is temporarily lost. This leaves the terminal growing point, the adventitious bud, which was dormant to be stimulated by the passing cytokinins and flushes, producing shoots, which like axillary buds become woody branches following incremental secondary growth.

Axillary buds, sometimes called endocormic sprouts (Latin for "within the stem," *endo* means "within") are considered by most to produce stronger limbs than those that have grown adventitiously having no bud trace. Both dormant buds and latent buds are strongly linked to the pipeline of water and nutrients found in the xylem and phloem through a connection called the bud trace. When a branch is cut or broken above a dormant or latent bud, a new shoot can quickly grow from the bud. Adventitious buds develop where no buds previously existed. These sometimes grow after a branch is wounded or cut back to mature tissue. These buds differ from latent buds because they develop close to the branch surface from deeper mature tissue and are not connected by a bud trace; consequently, the branches that develop from adventitious buds are not strongly connected to the trunk or main branches and can easily fail. Improper pruning techniques, such as cutting branches back to stubs, often activate dormant or latent buds or cause adventitious buds to form behind the stub. For several years following damage or injury, the young shoots that arise from these buds are weakly attached to the parent stems and are easily broken. It is not until new layers of wood form annual growth rings around the branches are they more strongly anchored.

FIGURE 2.3 Adventitious bud.

Determining whether a limb has grown from an axillary or adventitious bud is seldom possible in an advanced age tree without dissection. An adventitious shoot which grows sufficient increments of wood and generates a sufficient diameter in proportion to its height is known to be stable in the same way that a limb grown from an axillary bud is stable which is held in place by overlapping layers of wood. Trees are physically growing and adapting to the environment via the growth of axillary buds, adventitious buds, and epicormic/endocormic expansion. However, epicormic branch failure is common and is based on a lack of sufficient branch diameter to support height or length.

FIGURE 2.4 Vascular cambium is located just beneath the bark layer.

VASCULAR CAMBIUM

Trees have secondary or lateral meristems which allow tree components to increase in diameter. The first lateral meristematic tissue is the *cambium layer*. The cambium is a narrow ring of dividing cells producing the cells that will become the vascular system of the tree.

The cambium produces two kinds of tissue: xylem vessels to the inside and phloem vessels to the outside, beneath the bark layer. Every year the tree creates xylem (new wood) on the inside, and phloem (new inner bark) on the outside and originates from the vascular cambium. The xylem carries water and nutrients from the roots upward, while the phloem carries hormones and sugars from the leaves (source) to leaves or storage sites (sink).

The principal cambium is the vascular cambium, which forms the secondary vascular tissues (secondary xylem and secondary phloem) and causes an increase in thickness of the axis of the branch, stem, or root. This growth is usually accompanied by the activity of a cork cambium, or phellogen, which develops in the peripheral region of the expanding axis of the branch, stem, or root and gives rise to the periderm, a secondary protective tissue system replacing the epidermis layer.

In temperate climates, the cambium does not grow during the winter and a darkened line can be seen in the wood where cambial growth slowed at year's end. These are the annual growth rings visible in many woody species. In more tropical climates, growing seasons are longer, and distinct annual rings may not be visible. In some tree species, the rings are very noticeable, because wood produced in the spring (earlywood) is less dense than wood produced in the summer and fall (latewood). The larger vessels are important in the spring because they are more capable of transporting larger amounts of water. Trees have a somewhat reduced water requirement in the fall which allows for smaller rings.

VASCULAR ARRANGEMENT

Deciduous tree species have distinctly different earlywood and latewood rings which are composed of transport vessels. The composition of vessels relative to their size, number, and arrangement is important to arborists and plant health care professionals. There are three categories describing xylem vessel patterns. *Diffuse porous* trees include those with uniformly distributed xylem vessels across each growth ring. *Ring-porous* wood presents very large earlywood vessels and latewood vessels that may be up to 10 times smaller in a seasonal increment. *Semi-ring-porous* trees are like trees in that the vessels or pores are large in the earlywood and smaller toward the latewood, but without the distinct zoning seen in ring-porous trees. These terms are not applied to coniferous species due to a differing transport system. Coniferous trees don't have vessels and water is conducted in the living tree through the tracheid cells. Different conifers have different growth ring characteristics, however. In some evergreens, the rings are nondistinctive; however, some display clearly visible growth increments.

An important application for arborists is recognizing the level of transport activity as growing season develops. Species with larger conductive vessels, such as those found in the earlywood of ring-porous species, lose almost all their conductivity as these vessels become gas-filled and embolized during the first freeze–thaw cycle. The narrower latewood vessels (or tracheids depending on the species) retain their conductive ability during the cold, dormant period. Consequently, ring-porous trees must grow new earlywood vessels before leaves appear the following spring, and most of the water used by the crown is conducted up just this one ring of new wood. New xylem must be produced from stored carbohydrates from the previous growth season. This causes a pronounced decline in stored carbohydrates in some ring-porous tree species in early spring and tends to reduce the duration of the growing season as they expand their leaves later in the year compared to diffuse porous tree species.

Species with narrower vessels, typically diffuse porous, carry fewer embolized vessels into spring and, by using positive stem (or root) pressure, can refill these obstructed vessels. The advantages of refilling the vessels are that transport can resume prior to the development

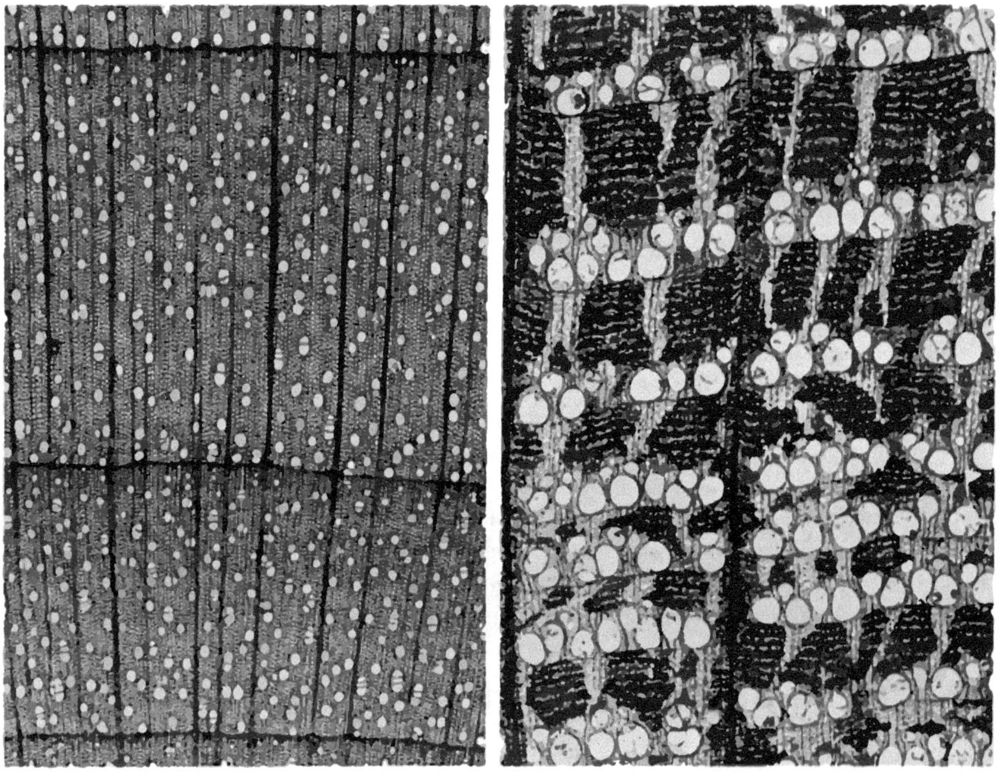

FIGURE 2.5 Ring-Porous and diffuse porous vascular system.

of new xylem growth in the new growing season. Typically, these trees experience earlier leaf expansion. This process can occur over a longer period and stored carbohydrates do not need to be used in new functional xylem development. The effect can be critical in the determination of good conductivity periods as well as predicting or diagnosing tree health. Trees producing positive pressure in the xylem vessels tend to "bleed" if they are wounded (pruned) during early spring. While, in most cases, this is not a real problem for the tree, it can cause rather unsightly sap that drips down the stem(s).

A certain amount of diameter growth also occurs through growth of the cork cambium, which produces the outer layer of bark. New cork cambium is produced each year; however, the outermost layer is shed so that the bark thickness of a mature tree remains nearly the same from year to year. While growth of the cork cambium may contribute greatly to diameter growth as a young tree develops a thick bark, diameter growth of a mature tree is mainly due to the production of wood by the vascular cambium, producing additional xylem vessels.

TREE ARMOR

Bark is the outer covering of a tree's branches, limbs, and stems. Structurally, it is a complex part of the tree found outside the cambium, and including the phloem vessels. There are several components to the bark layer. It is composed of protective tissues that moderate the temperature inside the stem, offer defense against injury, and reduce water loss. The outermost bark is composed of nonfunctional phloem and corky tissues which we typically see on trees.

It is the waterproof skin that seals the inside of the tree to provide anoxic moisture levels and helps prevent pests and pathogens entering the tree. Inside this layer is the inner bark that is dominated by the active, living phloem vessels that transports sugars and other compounds to all parts of trees. Most often these are sugars produced by the leaves being taken to where they are needed around the tree but can include many other compounds such as plant hormones.

FIGURE 2.6 Bark is the protective covering and is formed by dead phloem vessels.

TRANSPORTING VESSELS

The phloem vessels composed primarily of living cells are not considered structural tissues as is the xylem. The *phloem* moves carbohydrates (sugars) produced in the leaves throughout the plant for storage or consumption. Movement of sugar in the phloem is relatively slow and transport requires energy. The older phloem becomes crushed as the tree expands and is incorporated into the outer bark layers.

The *xylem* is comprised of basically nonliving cells and a small percentage of living cells and is responsible for four primary functions:

- the movement of water and dissolved minerals,
- supporting the weight of the tree,
- storing carbohydrate reserves, and
- providing defense mechanisms against decay.

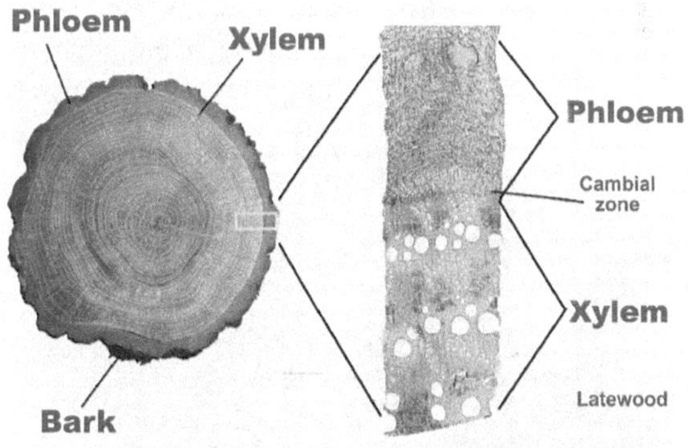

FIGURE 2.7 Phloem and xylem vessels are responsible for transporting food and water in the tree.

New xylem vessels or *sapwood* conducts water primarily as a connected pipeline from roots to leaves, providing strength but can also serving as storage for starches. The sapwood also contains living *parenchyma* cells. These cells are an important component in the tree and a method of storage and movement of *assimilates*. Parenchyma cells can run inward like spokes on a wheel (axially), and around the rings of the tree (radially) and are called the *rays* or radial parenchyma. These ray cells grow radially across the grain in small layers that extend across the growth increments of xylem and into the phloem. Rays transport carbohydrates and other compounds into sapwood, store carbohydrates, and assist in restricting decay in wood tissue through specialized chemical properties. These two systems of living cells connect to provide a complicated connectivity known as the *symplast*, that has roles in storage, transport, support, and defense.

The older, nonfunctioning xylem constitutes the inner wood of a tree and contributes to diameter increase. As the sapwood xylem vessels die each year, the tissue becomes nonconducting, nonliving *heartwood*. Typically, this is considered a storage area for carbohydrate reserves as well as mechanical support for the tree. The critical change between sapwood and heartwood is death of the ray cells and parenchyma cells. The reason this can be seen as critical is that these cells are responsible for creating chemical changes within the tree to help minimize the impact of decay. Various chemicals, such as phenols in angiosperms and resins and terpenes in gymnosperms, are increased in the rays in defense of decay.

However, some trees do not create heartwood such as Salix, Populus, and some Acer species. Many tree species maintain consistent living rings from 3 growth rings up to 20 and some species even more. As trees and tree parts age, there is a reduction of sapwood compared to the amount of heartwood present. However, there are several sapwood rings present for moving assimilates.

The structural components of the primary cell walls are a composition of *cellulose, lignin,* and *hemicellulose.* This combination provides the strength of the tree and allows it to stand under various loads. It is cellulose which plays a large role in providing flexibility within the tree fibers under dynamic loads such as wind. Lignin is a component of plant cells and, when formed in the cell walls of the wood, provides the strength and rigidity allowing trees to grow tall. Cellulose and lignin are critical components of wood strength; when affected by decay organisms and allowed to progress, health declines and strength loss results creating risk situations with potential tree failure. A visualization used to demonstrate the roles of these primary cell materials would be to imagine the cellulose fibers being much like the reinforcement bars in concrete providing flexibility to a structure. The lignin would be like the surrounding concrete which provides load-bearing strength for the tree. Together, they form a strong, sustainable biomechanical structure withstanding loads from the tree's own weight and bending strength from external loads such as wind.

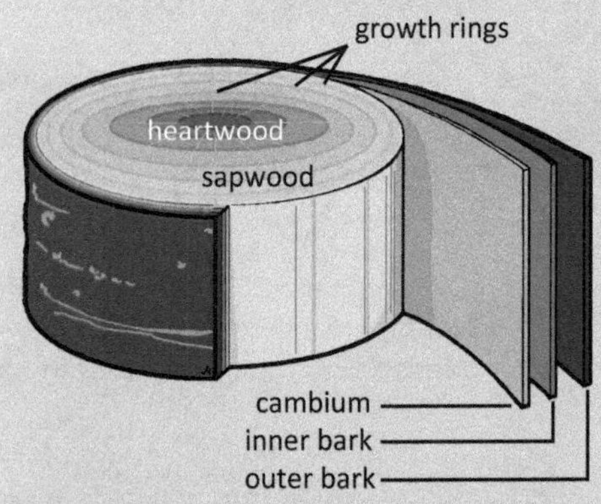

FIGURE 2.8 The basic parts of a tree.

The anatomy of a tree:

- **outer bark:** The outer bark is the tree's protective armor and varies with thickness and texture.
- **inner bark:** The inner bark is composed of phloem vessels and as it dies it becomes part of the protective outer bark.
- **cambium cell:** The cambium cell layer is a zone of intense activity giving rise to phloem layers and xylem vessels.
- **sapwood:** Sapwood is a general term for the active xylem vessels. It may be only one ring to two or three growth rings in depth. As new growth rings are created, the older rings turn to heartwood. Sapwood has active parenchyma cells which support wound wood development and decay resistance.
- **heartwood:** Heartwood is nonliving wood fiber formed from dying parenchyma and is responsible for storage.
- **growth rings:** Growth rings represent the vascular regions separated by earlywood and latewood growth.

TREE ROOTS: SUPPLY AND SUPPORT

Tree roots are supply lines and support components. Tree roots absorb water and nutrients from the soil, serve as storage for carbohydrates and form a structural system which supports the trunk and crown. The root system is considered a challenge for tree care and plant health care mainly due to the inability to inspect or investigate as a belowground component. A common misconception is that the root system is a "reflection" of the trunk and branches. This means many believe the tree roots grow as deep as the tree grows tall. When, in fact, the root system of urban trees is surprisingly shallow, dominated by long lateral roots spreading out just below the soil surface rather than a deep tap rooted system. Many urban tree roots fail to reach depths any further than 3 feet, but can extend deeper, depending on the soil health and texture. Root depth is often miscalculated; however, root spread can be sure to reach the dripline, extending outwards well beyond the branch spread, with no impediments.

ROOTS AND SOIL CONDITIONS

Soil conditions and physical barriers to root growth make determining location and spread difficult to identify. Root growth is opportunistic, growing only where the soil can sustain growth. Roots proliferate wherever they can find adequate moisture and oxygen, which is why the greatest root concentration is found close to the soil surface where the soil is loosest, and water, oxygen, and nutrients are readily available. However, there isn't any evidence to support identifying trees as "deep rooted" or "shallow rooted" species. Any trees can develop a deeper root system if conditions allow.

A healthy root system has the potential to pull water and mineral resources from the soil, so root development in its planted environment affects its effectiveness to obtain resources. The built environment produces soil conditions which are often harsh and restrictive to tree root growth. These include situations such as poorly aerated and nutrient-deficient conditions as well as compacted soils with increased bulk density. These circumstances prove to resist a developing root system. The root framework has four primary functions (as pictured in Figure 2.10).

The larger roots have many small, but often short-lived, branch roots. These are the absorbing roots which are small, fibrous, primary tissues growing at the ends of and along the main, woody roots. The absorbing roots have epidermal cells that may be modified into root hairs, aiding in the uptake of water and minerals. As with shoot tips in the crown, root tips contain a meristematic zone where the cells divide and elongate. The death and replacement of fine roots occur simultaneously. They are sensitive to soil moisture, nutritional conditions, and soil temperatures. Roots will always grow where adequate moisture and oxygen are available.

As mentioned earlier, rooting depth varies greatly among species and often shows little relation to the size of the plant above ground. Most plants tend to develop a high concentration of roots in the surface soil, perhaps because it is well aerated, contains a higher concentration of minerals than deeper soil horizons, and is sufficiently moist. Most absorbing roots are found in the upper 12–18 inches of urban soil.

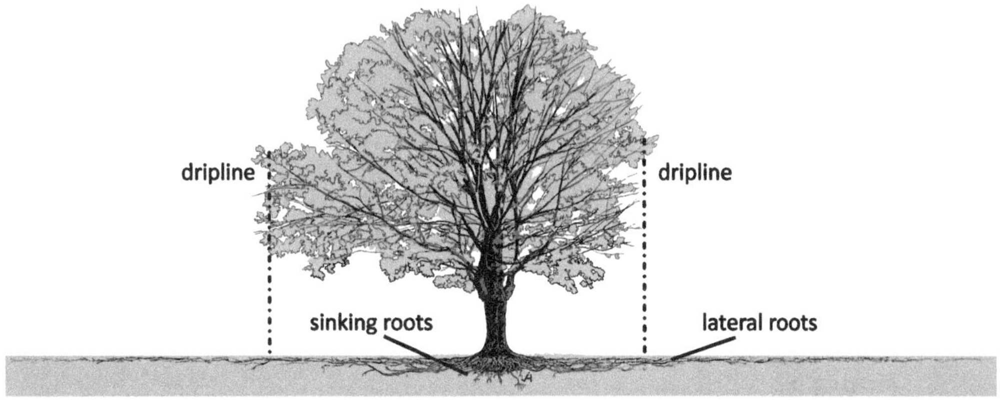

FIGURE 2.9 Basic root system and root types.

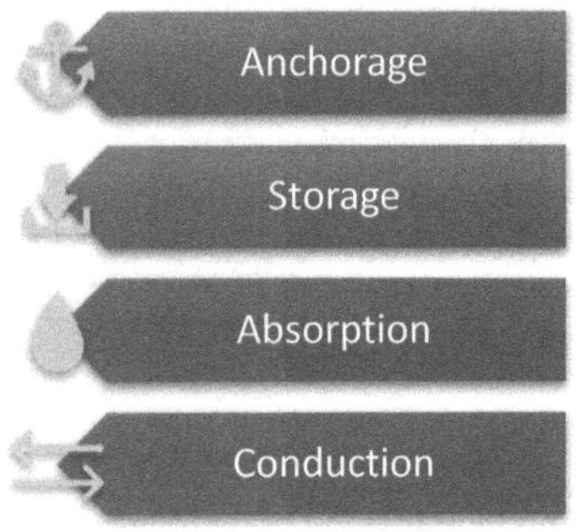

FIGURE 2.10 Root framework.

Although tree species may have the genetic potential for generating deep root systems of over 6 feet, rooting depth in urban situations is frequently restricted by dense or forbidding soil layers or by underground infrastructure. Lateral root extent is likewise subject to restriction by dense soils under hardscape, infrastructure such as curbs or walks, or lack of supplemental irrigation in dry areas.

Impediments to a healthy root system are frequently the underlying cause for a wide array of tree growth and health problems aboveground and belowground and are expressed in many ways. This may include decline and branch dieback as well as girdling roots. Deep roots provide important advantages including the gaining of additional water and mineral resources, improved hydraulic movement in the tree, and reduced risk issues. Root development and expansion, along with how it interacts with soil properties, influence tree stability and sustainability, which impacts longevity in the urban forest.

Roots will advance outward rather quickly into long, spreading, branching roots. Roots may extend laterally for considerable distances, depending on the tree and soil conditions. Roots of trees grown in the open often extend 2 to 3 times the radius of the crown. Variations

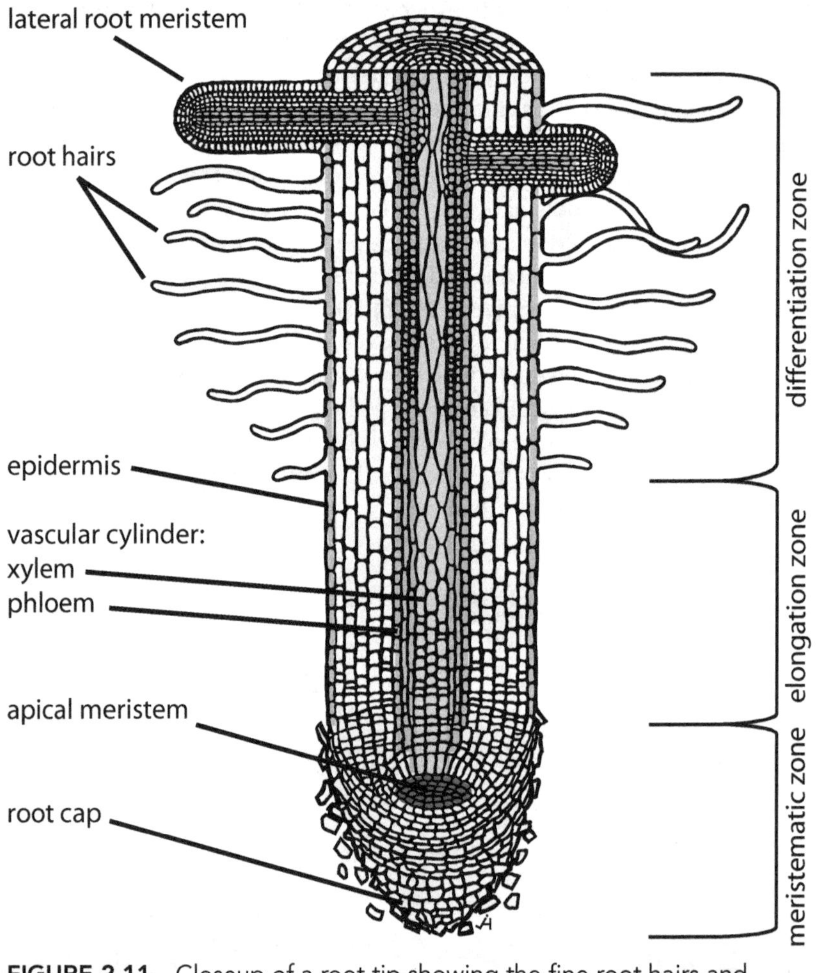

FIGURE 2.11 Closeup of a root tip showing the fine root hairs and root cap.

in the distribution and extent of roots are important because trees able to produce deeply penetrating and branching root systems absorb water and minerals from a larger soil volume than trees with more restricted root zones.

The transition zone where the roots join the main stem is known as the *root flare*. Natural movement is responsible for growth that reinforces potential biomechanical weaknesses within the tree. For the trunk and root transition zone, this results in additional wood at the base of the trunk and the production of a trunk flare. If trees do not experience limited crown movement, such as in a crowded forest area, then this adaptive growth does not occur, leaving the tree vulnerable to failure when exposed to winds.

Trees in the open grown areas form a swelling, where roots join the trunk that is visible at the soil surface. This root flare (also called trunk flare, root crown, or root collar) should be visible on trees in the landscape. This is a critical part of the tree to identify and for proper health and risk management. Tree stems are created to remain above ground as opposed to roots which are particularly suited for belowground environments. This is due primarily to the external covering of the tissue and its exposure to moist conditions. The buried stem or root flare increases vulnerability to decay due to elevated and prolonged moisture levels. It is important to identify the root-stem transition zone and make sure it is level with the existing soil grade.

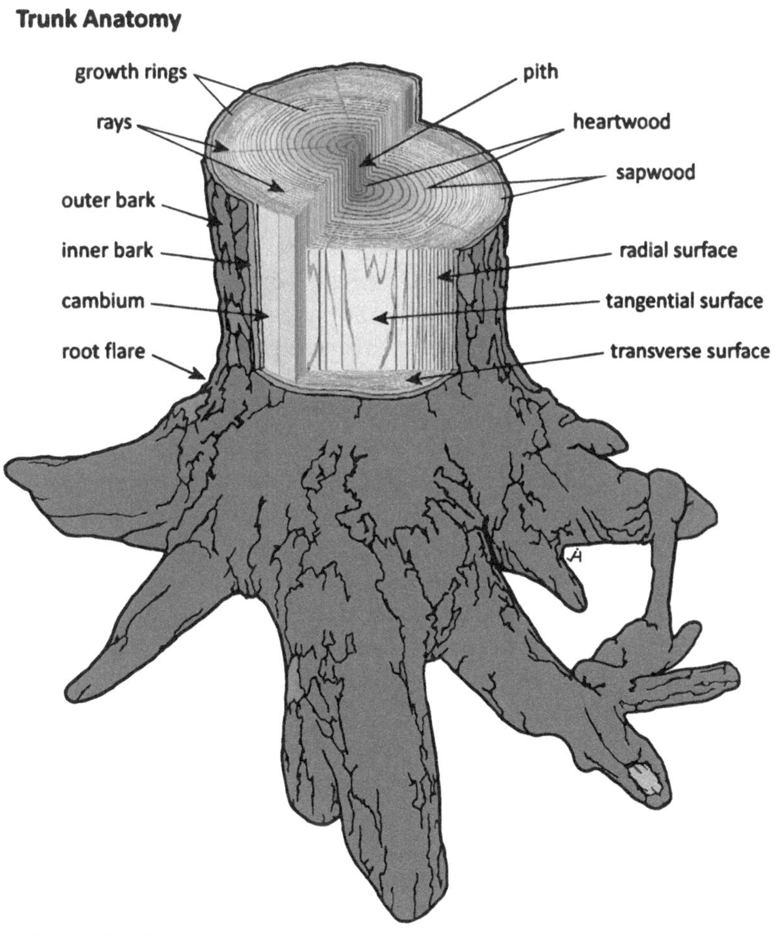

FIGURE 2.12 Tree trunk anatomy and roots.

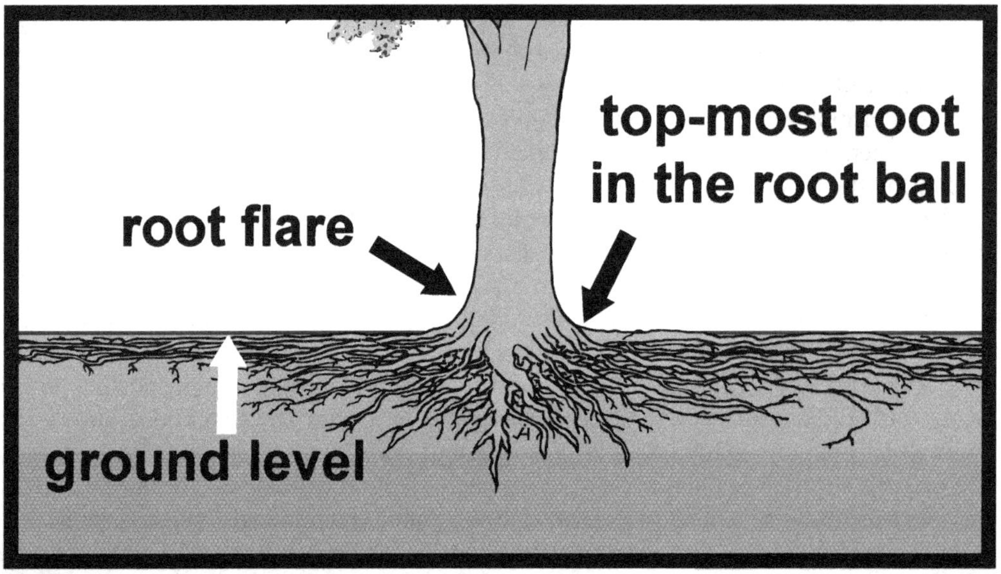

FIGURE 2.13 Root flare location.

LEAVES AND PLANT PROCESSES

Leaves serve as the primary photosynthetic structures and are the primary source of food production. Green, leafy areas and young stems contain a green pigment called chlorophyll. Any part of the tree that is green performs *photosynthesis,* using sunlight, water and carbon dioxide to produce sugar.

Sugar can either be used to provide energy where needed, or it can be transported to stem and roots, where it is either used or converted to starch for storage. Sugar and starch are sometimes called *photosynthates,* or *carbohydrates,* produced by photosynthesis. Trees convert the photosynthates as energy for maintenance and growth. Plant processes including photosynthesis, respiration, and transpiration are critical to creating, converting, and using food for energy. *Photosynthesis* (PSN) is the process by which green plants use light energy to build carbon molecules that make carbohydrates. Sugars are translocated in the roots, trunk, and branches in the phloem often at the rate of several feet per hour through specialized sieve tubes to areas of demand or sink areas.

Respiration (RESP) is the process by which the carbohydrates generated by photosynthesis and stored as starch are broken down into simpler compounds and used by the tree. The tree is constantly converting this food source into energy for vital processes. Approximately half of all photosynthates produced are used in plant processes, especially respiration, and the maintenance of tree components. This includes the energy needed to take up water and nutrients, the active transport in phloem, and the production of defense chemicals to name a few. Photosynthates are converted through the respiration process and used for growth of the roots, stems and branches. Just as the fine roots must be continuously replaced, the development of shoots must occur each year to produce new leaves, even in coniferous species that retain their leaves for several years. Respiration is a constant process and occurs at different levels of intensity throughout the year.

If growing conditions are favorable, enough sugar is produced for all maintenance, growth, and storage requirements, and any additional photosynthates can be applied for additional growth. This excess carbohydrate production often goes to secondary growth and can be measured as an increase in stem and trunk diameter. Therefore, the annual rings in wood often are wider during good growing years, and narrower during years with poor growth conditions.

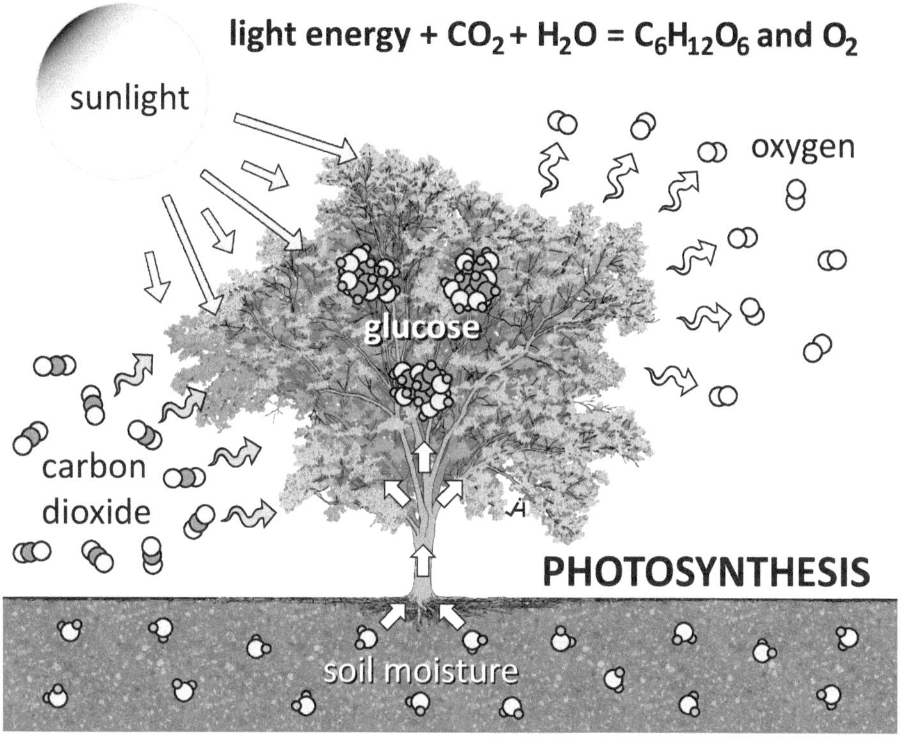

FIGURE 2.14 PSN diagram.

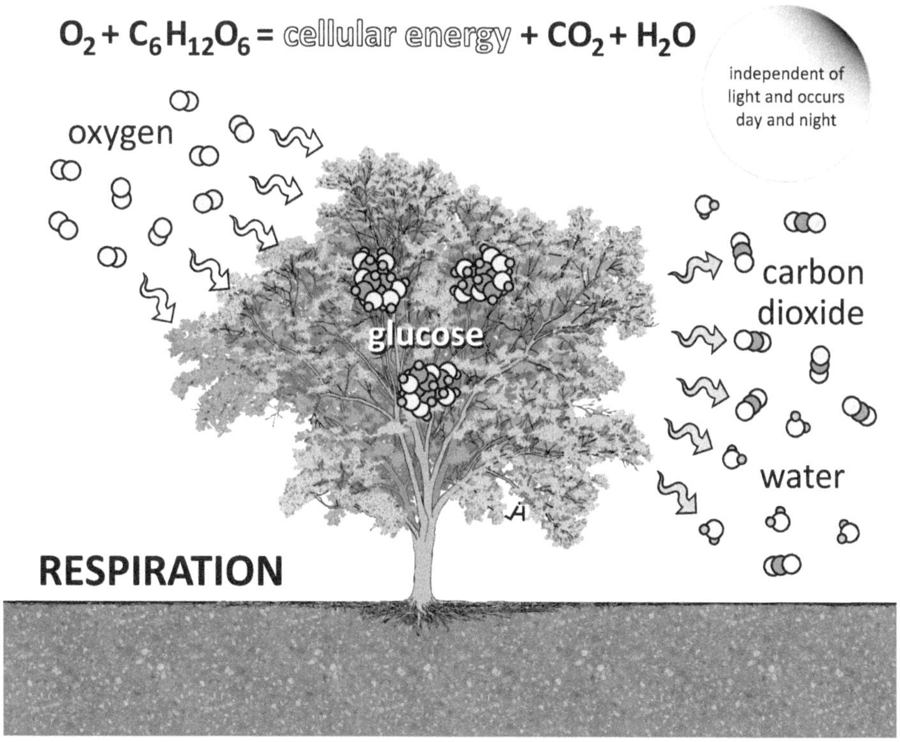

FIGURE 2.15 RESP diagram.

Any negative changes to the live, green tissue on a tree, whether environmentally induced or through arboricultural practices, will influence tree health and vitality. Healthy plants are stronger and more resistant to pests and decay organisms. Additionally, trees in good health have an increased recovery capacity from wounds. This includes a strategic system of defense against attacking wood-decaying organisms.

Leaves act as solar panels which collect light for energy; however, they are also the area of greatest water loss from the tree from exposure to environmental conditions and natural plant processes. Trees use or lose water through transpiration and evapotranspiration. It can be said that water is the single most limiting factor in tree growth and survival. It is a vital "nutrient" that must be available in adequate supply or trees decline and eventually die.

Transpiration is a physiological process where the tree responds to soil and atmospheric factors. Transpiration is the passive movement of water through the tree system which also allows columns of water to move great heights. Simply stated, it is water evaporation from foliage. First, water is taken up by tree roots from the soil and evaporated through the pores or *stomata* on the surface of leaves. This specialized organ on the leaf controls water loss through the movement of guard cells. Stomata, tiny pores in the leaf surface, provide the access between the internal leaf environment and the atmosphere. Stomata have two guard cells that act as gates controlling the size of the opening. When they are closed, they have sealed the entry and are resistant to water loss; when they are open, they expel water vapor as well as give access to carbon dioxide from the atmosphere necessary for photosynthesis, but in so doing they provide the gateway for water loss.

Water movement (hydraulics) through a tree is controlled by the tug-of-war between water availability and water movement in soil versus water loss from leaves. For example, water movement in a ring-porous tree like a red oak is 92 ft/hr, in a diffuse porous tree like a

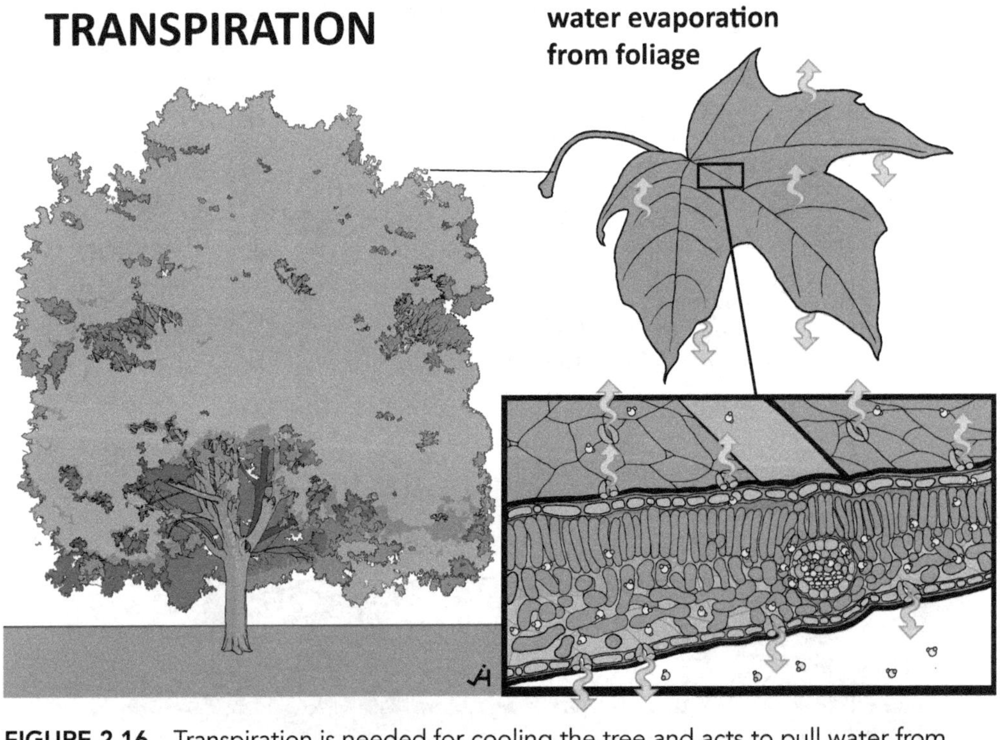

FIGURE 2.16 Transpiration is needed for cooling the tree and acts to pull water from the roots to the leaves in the xylem vessels.

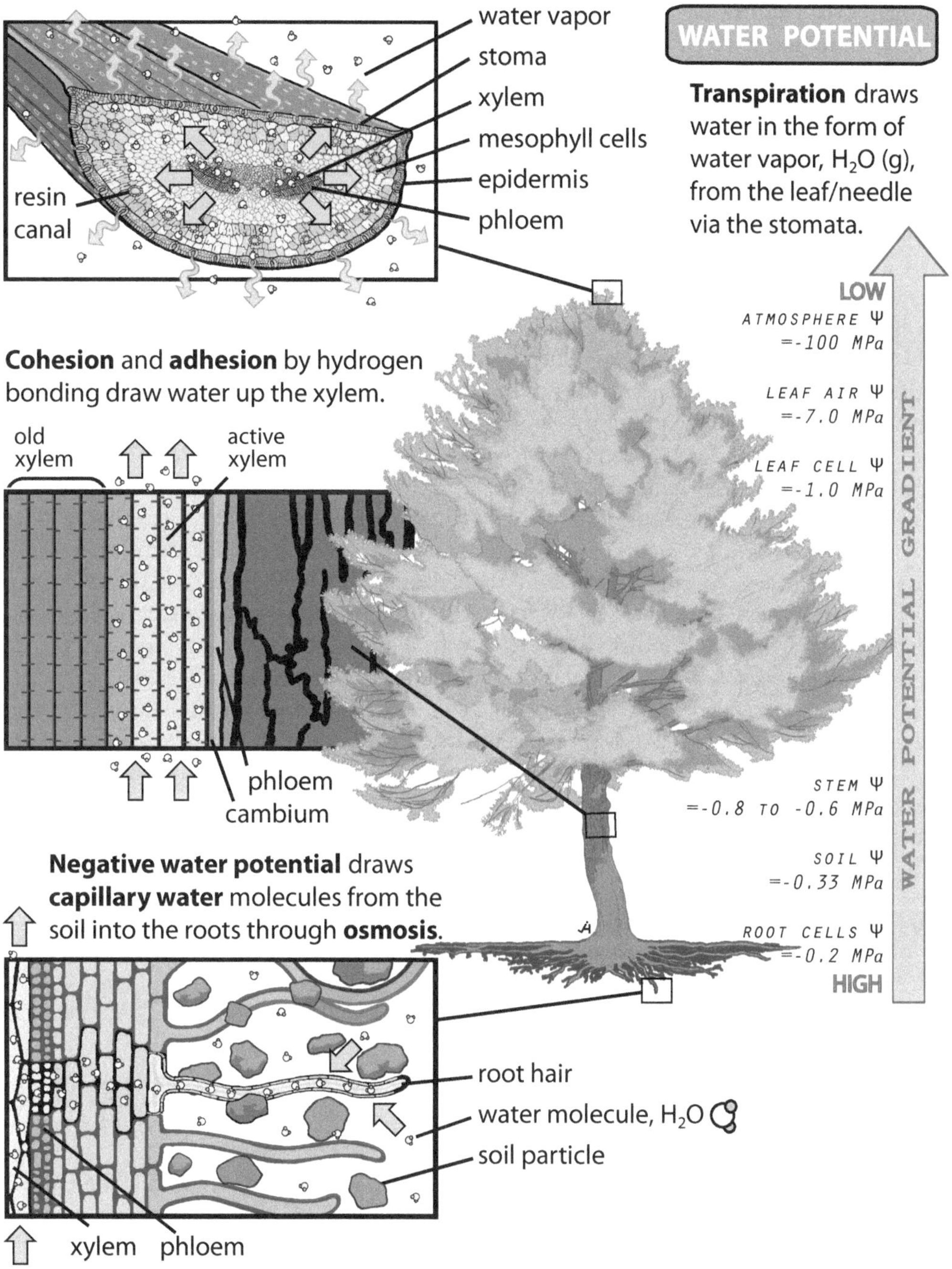

FIGURE 2.17 Water potential.

basswood is 11 ft/hr, and in a pine tree is 6 ft/hr. Trees can absorb between 10 and 150 gallons of water daily, yet of all the water absorbed by trees, less than 5% remains in the trees for growth. They rely on available water in the soil to "rehydrate" during the nighttime hours, replacing the water loss during the daytime hours.

This may seem like a negative event. However, this process is the very essence of mass flow because the evaporation of water from the leaves through the stomatal openings provides the pulling force that draws water and minerals up the tree from the soil. Simply stated, without transpiration (water loss) it wouldn't be possible to move water into the tree roots and throughout all parts of the tree system.

The second process is the interception of water by the surfaces of leaves, branches, and trunks during rainfall, and its following evaporation. Together, these two processes are often referred to as *evapotranspiration.* Evapotranspiration is a measure of total tree water use, including evaporation from the soil surface over the roots and transpiration from within the leaves. Both transpiration and evaporation are strongly affected by sunlight, temperature, humidity, and wind speed. As an example, if the relative humidity of the air surrounding the tree increases, transpiration rates decrease. It is easier for water to evaporate into dryer air than into more saturated air. Also, direct sunlight and increased wind will act to "draw" moisture out of the leaf, increasing the evapotranspiration rates or water loss. In certain conditions trees can release nearly 95% of the water they absorb.

Why is water so important for the tree? Well, nearly every tree process such as photosynthesis, respiration, and transpiration rely on water to function properly. Water is an essential nutrient and is as important, if not more important, than other nutrients. This is because water is required to put all our other elements into a form usable by the tree. Almost all essential elements are ionic forms dissolved in water, giving them the ability to move to stems, branches, and leaves for energy. The daily water requirement will depend on the leaf area of the tree and species, and how much water is lost through evapotranspiration. Mature landscape trees that have full crowns not excessively shaded or restricted by competing trees usually have a larger demand for water than similar trees in woodland. The goal of tree care is to avoid or reduce the negative impacts of water loss. If adequate soil moisture is available, water loss will go unnoticed as it is replaced naturally. During dry conditions, soil moisture content can be reduced to the point where tree roots can no longer pull the water molecules from the soil particles. This results in responses from the tree such as wilting, early fall color, scorching, and other symptoms.

WHY DO TREES DIE?

There is much speculation on why trees finally go past the point of no rescue treatment and die. How a tree dies may be described as a strategic physiological process. Despite decades of research on tree tolerance to environmental stressors, especially drought, the physiological mechanisms by which trees succumb are a continuous point of examination. Nonetheless, it could be argued the main reasons trees experience physiological deaths are carbon starvation and hydraulic failure, or simply a shortage of food and water.

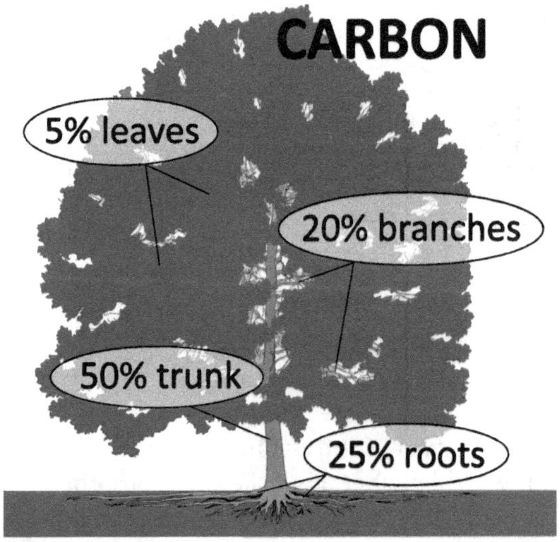

FIGURE 2.18 Carbon storage occurs in all parts of the tree.

Green plants rely on the process of photosynthesis to create sugar as energy. Sugar is broken down through the process of respiration, a key process for survival. It's a physiological process in which photosynthesis creates the energy source and relies on respiration to oxidize the carbohydrates into food. The tree then relies on translocation, which requires water and nutrient movement through the plant vascular system, allocating these resources throughout the plant.

FIGURE 2.19 Tree damage can lead to decline and pest infestation.

Carbon can be considered the currency for all plant life. When energy resources in the form of complex carbon molecules become exhausted, the tree can no longer support respiration to create energy. This can happen when the tree is depleted due to the inability to create and breakdown food products. If storage or sink areas which includes storage components, such as branches and stems are depleted of carbon storage as well, there are no energy reserves which the tree can draw upon. The result is decline. This situation can occur from several stressful growing seasons in which the tree cannot produce adequate and abundant energy resources which leads to something like an overdrawn account.

The process of respiration is constant, even in dormancy to some level. During the photosynthesis and respiration process, the production of carbohydrates must exceed its energy demands especially during the growing season. Without a surplus or storage of resources, decline begins when the process is imbalanced, and decline eventually follows, unless conditions improve. Proper maintenance and care means providing supplemental moisture and ensuring adequate nutrition to help maintain sustainable energy production, utilization, and reserves.

The next factor in why trees die can be attributed to hydraulic failure. This is the loss of conductivity or the inability to move water to the leaves and food in the form of carbohydrates throughout the tree. This means the tree cannot adequately translocate water and assimilates where needed in the tree. Water is an important component of many tree processes. PSN and RESP are no exception to this factor. The transport system must be capable of delivering water speedily and preferentially to those parts of the living crown which are most actively transpiring or losing water. Translocation is simply the movement of something from one place to another, in this case water and other soluble products. However, if that connective transport system is damaged to the point of ineffective water movement, system failure occurs. Tree components are subject to the impacts of environmental stress, especially temperature extremes and pests. This translocation of critical resources must occur, or the tree suffers the consequences.

WOUNDS AND RECOVERY

Throughout the life of a tree, wounds can occur such as simple chronic injuries from insects or birds to more complicated and acute harm like mechanical damage by man. When tree tissue is damaged or wounded, the newly uncovered cambium layer is exposed and that is when to expect damaging intruders. Injuries can result from pruning cuts, storm damage, animal foraging, temperature extremes, trunk abrasions, pest attacks, and people. Injury which opens up the cambium layer on the tree can attract pests. Insect pests are especially drawn to trees in distress, due to the phytochemicals dispersed from exposed tissue.

FIGURE 2.20 Decay results in strength loss within the tree.

Most diseases affecting trees are fungal organisms. These biotic agents of destruction introduce enzymes into tree cells, digesting the living tissue responsible for food and water translocation (phloem and xylem) or structural support resulting in unhealthy, unsightly, or unsafe trees.

After a tree has been injured, reactions are triggered prompting the tree to create boundaries or walls around the affected area. The boundary-setting process works to limit the loss of normal wood function and resist the spread of discoloration and decay. The process of wood decay is progressive. Basically, it begins with incipient decay because of an injury, then advances to more advanced stages of decay and fiber loss. Incipient decay may be classified as wood in the early stages of infection or invasion by a wood decay fungus, but the wood is not yet softened or broken down. This is proceeded by some discoloration due to the enzymatic breakdown of cell tissue. Discoloration is the response of the tree to microorganisms resulting in darkened wood but little strength loss. Decay is the complete breakdown of cellulose, lignin, and hemicellulose either alone or in combinations. The rate of the discoloration, decay, and tissue loss process depends on the severity of the wound, position in the tree, size of the wound, time of year, species, tree age, and the types of infecting microorganisms.

THE CODIT MODEL

Trees have a recovery system that can allow the tree to seal or heal the affected area. Compartmentalization of decay in trees (CODIT) is a physiological process which creates chemical and mechanical boundaries to resist wood decaying organisms such fungi. It creates a separation of healthy tissues and infected tissues by reaction and barrier zones. The traditional concept of wood decay in living trees was developed late in the 19th century by Robert Hartig, a German forester who pioneered modern forest pathology. The traditional view maintained that wood decay is caused by fungi entering through wounds in the bark. Once introduced into the tree, decay fungi will move unrestricted throughout the stem and older branches because wood is primarily dead wood fiber, with no active living tissue to help slow the progress. The current CODIT concept was relatively unknown by the arboricultural community until Dr. Alex L. Shigo, a tree pathologist with the United States Forest Service, and colleagues presented an expanded concept called CODIT. The current concept still maintains that wounds provide an entrance for wood decay organisms. Also, it considers tree response to wounding and infection where the affected tissues are compartmentalized or walled off from healthy tissue. Additionally, it is a model for arborists to easily understand the wound response strategy and provides a basis for better practices to improve tree survival and health. Effective compartmentalization preserves a tree's mechanical strength to help reduce the likelihood of failure. More currently, some practitioners use CODIT to indicate the compartmentalization of damage in trees, instead of decay. The concept of damage is considered more comprehensive than decay and includes all physiological changes after injury involved in the wound recovery process.

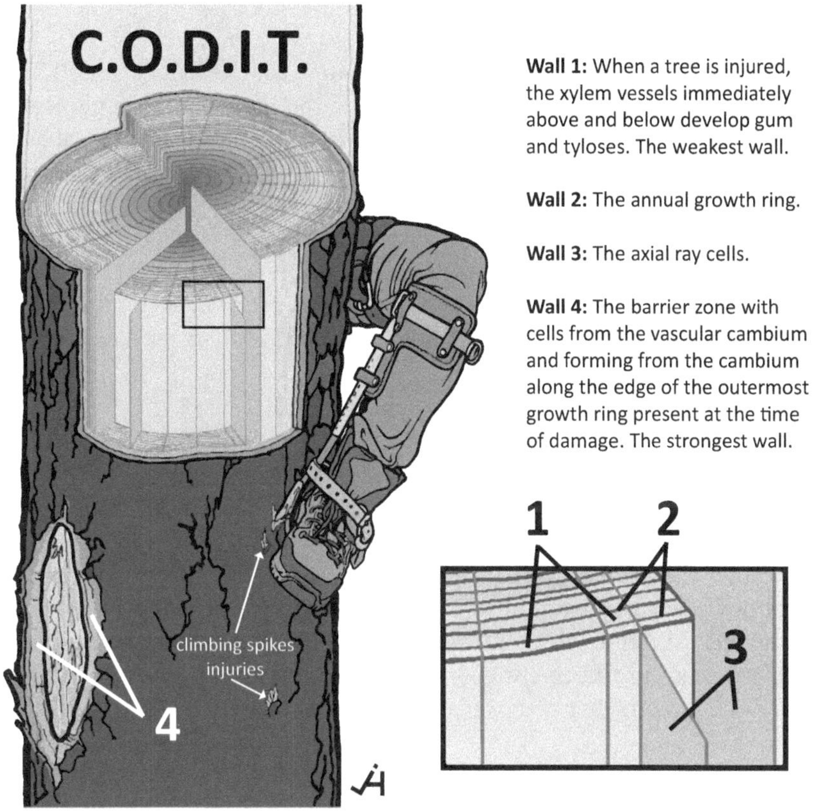

FIGURE 2.21 The CODIT model.

Tree wounds experienced during the growing season will hopefully stimulate the formation of callus tissue which develops at the injured area or parts of it. Callus tissue is an unorganized group of essential parenchyma cells. As the callus develops and grows, wound wood develops which is capable of covering the exposed injured area. Wound recovery rates vary widely for each tree and tree species. The speed of healing is greatly affected by the developmental age of the tree, environmental conditions, and vigor and health at the time of injury. Some trees may never completely close their wounds due to genetic capacity or inadequate resources. Tree wounds may close more slowly from poor pruning cuts. Other species may completely occlude a pruning wound in less than a year. Numerous studies reveal faster wound closure results in less decay and dysfunction in the tree.

FIGURE 2.22 Poor pruning cuts slow wound occlusion.

THE BOUNDARY WALLS

In Shigo's model, there are four walls or boundaries which exhibit specific responses to the invading organism. These are Wall 1, Wall 2, Wall 3, and Wall 4. Each form in a different manner and protect the tree in unique ways. The walls are numbered in increasing order of their successful ability to slow the movement of decay within the tree. For example, Wall 3 is stronger than Wall 1. Wall 1 may or may not be present at the time of wounding. Walls 2 and 3 are always present in the tree. Wall 4 forms in response to injury, outwardly.

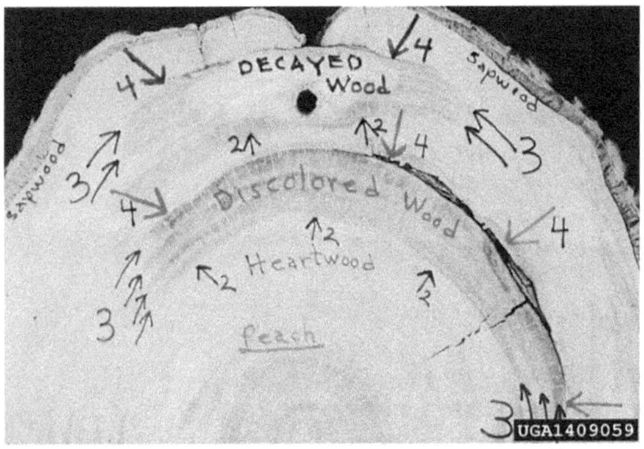

FIGURE 2.23 Boundary walls visible on this wood section.

Wall 1

When a tree is injured, the xylem vessels immediately above and below the injury develop gums and tyloses, forming Wall 1. It is considered a chemical barrier forming vascular obstructions which creates a weak boundary in some trees but is stronger in other. Because Wall 1 is weak, decay in some trees can advance rapidly up and down the trunk or along a branch from the injury resulting in long columns of decay and hollow branches and trunks. This wall is the weakest of the four walls.

Wall 2

The annual growth rings make up Wall 2. The transition from one growth ring to the next slows advancing decay organisms inward. Decay organisms often have a tougher time moving across growth rings spreading inward than they do through the vessels (Wall 1). The function of Wall 2 is considered a physical barrier and obvious when examining the cross section of a trunk or branch that was injured previously. A darkened area inside towards the heartwood often appears where it is attempting to slow advancement inward toward the center of the tree or branch at the boundary of one growth ring with the next ring.

Wall 3

This wall is identified by the axial ray cells that appear as comparable to spokes on a wheel. Rays are occupied with active parenchyma cells and are a vital component in resisting the lateral spread of decay. They have decay fighting capabilities, rich in stored energy. Discoloration and decay are increasingly obstructed moving across Wall 3 compared to Walls 1 and 2. The strength of Wall 3 can be recognized by viewing a cross section of a trunk or branch with prior injury. There is a clear delineation between darkened tissue and normal light-colored wood moving around the cross section in a circular direction. If Walls 2 and 3 fail and decay organisms break through, the affected trunk or branch can become hollow more quickly.

CODIT Walls 1, 2, and 3 are part of the reaction zone in the injured area. They are composed of existing wood undergoing changes in defense of the advancing pathogen moving through the wood fiber. The reaction zone is the beginning stage of compartmentalization, followed by initialization of the barrier zone, which includes the fourth and strongest boundary zone, Wall 4.

Wall 4

This wall is the barrier zone, containing cells from the vascular cambium and forming from the cambium along the edge of the outermost growth ring present at the time of damage. It begins at the point where the tree was injured, and it may extend to all or part way around the tree. Wall 4 is a structural barrier of new wood tissue separating the wood formed before the wounding from the wood formed after the wounding. It is the most resilient barrier preventing the spread of infection to the newly formed wound wood. This barrier zone is vital to continued growth and survival of the tree. Typically, Wall 4 is highly lignified with robust parenchyma cells present and suberized cell walls. The vascular cambium is protected by the compartmentalization process allowing the creation of new cells in the barrier zone.

Wall 4 remains in the same position within the tree but may extend further around or up and down the trunk over time. It does not move outward with the newly developed cambium layer and may eventually become encased by the subsequent growth rings as the tree grows and if the tree remains healthy. The benefit of Wall 4 is decelerating outward movement of decay; but there are two important weaknesses. Stored starch can get trapped in the rays located inside of Wall 4 and become available to feed decay organisms, keeping them active. The other issue is if a crack forms along Wall 4 by separation of the fibers, called a ring crack and may follow Wall 4 all or part way around the trunk. One or more secondary cracks, called radial cracks, can form from the ring crack along a ray. Another serious crack is the closure crack that occurs as the callus and wound wood attempt to grow over and close the wound. This crack often extends from the point of injury out to the current location of the bark. Sometimes this crack never closes. Even if it closes, the crack remains along with its associated weakness and can increase likelihood of failure.

FIGURE 2.24 Wall 4 developing to close the pruning wound.

Walls 1, 2, and 3 are preexisting walls, called the reaction zone. Wall 4 is formed after wounding with newly formed highly lignified cells with fungistatic properties called the reaction zone. The living parenchyma cells play a critical role in all four walls in relation to the CODIT model.

CODIT VARIABLES

The ability of trees to compartmentalize varies with each tree species. CODIT characteristics are weak in trees considered poor compartmentalizers of decay. Examples of weaker compartmentalizers can be found on trees that develop hollow trunks. Trees with less decay and smaller cavities exhibit stronger boundary formations and are considered good compartmentalizers. The CODIT process is dependent on key factors including available resources, developmental age, and growing conditions. A healthy tree, which is considered a good compartmentalizer, will generally seal wounds faster, a trait often observed in younger trees as well. Trees that are planted in quality soils, containing

TABLE 2.1 CODIT Boundary Activity

CODIT Boundary Activity	
Wall 1	resists spread up the vascular system
Wall 2	resists spread to the heartwood
Wall 3	resists spread around the growth rings
Wall 4	prevents decay moving into new wood

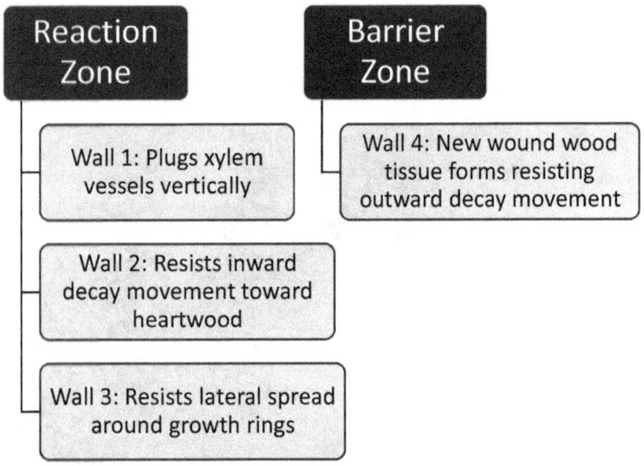

FIGURE 2.25 Zone characteristics.

adequate nutrition levels, grow in a way that favors the healing process. The tree must have adequate energy resources to fund the CODIT process, developing these internal changes. The trees' ability to recover from injuries will also depend on its development age which is primarily a function of its vitality. The term vitality typically refers to the response of the tree to environmental stressors. Tree vigor is more accurately used as a determinant based on the annual growth of wood within the tree, and overall visual assessment of crown and stem. Tree vitality and vigor share many attributes but are not interchangeable. Vitality is more about tree survival and outcomes of vigor from a genetic perspective as well. Finally, adequate water and nutrition are important to facilitate the processes which supply the resources which support the CODIT process such as photosynthesis and respiration. Any shortage or shortcomings of these factors reduces the rate of recovery. Trees will heal, occluding a wounded area if they have the time, ability, and resources to do so. Predicting CODIT potential is challenging because of the inherent variability.

There are few ways wound closure can be hastened and less inhibited. First, it is essential to avoid limiting oxygen availability to the wounded tissues. Oxygen is necessary for

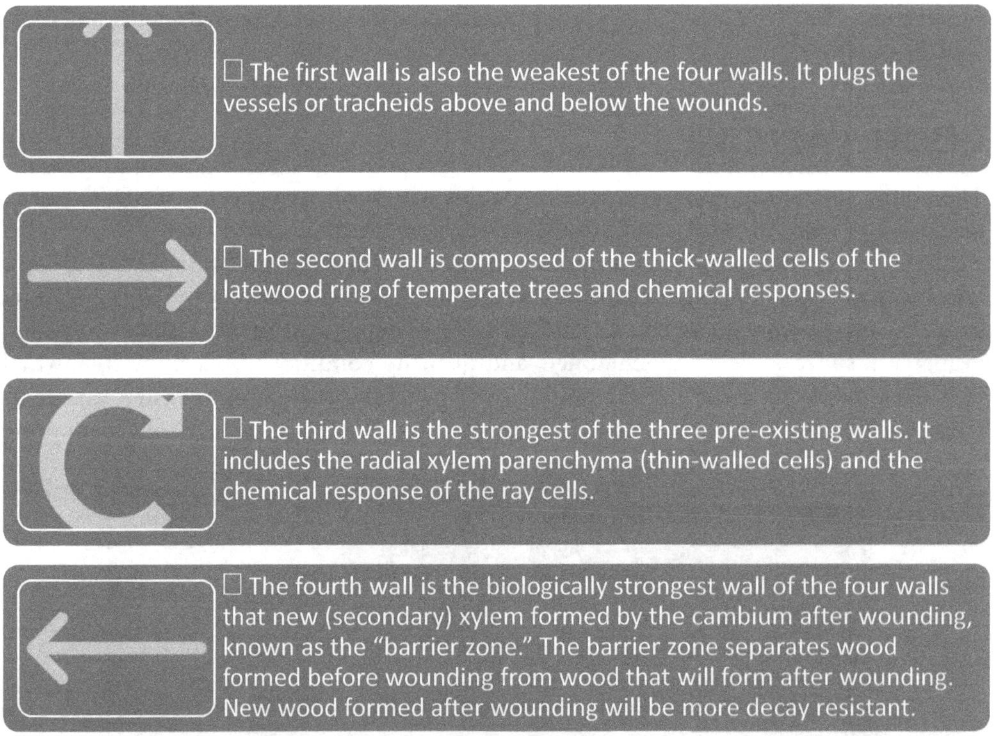

FIGURE 2.26 Mode of action in the CODIT process.

proper recovery. For example, painting a wound with any kind of material that interferes or impedes oxygen exposure will slow or even prevent wound closure with poor callus formation. Wound treatment with petroleum-based products is prohibited by best management practices and arboricultural standards. In fact, research indicates any type of wound dressing can essentially slow the healing process. There is one exception for the application of any covering for treating wounds. This is in areas where infectious disease occurs such as oak wilt. Wound paints which are water based may be useful in preventing insect spread of the oak wilt fungal pathogen through vectors.

SEASONAL INFLUENCES ON TREE FUNCTIONS

Trees must acclimate to survive the cold, frozen conditions of winter; they must rely on physiological and structural adaptations for their existence. The increasing height of mature trees is critical for getting those leaves up to light absorbing elevations, but it can be a liability in the winter, as live, green tissues are exposed to the weather. There are some strategic moves trees use to overcome and overwinter until the warmer growing season appears.

DORMANCY AND FOLIAGE

Deciduous trees shed their leaves when daylength shortens and temperatures begin to fall. This foliage shedding avoids the problem of maintaining live tissue when water is unavailable to the tree. Coniferous trees, needled evergreens, retain foliage and have special adaptations to protect themselves from drying out, or desiccation. Early spring growth, when buds swell and begin to break, requires tremendous amounts of water and nutrients. Trees must produce and store a sufficient amount of energy reserves during the growing season to meet this coming demand. In this effort, most growth slows and ceases in this preparation mode.

The shorter days modify the production rates of phytochemicals especially hormones. The most prevalent is the increase in abscisic acid (AA). This special chemical messenger does a couple of things. It slows protein and ribonucleic acid (RNA) production, which are critical in cell growth and increasing cell membrane permeability, which is important in the acclimation process.

Deciduous tree species drop their leaves at various times and are influenced by environmental, developmental age, and physical conditions. Temperate trees include a large number of woody species that shed their leaves in autumn and are completely leafless during winter months. Leaf senescence in deciduous trees is a complex process typically involving the physiological formation of an abscission layer causing the petiole to mechanically detach from the stem. This natural trigger mechanism creates abscission layers which form between the leaf petiole and the tree branch. Cells along this line expand at different rates, and enzymes break down these cells into scar tissue. The rigid, tough area observed at the base of a fallen leaf is the abscission zone. When tree cells are abandoned, they leave behind *suberin*, from the Latin word for cork. It's a conversion of living cells into a form of protective sealant. This allows the tree to drop their leaves to maintain the overall tree moisture inside or to prevent the area on the tree from drying out. They pull in all the nutrients they can from the leaves and cut the leaves off from their stems.

There are several deciduous species which actually retain senesced leaves on their stems and branches through much of the winter. This incidence is called marcescence. Marcescent leaves are a little researched and less understood part of the dormancy process.

The strategy has been explained by botanists with several proposed hypotheses, including:

- marcescence has no adaptive function but rather, an evolutionary byproduct;
- marcescent leaves deter winter browsing herbivores;
- leaf retention through winter improves nutrient resorption during autumn senescence;
- prolonged leaf shedding into spring minimizes nutrient leaching and promotes decomposition;
- marcescent leaves protect overwintering buds from frost or desiccation as well as smooth bark trees from frost cracking; and
- marcescent canopies provide winter cover for animals (including insects, birds, bats), thereby affecting tree nutrient availability via excrement.

This phenomena is common and easily identified during the winter, but much remains to be understood on the physiology, evolution, function, and ecological implications of leaf marcescence.

Prior to the leaf shedding process, the autumn season welcomes a pleasurable change in the weather and temperate trees. But why do leaves change color when fall weather appears rather than just drop? Once again, it is about daylength and temperature. The important thing is not that the amount of sunlight has *decreased* but rather the amount of dark has *increased.*

Dormancy begins by receiving signals on the length of nights, which is a response called *photoperiodism*, a Greek derivative for light and time. As the amount of light per day goes down, plants produce ethylene and abscisic acid leading to a change in the ratio of auxin and cytokinin. Plant biologists define *auxin* as the hormone that induces cells to elongate or grow, and *cytokinin* is involved in cell division.

The process to discontinue chlorophyll production for the winter is a chemically complicated biochemical process. When there are less daylight hours and temperatures become cooler, photosynthesis slows down and there is less chlorophyll production. This reduction reveals yellow and orange, pigments called carotenoids, which are usually hidden by the abundance of chlorophyll present in leaves during the growing season. Unlike chlorophyll and carotenoids, which are present in leaf cells throughout the growing season, anthocyanins are produced mainly in the fall. This natural chemical gives color to familiar fruits such as cranberries, red apples, cherries, and plums. The complex compounds in leaf cells react with excess stored plant sugars and exposure to sunlight creating vivid pink, red, and purple leaves. A mixture of red anthocyanin pigment and beta carotene often results in the bright orange color seen in the leaves of some tree species. Also, bright sunlight striking the leaves allows us to see a variety of leaf colors throughout the year. Leaf pigments play a crucial role in the range of colors seen by the human eye. Soil moisture levels have an impact on the ability to produce good fall color as well. A summer drought can delay color change for a few weeks as well as reduce foliage intensity. The ideal conditions for producing the best colors are good summer weather, with timely rainfall, and sunny fall days with cool night temperatures.

Conifers or needled evergreens also drop some of, but not all of, their "leaves." Most retain needles for 2 to 3 years before shedding them. Evergreen needles have a waxy layer on the epidermal cells called cutin, which reduces water loss. Needles also have much tighter stomatal closures which help prevents desiccation. This retention of needles allows for the extension of the photosynthetic season and taking advantage of winter thaws and permitting slow rates of photosynthesis during cold weather. However, this can also compromise survival in terms of water loss and winter damage from desiccation.

IMPORTANCE OF WINTER WATER RESOURCES

Winter conditions make finding moisture and the translocation of nutrients and water a challenge. Trees need water to survive, and keeping tissue and cells hydrated is critical. One of the main types of plant damage and loss is from winter desiccation, especially on evergreens. However, trees are resourceful, finding ways to obtain water either from the ground, within the tree, or from the subnivean zone. Above-ground tree components such as twigs and leaf buds are watertight and have a reduced water demand and less moisture loss. Conifers, however, have a much larger surface area of living green needles. Any photosynthesis that might occur will increase water demands and risk of damage. Water vapor moves from areas of high concentration to low concentration. These concentrations are usually higher inside needles, so the tendency is for water to be lost from foliage; however, conifers have advanced structures creating somewhat of a barrier to evapotranspiration.

Warm, sunny days during the winter months present the greatest challenges for evergreen foliage respective to water loss. Dark coloration of leaves and needles readily absorbs heat and raises needle temperatures significantly above ambient air temperatures. Metabolic rates rise and internal vapor pressure increases. Despite thicker air boundary layers, the net effect is greater water loss and potential desiccation. Cloudy, breezy days are better for conifers. Clouds block warming solar energy and wind readily removes heat from the needles, reducing the vapor pressure gradient.

Deciduous trees need some supply of water, even in the winter season. There are three potential sources of water, which includes the soil, internal tree reservoirs, and subnivean vapor absorption. The insulating effects of snow can result in ground thaws or prevent freezing in the first place. This means that the liquid form of water is available. The sapwood of trees and branches contains water. Often this water is frozen and unavailable. However, solar radiation energy and winter thaws can melt the sapwood water, making it available for transport in the tree.

Conifers have some fascinating adaptations that overcome the problem of broken water columns. Within the transport tubes, conifers have tiny "check valves" between each tracheid. The tiny check valves between each tracheid in conifers are these tori. When ice forms and volume expansion increase pressure within a water column, the "float" within the check valve, called a "torus," seals the ends of each tracheid making up a tube. This mechanism helps to prevent the spread of air bubbles (embolisms) that can disrupt water transport, thereby enhancing the tree's survival under drought conditions, winter and summer.

WINTER ACCLIMATION

Trees don't die during the cold; they simply go into a dormant phase, much like animals hibernate in the winter and undergo changes that allow them to survive the harsh weather. However, all trees do have a "killing temperature," where ice crystals form within cells, resulting in cell death. Killing temperatures vary among different species between populations of the same species, and even among different tissues. For example, root killing temperatures in moist soils can range between 0 and 24 °F. Some species of spruce and withstand below 0 °F temperatures. Many variables influence these thresholds such as stage of dormancy and developmental age of the tree.

Plant acclimation is prompted by changes in the photoperiod, just like leaf drop. Abscisic acid (AA), once again, plays a key role. Physiological changes include an increase in AA, and lipids (soluble fats). Also, proteins change as they depolymerize and cell membrane permeability increases. Solute concentrations within cells increase, slightly lowering the internal freezing point. Therefore, as temperatures drop, water outside cells freezes first. When freezing occurs, a specific heat is created, and small amounts of energy help internal cell fluids remain unfrozen. Twig temperatures actually rise several degrees during this process. Water moves out of cells, attracted to the ice crystals in the pore spaces. This process effectively reduces the freezing point of cell water to the killing temperature.

OTHER TREE SURVIVAL CHALLENGES

There are other challenges for trees besides the physiological issues each winter. Snow accumulation and ice storms can be a major issue, especially for evergreen species. To offset this weight-loading problem, conifers grow in an excurrent shape and display alternative growth and branching patterns, including unique angled branching patterns to help shed snow. Also, conifers possess longer wood fibers, which provides more elasticity. It doesn't help the tree when attempting to remove snow or ice from trees, allowing it to dissipate naturally.

Many of our furry forest mammals such as deer, mice, and other rodents find sustenance in the living bark of trees. Squirrels will often seek moisture and nesting materials from the soft bark of some trees, completely removing the bark in some cases. If the bark is chewed and removed all the way around the stem, the girdling can kill the tree. Also, heavy browsing of the flower buds and shoots of deciduous trees can cause severe damage to young trees.

The final winter challenge for trees is caused by humans. Road salt, thawing chemicals, and vehicle exhaust can be damaging and lethal to trees, especially evergreen species. Some plants are more resistant to these pollutants and will survive the pollutants. The more vulnerable species include white pine, red pine, hemlock, basswood, ironwood, sugar maple, and red maple.

THE ARRIVAL OF SPRING: BUD RELEASE BREAKING DORMANCY

Dormancy involves a temporary suspension of growth. To establish dormancy before the winter season or colder weather arrives, trees must sense and process regular and reliable seasonal environmental signals. The complicated and often confusing part of this tree process is that the same temperature routine that induces dormancy in these species also controls its release. number of chill units has been accumulated, heat units speed up budbreak.

Growing points for leaves and flowers or buds are typically present in the tree throughout the year. However, in the cold months, trees make that bud dormant as well. The chemically induced hardened bud protects the growing tissues underneath from the damaging temperatures. They lie dormant until chilling requirements are met and begin releasing at the proper time to reduce the potential for cold damage. The process, known as a "cold requirement" or "chilling requirement," is critical for keeping buds dormant until the proper timing for bud break, which involves a minimum number of days of cold temperature. After this critical number of cold days has passed, and provided the trees are warmed to an adequate temperature, they will break or expand their buds in the spring. Both factors of cold days and adequate warming temperatures are essential for the process.

This explains why trees from lower, warmer latitudes don't do well in cold climates. Their need for cold temperature or reduced water can be quickly fulfilled on a very warm, sunny day in late winter, leading to premature budbreak. Subsequent drops in temperature will result in winter injury.

This occasionally occurs in regions with lower latitudes, where trees have their minimum requirement for cold days fulfilled early in the dormancy phase, and temperatures rise relatively early in the spring. The trees then release their buds, which begin to swell and break, creating tender, new leaves. These newly formed leaves are quite vulnerable to lower temperatures, which can result in catastrophic tree losses. This vulnerability arises from the fulfillment of the cold requirement, followed by a warm period and then a late spring frost.

The information provided in this chapter is based on the relevance and application of the science to arboricultural practices. There is an infinite amount of tree biology and physiology available and noted in the references, however, not always directly applicable to common practice. The selected topics discussed here are considered frequent and common areas of interest for the working arborist.

HOW OLD IS MY TREE?

Everyone loves big, veteran trees. They have a certain distinction and royalty in our forests and landscapes. The first question asked is, "I wonder how old that tree is?" Determination of the actual age of a standing tree is difficult. There are several methods and techniques. Some are fairly accurate and some are just good guesses.

If you know when the tree was planted and the age of the tree at the time of planting, obviously, you can easily and accurately determine its age. Most trees are 5–10 years old when they come out of the nursery, so add that to the time planted. The second most accurate way to estimate tree age is to count the annual rings of wood growth.

Annual rings can be counted using an increment borer. An increment borer is a tool used to extract a core sample from a tree, which can then be used to determine the tree's age by counting the annual growth rings. To use the borer for this purpose:

- Determine the radius: Determine the radius of the tree (which is half of the diameter). This will help you know how far to insert the increment borer into the trunk, the halfway point.
- Insert the borer: Insert the increment borer into the trunk at standard height which is typically 4.5 feet above the ground.
- Rotate the borer: Turn the handle of the increment borer clockwise into the tree. Be sure the specialized bit is long enough to penetrate to the halfway point of the diameter.
- Extract the core: Once the bit has reached the halfway through the trunk's diameter, use the extractor to retrieve the core sample.
- Count the rings: The resulting core sample extracted from the tree reveals dark and light increments which are the tree's annual rings. The number of rings indicate the age of the tree.

There are detailed instructions and procedures for accurate and effective core sampling. Be sure to understand the implications of the process. This includes proper care and use of the instrument. Also, this process does wound the tree which can introduce decay into the tree.

FIGURE 2.27 An increment borer can be used to determine tree age.

In the absence of technical tools used by a forester, an educated guess can be made with the following method. By following these easy steps, you can get a rough estimate of a live, standing tree's age, without knowing when the tree was planted.

Step 1: Measure the circumference (c) of the tree trunk using a measuring tape that measures in feet and inches. This should be done at 4.5 feet above the ground or slightly below shoulder height. This is known to arborists as the diameter at breast height (DBH).

Example:

A white oak tree measures 5 feet, 10 inches in circumference.
c = 5 feet, 10 inches (70 inches)

Step 2: Calculate the diameter (d). Divide the circumference by 3.14, a constant known as "pi." Formula: c/3.14 = d. For the white oak tree: 70 inches/3.14 = 22 inches DBH.
Step 3: Multiply the diameter of the tree by the *growth factor* as determined by species.

This is where we have so many variables that affect the accuracy of our answer. Growth factor tables assume a consistent or linear relationship of diameter increment to years of growth. This assumes little variation in the many problems which can affect tree growth.

Growth factors are variable and relative to their growing environment, genetics, and available resources. The factors provided are generally speaking "ideal" growing conditions. Growth factor tables assume a consistent linear or straight-line relationship of diameter increment to years of growth. The tables assume small year-to-year variation will average out both within an individual tree and among groups of trees as well. Also, woodland trees and urban trees grow quite differently. Trees in our neighborhoods, along streets and in parks, are often under more stress and grow more slowly. Natural woodland trees on undisturbed sites with less pressure may grow more rapidly. Also, trees growing in a woodland environment typically have a restricted crown and therefore increase in circumference at about half of the rate of full, open-grown trees found in a park or residential landscape. History tells us there are very few trees much older than about 250 years in the Midwest due to the early settlers clearing our forests for farms. Tree growth rates are affected tremendously by conditions such as water availability, climate, soil conditions, root stress, competition for light, and overall plant vigor. Further, the growth rates of species within genera can vary significantly. A white oak growing in a moist, well drained site will grow faster and be younger than a similar white oak in a dry, stressed tree lawn. So, only use this formula as a very rough estimate of a tree's age. However, in my experience, many trees in urban areas are a factor of approximately 1.8–2.75 years per inch at DBH.

Back to our white oak tree. We determined that the tree has a 22-inch DBH, so you would then multiply it by the growth factor of 5.0 (refer to table below), and our answer is 110 years old. It may be more accurate to apply my "urban forest factor" of deducting 25%–50% from the age calculation. This is an anecdotal deduction based on experience with aging trees in different environmental situations. If this factor is applied, the tree is aged at about 83 years. So, somewhere in that range would be a good guess.

Formula for approximate tree age: DBH x growth factor
DBH = c/3.14

TABLE 2.2 Growth Factor Suggestions for Various Tree Species

Tree Species	Growth Factor	Tree Species	Growth Factor
Red Maple	4.5	White Oak	5.0
Silver Maple	3.0	Red Oak	4.0
Sugar Maple	5.0	Pin Oak	3.0
River Birch	3.5	Linden or Basswood	3.0
White Birch	5.0	American Elm	4.0
Shagbark Hickory	7.5	Ironwood	7.0
Green Ash	4.0	Cottonwood	2.0
Black Walnut	4.5	Dogwood	7.0
Black Cherry	5.0	Redbud	7.0

FIGURE 2.28 Arborists use a diameter tape (d-tape) to accurately determine tree diameter

CHAPTER

3 THE BASIC NEEDS OF TREES

LEARNING OBJECTIVES

- Distinguish between soil texture and soil structure and their impact on soil characteristics.
- Recognize the importance of soil chemistry including pH, cation exchange capacity (CEC), and buffering capacity of soils and the impacts on tree health care and nutrition.
- Define bulk density and how it is affects tree health and growth.
- Explain the importance of water on absorption of essential elements and movement of phytochemicals.
- Describe the connection between gravitational water, field capacity, permanent wilting point, and infiltration rate.
- Describe the principles of water uptake, transpiration, and the movement of water in the soil.
- Recognize drought stress in trees and how trees manage water deficits.

Trees have basic needs to be healthy, stable, and sustainable. These include air, water, soil, nutrients, and space. Of course, these are very generalized and simply stated; however, just as humans have a hierarchy of needs, so do trees. The challenge that arborists, tree managers, and owners have is to balance and maintain these basic needs in the urban environment. Typically, if there is an issue with one or more of these basic needs, stress develops and may continue to decline unless interrupted. Tree life requires the critical elements of air and water in adequate amounts and quality. Space is reference to adequate room to grow and expand into its location both aboveground and belowground. The trees and plants cultivated in urban landscapes also require suitable soil to

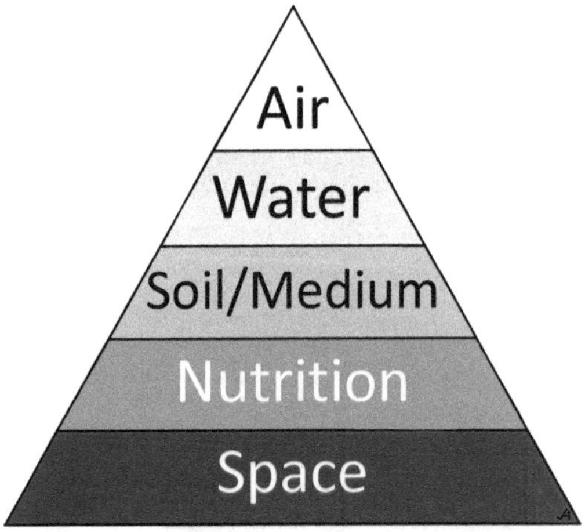

FIGURE 3.1 The basic needs of a tree.

sustain growth. The quality of soil lies primarily on its consistency and composition as well as its ability to supply nutritional necessities. All these basic needs play a part in the success of any tree community or system. In order to manage trees, the arborist must be capable of investigating, analyzing, and evaluating all of these conditions and act with appropriate interventions.

URBAN SOILS

Soil, along with water and air, provides the basis for plant life and human existence. As a dynamic natural body composed of solids, liquids, and gases on the Earth's surface, soil contains living matter and is capable of supporting plant life. It has either formed in place, naturally, or has been transferred to its present location by wind, water, ice, gravity. Now more commonly, delivered to the site by truck or other conveyance as recent as an hour ago, a week ago, or several months ago. Most soils we encounter as arborists and tree care professionals are not endemic to the site. They are different primarily from natural soils formed organically from deposits and the underlying bedrock weathering process. The "unnatural" soils occur from the mixing, filling, or contamination of land surface in urban and suburban areas. These soils can be any combination of native soils along with construction debris, solid waste, or other fillers. These facts illustrate why soils are very simple by nature, but quite complex in the built environment. Urban soils are an ecosystem unlike any other because they are dynamic, fragmented, highly pressured, and constantly under siege. There is continual destruction and disruption to the soil structure as opposed to a more natural, wooded ecosystem which remains relatively undisturbed.

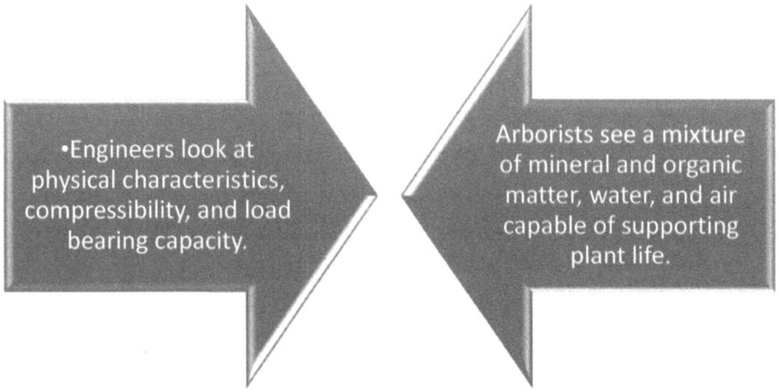

FIGURE 3.2 differences between needs of engineers and arborists.

SOIL CHARACTERISTICS

The definition of soil can be different depending on perspective and application. Urban soils are created by the truckload, literally, with tremendous variation in each load oftentimes. New development typically necessitates removing, relocating, and adding soil for engineering purposes. However, there is a distinctive disparity between mechanical and biological needs.

There is an antagonistic situation between the needs of engineers and arborists. Soil preparation depends on the application. Developers, engineers, and builders look at soil as a material with physical characteristics important for compressibility and load-bearing capacity. Whereas arborists and landscapers see soil as a consolidated cover of mineral and organic compounds, water, and air capable of supporting plant life. Soil management on a landscaped site couldn't be any more disparate in the built environment. Comprehensive tree care includes managing the roots which entails managing the belowground environment.

FIGURE 3.3 Construction leads to compaction without precautions.

PHYSICAL SOIL PROPERTIES

All soil is made up of air, water, organic matter (numerous kinds of living or dead organisms), and mineral matter (sand, silt, and clay). In the built environment, where there are businesses, buildings, parking lots, and other densely populated areas, the amount of each of these soil components varies greatly regionally, locally, and even on the same site. Also, the type and texture of soil present on a site can vary dramatically within distances of only a few feet in the same landscape. Soils form through a group of processes. The soil-forming factors are parent material, climate, living and dead organisms, time, and landscape position. When all soil-forming factors are similar, a similar soil is produced. If we change one or more of the soil-forming factors significantly, then a different soil is produced.

The soil profile in mostly undisturbed locations consists of different layers or horizons that form as the result of the soil-forming factors. The major soil horizons arborists and landscape professionals encounter are designated as O, A, B, and C horizons. All these horizons, also known as layers, are not always evident in every soil profile. For example, most agricultural soils or urban soils do not have an O horizon because organic horizons are usually mixed with A horizons during tillage or other mechanical disturbances. Occasionally, in historic residential districts of cities, an arborist may find areas that have revitalized a visible O layer over the course of many years. These are usually small areas or specific properties previously untouched or properly managed by arborists or horticulturalists for decades.

O Horizons

When present in unmanipulated soils, O horizons are the uppermost layers and form on top of A horizons where they occur. They are formed by the accumulation of fresh and decaying plant parts, such as leaves grass, needles, and woody debris. O horizons are dark colored (mainly black or brown) due to the production of humus by decomposing plant and animal materials. They are much more prevalent and prolific in forested or wet areas. Organic

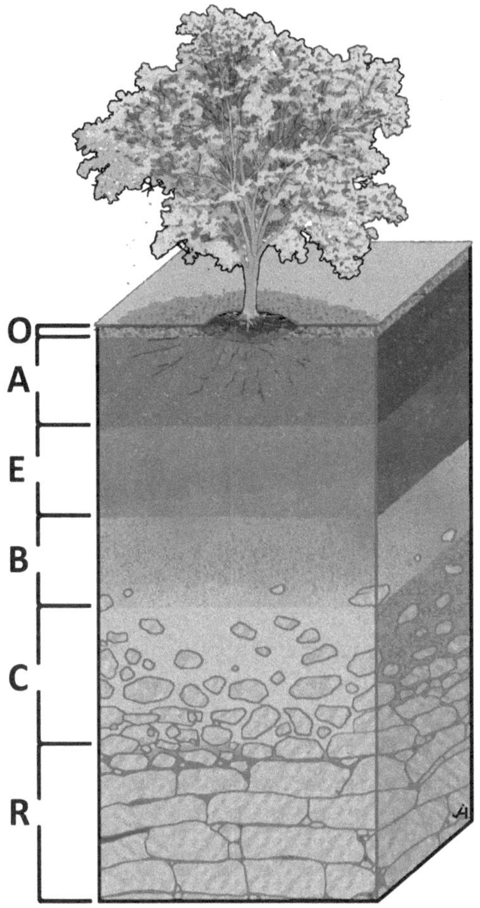

FIGURE 3.4 Typical soil profile on undisturbed sites.

matter is considered by many soil scientists as the most critical component of soil health. Decomposition of the various plant, animal, and microbial tissues provides significant benefits to trees and plants of all kinds.

Organic Soil Matter

Organic soil matter (OM) is comprised of mostly humus, stabilized organic matter, and actively decomposing organic matter with a small fraction of living organisms and microbial activity. Some of the key functions of OM include improved soil tilth. This term refers to the physical condition of soil for growing plants. The factors that determine soil tilth include water drainage, rate of water infiltration, aeration, moisture content, stability, and the formation of the aggregated particles in the soil. So, organic matter acts as the "glue" which allows the formation of soil aggregates. Also, OM absorbs and holds nutrients and soil water in place for plants, is a source of plant nutrients, and acts as energy and food for the living soil organisms. Organic matter could figuratively be considered the engine that drives plant and soil life. Without the process of decomposition into organic matter and nutrient cycling, soils would not have the ability to continue feeding plant life long-term.

A Horizons

The upper layer of the soil profile is the A horizon, below any organic layer present, and is made up mostly of mineral material. The A horizon can be rich in organic matter, especially if it has been manipulated during construction or other disturbances. Typically, it is a loose, dynamic mixture of bacteria, fungi, and even small animals and invertebrates in the soil where roots are found. Often referred to as topsoil, this is the location where most plants find nutrients, air, and water for survival. The A horizon is usually darker in color than the other layers of soil and is typically 2–8 inches deep.

B Horizons

The next layer down is called the B horizon, representing the subsoil. Subsoil is the layer of soil immediately underneath the surface topsoil. It is the weathered soil layer extending between the natural topsoil and the unweathered parent material below. Typically, subsoil consists of the same variable mixture of minerals and small particles (e.g., sand, silt, clay) as the topsoil, but it has a much lower percentage of organic matter and humus (fine organic matter derived from the decomposition of plant and animal substances). Subsoil tends to be paler than its overlying topsoil due to the lack of dark humus and varies in color from red to yellow. Very little living matter or organic matter is present in this layer. Few, if any, roots actually penetrate this level due to a lack of oxygen and impervious qualities; however, some trees may find additional physical support in this deeper layer. Nutrients are often deficient in this horizon as well due to a lack of soluble mineral elements required for health growth. Poor aeration and moisture content is common in this less friable, hardpan soil layer.

It is important for the tree care professional to make distinctions in the A and B horizons for plant health care reasons. Most site development activity leaves conditions lacking topsoil and finding primarily the B horizon soil layer. Characteristics of soil in the built environment depend largely on how deep the site has been excavated during development and if new materials were brought in and mixed with the original soil constituents. Many times, topsoil is removed from the site prior to construction and may or may not be returned to the site. After excavation, subsoil may be placed as fill over topsoil. Changing the order of the soil layers or mixing the topsoil and subsoil can alter soil properties greatly. These variables make predicting soil behavior such as pH and CEC values difficult in tree planting areas. Unfortunately, in most urban and peri-urban locations, the B horizon or a fragmented mixture is what is available for a growing environment, leaving a challenging, damaged soil situation to grow trees. Since nutrient availability to plant roots relies heavily on soil microbes, the lack of both diversity and total populations of soil microbes reduces the amount of available nutrients naturally, and the ability to successfully fertilize trees with synthetic nutrients via fertilization.

The last horizon encountered is the C layer or horizon, which consist of primarily parent material and bedrock, from where the upper horizons develop. This layer of earth material is often composed of unconsolidated parent material from which the A and B horizons have formed. These horizons host very little biological activity; however, knowing the composition of these parent layers can help determine the type of soils present on a site.

PHYSICAL SOIL PROPERTIES

Soil is a mixture of solids in the form of minerals and organic material, air, and water. Ideally, soils would contain about 50% solids, 25% air, and 25% water in the pore space. The texture of a mineral soil is based on the amounts of sand, silt, and clay in the soil. Sand, silt, and clay is determined based on the size of each individual soil particle. These size relationships can be demonstrated by imagining that a sand particle is the size of a football, a silt particle is the size of a softball, and a clay particle is the size of a marble.

Soil texture affects water and air movement through the soil as particles of different sizes pack together and thus determine the size and spacing of pores and channels. Sandy soils particles have the largest pore space allowing water to drain through the pores more freely. Silty soils have smaller pore spaces, so water moves through them more slowly. Clay dominant soils have very small pore space, and so they tend to hold more water. The mixture of particle sizes affects water, nutrient, and other chemicals' absorption rates. The *soil triangle* can be used to determine soil texture by comparison with the amounts of particles of any two sizes. For example, a sandy loam texture designation would include 65% sand, 25% silt and 10% clay when look at the soil texture triangle.

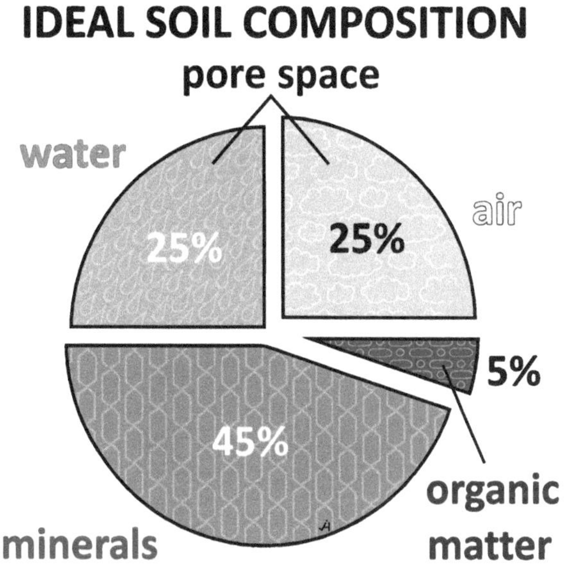

FIGURE 3.5 Ideal soil contains the appropriate combination of solids, liquids, and air space.

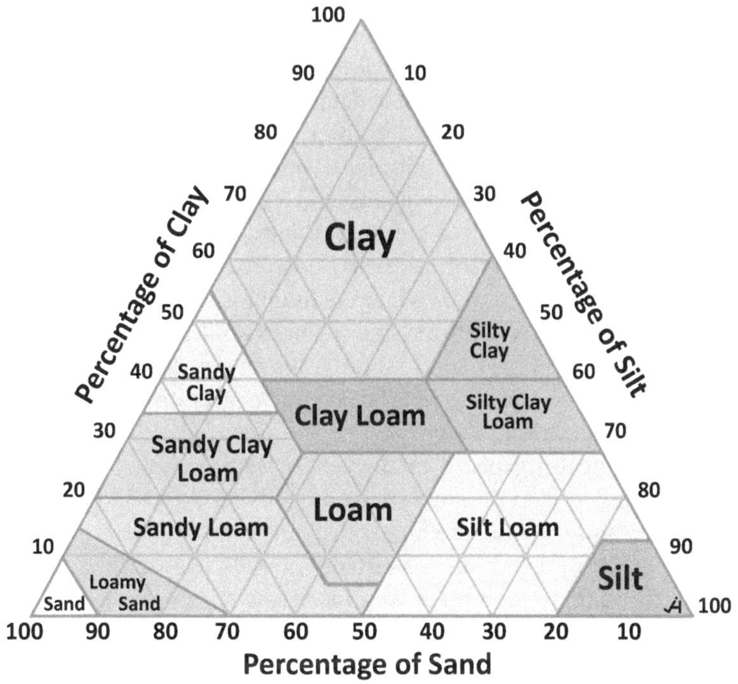

FIGURE 3.6 Soil triangle.

TABLE 3.1 Soil Diagnostics: Soil Color Can Indicate Soil Conditions

Soil Pigmentation	Potential Cause
Darker brown or black	High organic matter content
Red or yellow	Iron oxide accumulation
Gray to green hues	No organic matter, anaerobic conditions
Tan hues	Higher gypsum content

Soil Color

Soil color differences in a profile reflect soil-forming processes and can be an indicator of soil moisture in some cases. These differences help to distinguish fill soil from natural soil. Important coloring agents in soil include parent material, soil moisture, extent of leaching, content of organic matter, and the iron content. Organic matter darkens the soil to a degree, depending on the content and the extent of decomposition. Iron gives soil a brown, yellow, or red color. Shades of blue or green may also appear, depending on iron amount, oxidation state, and hydration state. When soil is saturated, iron can become soluble and can be removed, leaving the soil with "mottled" brown and gray colors or completely gray colors, depending on the extent of the moisture.

Helpful Fungi and Organisms

Healthy soils are a vast diversity of organisms. Microorganisms and beneficial fungi are present wherever a suitable environment exists. Endophytes are also present, which act to enhance nutrient acquisition, enhance growth, decrease biotic stress, and provide other benefits. However, colonization is quite dependent on soil health and conditions. Tree health relies on soil health, which allows appropriate nutrient cycling through decomposition and decay. When this edaphic environment is damaged through compression, contamination, or other disturbances, the result is a loss of active soil carbon, reduced plant and animal decomposition, and reduced microbial activity. Dead microbes = no carbon production.

The ancient symbioses between plants and certain types of soil fungi provide well-documented benefits. Mycorrhizae create symbiotic associations between many plants and are the beneficial fungi that colonize their roots. The term means fungus roots, and they gather nutrition from the tree roots while the fungi assist plant absorption of water and nutrients. Many green industry professionals are often unaware of these relationships and may inadvertently injure or kill the beneficial fungi through normal arboricultural and horticultural activities, including excessive fertilization and aggregate destruction through construction activities.

These helpful fungal organisms are present at varying levels in all soils. Inaccurate assessment of soils is common thinking that poor soils need and addition of the fungus. However, their activity and presence are highly dependent on soil health. Mycorrhizae inoculum is present in adequate quantities in most urban soils. However, the conditions are often unsuitable for the fungi to proliferate. As a result, the response is to add commercially available inoculants to build population. However, studies on gardens and landscapes find that mycorrhizal amendments are generally ineffective and unnecessary. Given the widespread presence of fungal spores already in the landscape, plants quickly become colonized by native mycorrhizal species. The preferred option is to focus on improving soil conditions, applying cultural practices, and improving colonization. One example of this would be to construct a large mulch ring around the tree, then add organic amendments like compost or biochar to increase soil activity. Plant health care practitioners have shown significant increases in microbial biomass in soils receiving liquid humates and biochar additives. The fungal to bacterial ratio increased as well in these applications resulting in increased biomass and favorable fungal populations. Additional information on modern organic amendments can be found in the appendix.

FIGURE 3.7 Soil supplements.

Pore Space

Soil properties can be dramatically changed by forces acting on the soil surface. This can be from pedestrian or vehicular traffic, lawn equipment, or any other loads. These loading impacts are especially damaging when the soil is wet. The factors most readily changed are the amounts of air and water space in the soil. This is called *pore space*. Compression on soils displaces air from the pore spaces between the soil particles leaving little to no room for air and water. Figure 3.7 demonstrates the lack of pore space on compacted soil.

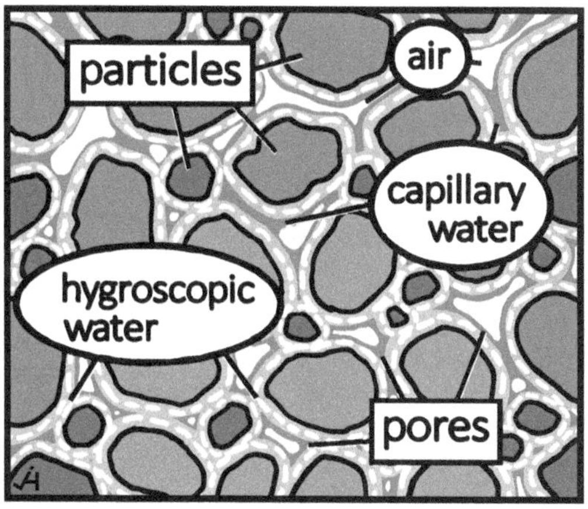

FIGURE 3.8 Pore space is critical for tree health, allowing air and water for the roots.

Measuring Compaction

Bulk density is defined as the mass of the soil particles divided by the total volume it would occupy. This condition may also be thought of as how closely packed the particles are in the soil. It is an indicator of the common soil health problem of compaction often encountered by plant health professionals. The extent of compaction in soils can be determined in a lab or in the field.

Engineers often measure compaction as a percentage of maximum dry density that may be obtained under conditions of optimum moisture content as determined by the Proctor density test. Then a nuclear densometer (soil density gauge) is used onsite to complete the process. A Proctor test is most often used to determine whether a soil is compacted to the minimum level to support structures. For engineering applications, 90% is a standard for many residential sidewalk applications and 95% and greater for structures. For planting soil, 85% is often quoted as the proper density, yet research indicates this value is root-limiting for many plants. A measure below 80% is more acceptable for planting soil. The Proctor test is a very reliable test but requires specialized equipment and geotechnical lab facilities for testing. This makes it a less practical tool for consulting arborists to apply in the field.

FIGURE 3.9 Cone penetrometer readings.

Penetration resistance is another method of testing compaction. This requires a tool called a penetrometer, which reflects the difficulty of pushing the tool's probe into the soil, measuring resistance. A penetrometer is a pointed steel rod with a gauge that records the pressure required to penetrate the soil. It is most useful for checking the relative compaction of planting soil installations where the soil type and moisture are relatively consistent. Penetration resistances are highly dependent on soil moisture and the speed it is pushed into the soil. A dry soil will have greater resistance than a soil at field capacity, which is ideal moisture conditions for a given soil. Therefore, caution should be exercised when using a penetrometer—it is most reliable when the soil profile is moist.

FIGURE 3.10 Soil penetrometers can be useful field tools for determining compaction levels.

Compaction measurement using bulk density values is easier than the Proctor test and more accurate than penetration testing. This requires samples to be taken properly and sent to a soil lab for preparation. This method is most effective and informative during the soil analysis phase on construction sites and in areas that have been mechanically disturbed. Calculating bulk density is the best diagnostic tool for existing soil and should be included as a part of an overall soil analysis.

Effects of Compaction

Compaction affects infiltration, rooting depth, water holding capacity, porosity, the availability of plant nutrients, and the activity of soil microbes. The total volume includes particle volume, interparticle void volume, and internal pore volume. Bulk density is not a fundamental property of the soil; it can change depending on how the soil is handled. For example, a soil can have a particular bulk density measurement of 1.3 g/cm^3, which is an acceptable metric for a soil with about 50% pore space. Later, the soil may be compressed by construction equipment in preparation for a new building. The arborist determines that the bulk density has increased to 2.1 g/cm^3, which is highly compressed and very limited pore space. This change demonstrates compression which moved the soil particles more closely together, occupying the once open pore space, and displacing air and water space with solids.

Note the differences in soil air and soil water in Figure 4.10, which compares compressed and uncompressed soils. The figure illustrates the preferred composition of healthy soil and the composition of a soil that has experienced compression by heavy equipment or vehicles.

As soil particles are squeezed or compressed together, pore space for air and water is minimized. The reduced pore space creates an unfavorable growing medium for any plant, including trees. This is called *compaction.* A key indicator of soil compaction is expressed by symptoms of decline, usually in the upper crown. When soil is compacted, there isn't enough room for water to easily penetrate the ground. This means even if there has been lots of rain, the moisture likely hasn't been able to reach the roots and the pore space. This level of compaction will cause trees to start showing drought symptoms such as wilting and curling leaves, yellowing leaves, decline in vigor and leaf scorch (brown outside edges or browning between veins). When symptoms are expressed in the upper portion of the crown, rooting and soil issues are prime suspects for compromising tree health. This can be seen from Figure 3.9. Bulk density is a useful metric to help evaluate soil conditions and diagnose tree health issues. The greater the bulk density, the more restrictive root growth and function. Recognizing soil texture can help speculate bulk density measurements and an idea of what to expect for the situation. More information on soil compaction and mitigation can be found in the next chapter.

Cation Exchange Capacity

The specifics of soil properties are measured using laboratory analysis. The properties of particular concern to arborists and other green professionals, generally measured in the laboratory, include a measure of pH, *cation exchange capacity* (CEC), and nutrient content.

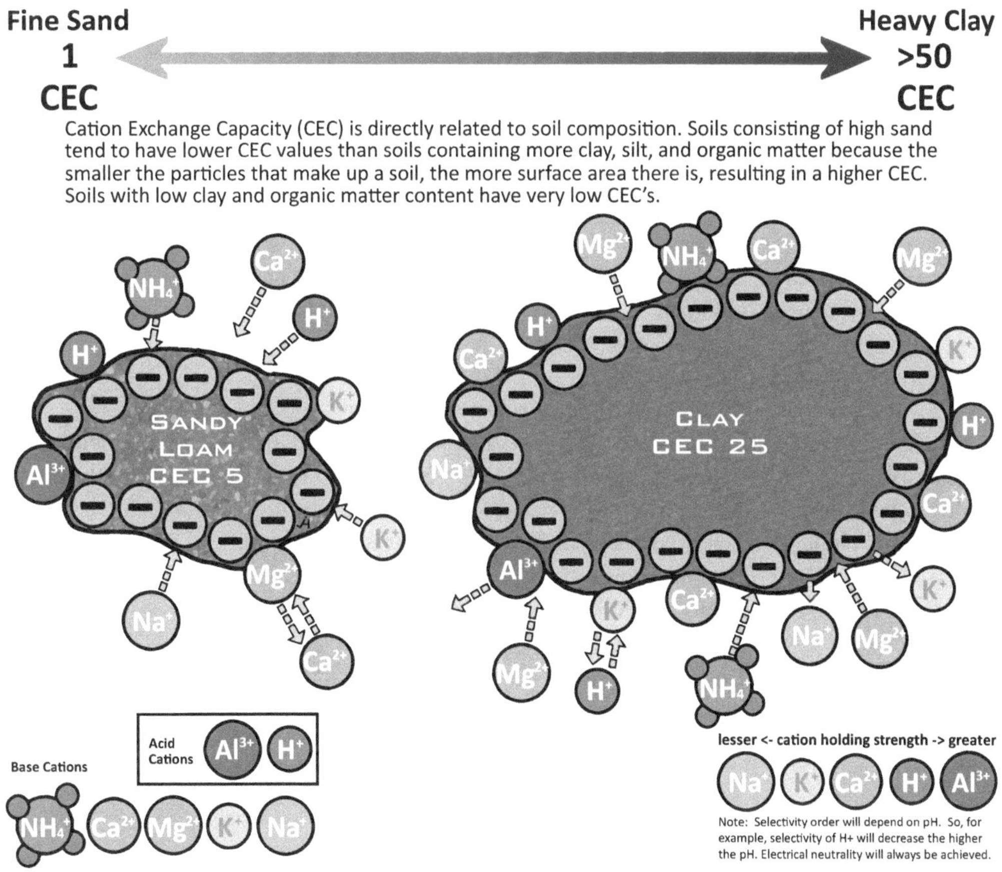

FIGURE 3.11 Cation exchange capacity.

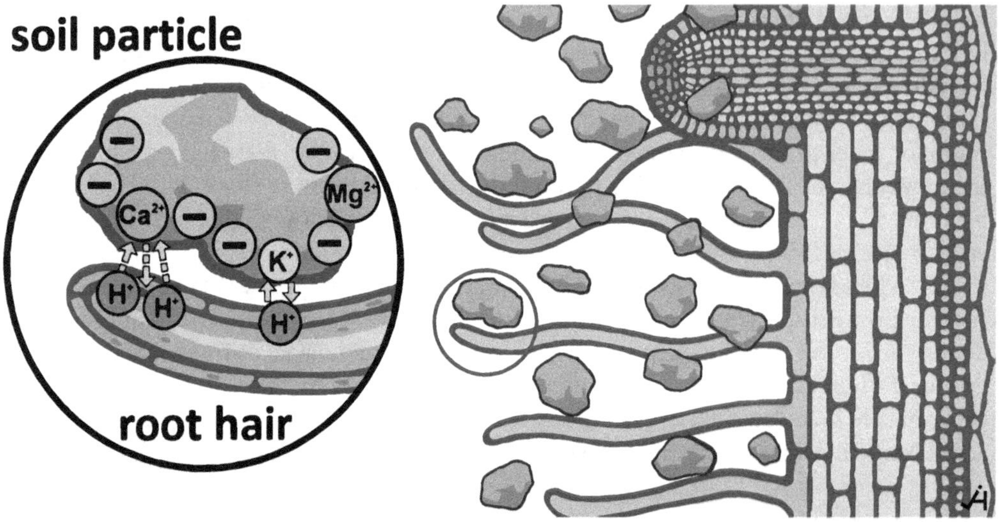

FIGURE 3.12 Cation exchange capacity.

Tree development and nutrition for healthy growth are linked to soil properties. The primary determinant of nutrient availability is dependent on pH and CEC, which is the ability of soil particles to hold and release nutrients for plants and microorganisms to use. Imagine that the soil particles that make up the consolidated soil environment are like magnets for soil elements and nutrients. Cations in the soil are positively charged nutrients, including nitrogen, sodium, calcium, and potassium. Whereas soil particles are typically negatively charged. Clay and organic matter in the soil have a negative charge. So naturally they attract positively charged nutrients and repel negatively charged nutrients. This explains why cations, the positively charged nutrients, find an easy home in the soil while anions, the negatively charged nutrients, are repelled, and are easily *leached* out of the soil. The amount of charge on a soil particle is constant such as clay, so if one cation is removed, other cations in the soil solution with the same charge must be added. Exchangeable cations are a major source of plant nutrients. For example, when a tree root takes up a calcium ion which has a +2 charge, two hydrogen ions with a +1 charge, which is furnished by the plant root, replaces the Ca^{+2} cation on the soil particle. The cation exchange capacity of a soil indicates the quantity of positively charged ions or *cations* the soil can hold on to its surface. CEC, or the ability to make this exchange, can have a significant effect on fertility.

Different plants and soil microbes require different kinds and amounts of nutrients. Various kinds of soils have different CEC values. For example, clay in the soil can attract and hold cations very tightly onto negatively charged parts of their surfaces. As a result, it is more difficult for plants to obtain the necessary nutrients from the soil solution. Soils with higher organic matter content have many active sites binding chemicals like the manner in which clay particles bind the chemicals in the soil. Clays and other soil materials are mixed with the organic matter in each soil layer to form a chemical system. Micronutrients are important food elements in the soil. Generally, they are used in very small amounts but are equally as important. Such micronutrients include metals like iron and manganese, which are necessary for plant growth and help plants to fight diseases.

Soil Chemistry and Nutrients

Soil chemistry information is important for proper management of tree nutrition and health. This should be a common request when samples are submitted to the testing facility. Arborists

TABLE 3.2 Recommended Cec Ranges Based on Soil Texture

Cation Exchange Capacity Ranges (meq/100 g soil)	Soil Texture Approximation
3–8	Sand
8–11	Loamy Sand
11–15	Loam
15–25	Silty Loam
25–40	Clay Loam
>40	Clay

can use CEC metrics of the soil to plan how much and how often to apply fertilizer. For example, soils with low CECs, such as sandy soils, require small quantities of fertilizer on a frequent basis. Excess fertilizer cannot be held by the sandy soil because of the low exchange capacity. Therefore, be careful not to overapply and aim to frequently apply as soil tests indicate. On the other hand, heavy textured soil with higher clay contents can go for longer periods of time without fertilizer because of the larger reserves stored on the soil exchange sites. When heavy textured soils are found to be deficient, however, it takes considerably more fertilizer to correct the deficit, and more exchange sites to fill. That's why recommendations are always higher for soils with high CEC such as clay dominant textures. This capability determines which nutrients stay in solution and are available for uptake by plant roots and which nutrients pass through the soil and are not available for plant and microbial use.

The other determinant of soil chemistry that impacts nutrient availability and growth is pH or potential of hydrogen. This is another common request for soil test samples submitted to the laboratory. This value is a figure expressing the acidity or alkalinity of a solution on a logarithmic scale of 1–14 on which 7 is neutral, lower values are more acid, and higher values more alkaline.

Soil pH

The pH scale (1–14) is a common measure of soil reaction. The lower the number, the more acidic. The midpoint of the pH scale is neutral (7.0), a good level for the growth of most plants. Changes in soil reaction, change in the pH, can have significant effects on the metal elements such as copper, zinc, and iron in soil. Metal toxicity to plants increases in strongly acid soils with a low pH (4). Metals in these soils are released from negative sites back into soil solution. At a higher pH (7.8), the metals often are bound in the soil. This indicates that the positively charged metal ions are held tightly to negatively charged sites in the soil. These sites may be on clays or organic matter. These strong, tight bonds restrict the availability of important metals for plant uptake.

Soil pH values are influenced by the soil microbe populations and diversity. The formation of the parent material plays a major role in pH determination as well because the mineral soil dictates the microbe composition. Also, it is important to recognize organic matter inputs often "buffer" pH values. OM inputs can increase microbe diversity, as well as "load" the soil with increased CEC sites for nutrient exchange. This improves soil and tree health without shifting pH values. This is one of the primary reasons to consider organic fertilizer which stimulates and supports the natural plant processes.

Industry Recommendations: Do not use more than 20 lbs. of sulfur per 1000 sq. ft. at any one time. If greater acidification is required, use split applications of 20 lb. and apply in succeeding years (as long as there are no other plants or turf).

Soil pH values are very challenging to manipulate in an established landscape. It can be difficult, if not impossible, to acidify large areas of soil such as found within a tree's root system. Because pH is logarithmic, a pH of 7 is 10x more acidic than an 8. Therefore, moving the pH needle from a pH of 8 to 6 is a factor of 100x. Most urban landscape soils would require between 35 and 56 lbs. of sulfur/1000 sq. ft. just to lower the soil pH by 0.5. This would be highly impractical and most likely lethal to any adjacent landscape plants.

Water Management

Water is undeniably one of the most influential factors affecting the establishment and sustainability of trees and other plants on a site. It is arguably the most limiting environmental factor for tree survival. It is also one of the most challenging resources to manage as well. There are many factors influencing water relations on the site that universal supplemental irrigation recommendations are impossible to render. Variations in soil texture, soil depth, environmental conditions, plant type, and more create unique situations for provisions.

A basic understanding of soil characteristics and water demands is important for diagnosing tree issues and sustaining growth. The primary factors to consider in the process is the water holding capacity (WHC) of the soil, water percolation rates, and plant requirements. The WHC of the soil, and the plant's use rate, will determine if supplemental irrigation is required in addition to natural rainfall and the amount to be applied at each irrigation. Imagine that the soil acts as a canteen, and its water supply must be refilled often enough to keep water available for the plant's withdrawal as needed for optimum growth and production. The WHC of the soil is determined by a combination of texture, structure, and amount of organic matter.

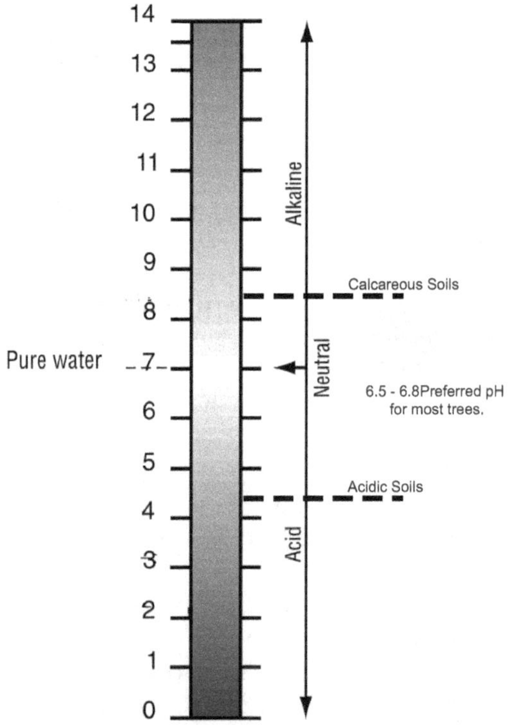

FIGURE 3.13 pH scale.

Supplemental Water by the Numbers

There are various methods for supplementing natural rainfall, including watering trees manually with a hose or other delivery device. A shared recommendation is to provide trees with at least one inch of water per week. This is considered a minimum amount for most trees and is subject to many variables. Environmental influences, evapotranspiration rates (ET), soil texture, and tree size all determine the amount of water a tree needs to thrive.

Additionally, it is important to understand water volumes to supplement natural precipitation more accurately, especially during drought conditions. The determination of the exact area to be irrigated and volume to supply is our goal. Typically, this would be all area inside the dripline of the tree. An inch of water applied would equal 0.62 gallons per square foot of mulch area or dripline area. In application, determine the area in square feet and then multiply the total area by 0.62 gallons. This would be the amount of water to apply for the 1 in. of water recommendation for that tree.

Example: If a mulch ring (36 in. diameter) is calculated to be about 1,000 sq. ft., multiplied by 0.62 gallons, our water requirement would be 620 gallons of water. This would be the amount of water necessary to provide 1 in. of water to the tree. (Area of a circle: $A = \pi r^2$)

The ability to measure delivery volume is the next step for supplemental irrigation. This requires timing output of the application device in gallons per minute for most accurate measurement. Other simple methods include using a catch device under sprinkler systems to determine rates. Water rates are important to ensure responsible water use and application of adequate water. Underwatering is not helpful and wasteful, while overwatering can be damaging to the plant and wasteful as well.

FIGURE 3.14 Be sure to recognize the factors that influence tree water loss, requiring supplemental irrigation.

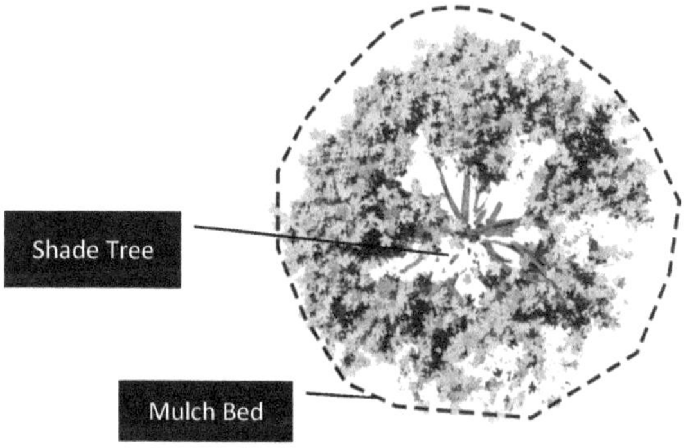

FIGURE 3.15 Efficient irrigation includes the entire dripline area of the tree.

Soil Texture

- The relative proportions of various sized particles in the profile; sand silt and clay

Soil Structure

- The manner in which the different sizes of mineral particles are arranged in groups aggregated within the growing depth of the soil.

Organic Matter

- Includes non-mineral substances incorporated into the soil profile and can hold several times its weight in water.

FIGURE 3.16 Factors affecting absorption rates

FIGURE 3.17 Prolonged wetness can create issues such as root diseases and lack of stability.

Percolation Rates

Another important soil characteristic is drainage or the rate at which water moves away from the root system. This is often considered the percolation rate of water. Percolations tests or Perc tests are one of the simplest methods for determining planting protocols and tree selection with poorly drained soils. This measures how long it takes water to drop one inch in a saturated hole dug in the soil.

A soil's percolation rate is determined by several factors and combinations of soil characteristics. Remember the percolation rate of any soil can vary with factors such as soil texture, structure, and the amount of OM. Generally, the amount of moisture taken in by the soil will be determined by its moisture content. Water will be taken in faster by a drier soil. The rate of intake slows as the soil profile approaches saturation. Soil texture is influential as well. For example, clay is the slowest and coarse sand and gravel being the fastest. If the soil profile is made up of layers with many textural characteristics, the percolation rate will be determined by the horizon having the lowest water diffusion rate, whether it be on the surface or in the lower part of the root zone. Additionally, compaction can reduce infiltration rates into the soil profile as well.

There are different ways to determine the percolation rate of soil. This test can be crucial in preparation for planting. Typically, the test is conducted as follows:

- Dig a 12 to 18 in. deep hole in the vicinity of the tree to be installed or the actual location.
- Place a ruler (or stick marked in inches) in the bottom of the hole. The measuring device should reach the top of the hole.
- Fill the hole with water. Note the speed of percolation.

When the hole is empty, note the time and calculate the time needed to drain the hole.

Trees and shrubs vary widely in their tolerance of flooding or poor drainage. Many riparian species, such as cypress, cottonwoods, sycamores, and most maples tolerate seasonal flooding or soils with prolonged wetness. Other species, such as hickory, oaks, and many

TABLE 3.3 Drainage Rate Table

<1/2 in. per hour	Poor drainage, likely compacted.
1–2 in. per hour	Good percolation and drainage rate.
>3 in. per hour	Rapid drainage, amendments may be required.
6–8 hours	Excavated hole is void of water, acceptable drainage for planting, overall.
8 hours	Unacceptable planting conditions if water is still present.

Note. Considerations include dry soils and moist soils will drain at different rates. Compacted soil and frozen or nearly frozen soils are slower to drain.

pines, will grow poorly, or will not survive at all, when planted on sites that flood or have poor drainage. For example, oak trees are generally tolerant of a wide range of soil conditions, but they prefer well-drained soils. A well-drained site with a percolation rate of an inch per hour is suitable for all species including sensitive species. If the site has poor drainage, you may need to improve the drainage rates with tile installation or select a different tree species that is adapted to wet conditions. In large part, if water has drained from the site or planting area within 6–8 hours, drainage would be considered fair for most tree species.

Water Availability

How much water is available to tree roots in soil? Available water capacity (AWC) is the soil moisture (water) that can be extracted and used by plants and is often referred to as capillary water. Capillary water is the soil water that is available to plants as it is trapped in the soil solution right next to the roots of the plant. Field capacity is the soil moisture that exists when the maximum amount of water is held against gravity in the soil profile. It may take between a couple of days to several weeks for all gravitational water to drain through the subsoil. *Gravitational water* refers to the water that drains freely from the soil by gravity after it has reached field capacity and moves further into the aquifers.

For practical purposes, the *field capacity* of well-drained soil can be taken as the moisture condition when the drainage losses from the root zone are minor. The soil water above field capacity is typically more available to plants than that at or below field capacity. Tree roots will use some of this free water. The amount the tree will use depends on how fast the water drains down to field capacity.

A very serious condition for trees and plants of any kind is the permanent wilting point (PWP) to designate the acronym for permanent wilting point. This condition measures the soil moisture that exists when plants can no longer obtain sufficient moisture to satisfy transpiration requirements and the plants wilt and remain wilted. As water continues to be removed from the soil through evapotranspiration, more of the pore space will void of water. As this process continues, only the tightly held water next to the soil particles remains. *Hygroscopic water* forms a very thin film surrounding soil particles and is not available to the plant. This type of soil water is bound so tightly to the soil by adhesion properties that it can't be pulled away up by plant roots. Since hygroscopic water is found on the soil particles and not in the pores, certain types of soils like heavy clay soils will contain a higher percentage of it, increasing WHC, but not necessarily helping the plants. This is the true *wilting point*. Tree leaves may wilt on days where they need a lot of water, but usually catch up at night when transpiration slows. When a tree can pull no more water it hits the wilting point and if prolonged can cause irreparable damage. These conditions are considered moderate to severe drought.

> Drought can have a major impact on tree health and survival by:
> - slowing and reducing growth,
> - reducing carbohydrate production,
> - lowering energy reserves for survival through winter,
> - inhibiting production of defense chemicals against pests, and/or
> - causing death to a portion or even the entire tree.

Drought Impacts

What is drought? It is an extended period without normal rainfall or precipitation, simply speaking. Specifically for trees, it means moisture in the ground is so limited they cannot extract enough water for biological processes.

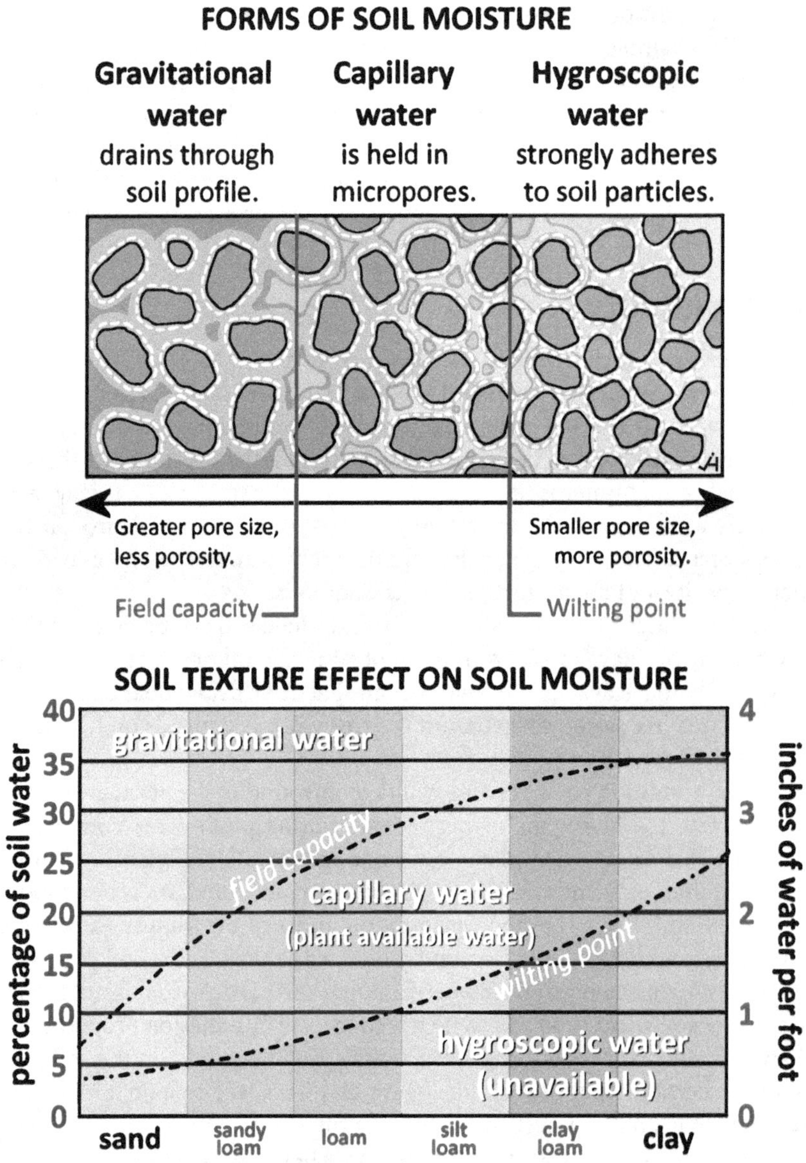

FIGURE 3.18 Soil moisture and soil texture effects on water holding capacity.

Our urban landscapes are suffering from worsening drought conditions, and trees are slowly declining and dying. This seemingly endless pattern of dry weather affects crops and plants of all kinds. It is critical for the arborist to identify these environmentally stressful conditions to create strategies to reduce the impacts on trees.

In many situations, prolonged drought conditions weaken trees, causing them to become more vulnerable to pests. Insects and diseases enter, weaken, and can kill all or part of the tree, depending on how badly the tree is predisposed to attack in this declining condition.

It is common to see an increase in borer insect activity, infestation of scale, and diseases such as wilt, canker, and root decay during periods of extreme dryness. At this point in the environmentally induced decline, we can expect serious, and often permanent, damage to our urban forests.

Drought stress promotes diseases through several different mechanisms:

- it may alter the plant's physiology, making trees more susceptible;
- it may reduce the ability of the plant to produce defensive chemicals or to "outgrow" the disease; and
- more severe drought stress can cause physical injury to tissues through drying and damaged tissues are then easily invaded by otherwise "weak" pathogens.

Legacy Effects of Drought

How a tree responds to prolonged or extended drought conditions depends greatly on health and vitality before the seasonal distress. A biological lag effect is common in trees where environmental conditions are challenging and prolonged. There will likely be a noticeable decrease in the number of new leaves formed in the buds, and the new stem segments (internodes) present. Drought influences the number of leaves, leaf size, and twig extension the following year when buds expand. This can mean fewer, smaller leaves and shorter growth between the leaf buds. The results of prolonged dry conditions might not inhibit the first growth flush but may decrease the number of stem units formed in the new bud that will expand during the second (or third, etc.) flush of growth. If drought continues, all growth will be affected. Thus, tree growth the next year will be atypical and, again, create predisposed conditions for diseases and insects if not monitored and managed. These conditions are only made worse in the urban forest, specifically where the available rooting area is already reduced with likely compacted soils low in organic matter.

Managing Drought Conditions

Trees and plants initiate biological responses to dry conditions. An initial response begins with stomatal regulation. The stomata are tiny openings or pores in leaves allowing gas exchange. Specialized guard cells surround stomata and function to open and close stomatal pores in response to current conditions. Stomata allow a plant to take in carbon dioxide, which is needed for photosynthesis, and help to reduce water loss by closing when conditions are hot or dry. When closed, this forces trees to rely on stored sugars and starches. However, if they run out of energy sources before the drought is over, they can die from *carbon starvation*. Water stress is known to cause xylem cavitation in the leaves, roots, and stems of plants. This is the formation of gas bubbles in xylem which interrupts the continuous stream of water along the column of vessels. Water flow through the tree is interrupted until the column can be reestablished with adequate resources. When trees lose too much water too quickly during a drought, these air

FIGURE 3.19 Removing turf and extending mulch rings helps reduce drought stress.

bubbles form in xylem vessels, preventing the transportation of water from the roots to the leaves, a process that can also result in decline and death.

There is direct, acoustic evidence of cavitation in trees which were first heard by Dr. John Milburn. He termed them as clicks audible within the plant. It is possible to hear a quiet popping sound that is produced by the water passing through the cells of the xylem tubes and cavitating as it mixes with air on its way upwards with enhanced listening devices. The process can be related to sucking on a straw in a near empty glass and taking up air with little liquid.

The first line of defense against any environmental or biological challenge is improved cultural practices. Strategies to help make trees more drought-resistant and reduce the long-term effects of prolonged dry conditions are necessary to reduce tree loss. Some of these strategies include protection and prevention. Always protect tree trunks, especially young trees, from mechanical damage from string trimmers, lawn mowers, and other equipment. Preventing damage to the bark and wood at the base of the tree maintains a continuous ring of water- and food-transporting tissues. Reduce competition for available moisture with other plant materials such as turf, shrubs, and groundcovers, where feasible, by removing plants and adding mulch. Extend the mulch beds and tree rings to the edge of the tree canopy (drip line) to protect those fine roots from drying out and increasing the soil's ability to retain water.

Supplemental Water Rates

It is of even greater importance to remember newly planted landscape plants may need additional help during the drier times. Considering that over 85% of the roots can be removed during the harvesting process on balled and burlapped plants—and that containerized plants are likely dysfunctional from root restrictions such as circling and girdling roots—there is very little means for the roots to draw much-needed moisture. Typically, trees should not be considered established for at least two growing seasons after installation. However, the establishment period depends upon species, health, and maintenance.

FIGURE 3.20 Apply water at the trunk flare for improved absorption.

Roots from balled and burlapped or container-grown plants may not grow out of the soil ball and into the surrounding backfill soil for many weeks after installation. The plant must depend on moisture in the original soil ball to sustain it until roots extend into its new soil environment. Since water will generally not move from the backfill into the soil ball, the moisture content of the surrounding backfill soil is not critical to the transplanted tree. The original soil ball must be replenished frequently with the correct amounts of water.

FIGURE 3.21 Probing the soil near the trunk of newly planted trees helps determine moisture conditions.

Let's look at a scenario for watering a newly planted tree. According to nursery standards, a 2-inch caliper red maple tree should have a soil ball about 24 inches in diameter with a depth of about 16 inches. The total volume of such a soil ball would be equivalent to the volume of about 20 gallons of water. Assuming that the soil in the ball is a clay loam, the volume of water that could be held by the soil ball is about 25% of the total soil volume (about 5 gallons). On a single windy, 95-degree day in the summer, the leaves on a reasonably healthy, newly transplanted red maple tree may lose 2 or 3 gallons of water through their leaves each day. Considering the limited root growth into the backfill soil outside the root ball, the tree may be suffering from drought stress in only a couple of days. Soil type is important and has an impact on moisture levels as well. Sandy soils usually require more frequent irrigation than the heavier clay content soils. Watering intensity and frequency is relative to the soil texture. Sandy soils may need more frequent applications than clay soils.

There are several methods for watering newly planted trees. More common methods include applying water by hand and hose to the planting pit. Another method is a dripping hose into the plant pit with the nozzle nestled up to the base of the tree. Also, drip hoses with perforations apply low rates over time. However, these aren't the most reliable and accurate methods of delivery. First, the person watering usually has no idea how much water is being applied. Second, there is no guarantee that the water applied goes into the soil ball as opposed to the surrounding backfill. In these situations, it is possible that much of the water applied may run off onto nontarget areas. In other cases, the planting hole fills with water and the soil ball stays saturated and too wet for an extended time. Drainage is a must! Prolonged saturated soils can be just as detrimental to the tree as prolonged dryness.

Other than estimating water needs by observing the plant for wilt, a more effective and proactive way to determine soil moisture is to use a soil probe. Remember, excessive water can kill a tree as well. Similar symptoms are expressed with too much water, such as wilting and chlorosis. Excess moisture reduces aeration properties, creating a lack of oxygen. Soil probes are great tools for water management decisions. The key is to deliver the proper amounts of supplemental water while allowing adequate drainage between deliveries.

Shallow rooting, a condition where roots are unable to grow to favorable soil depths, is typically from compaction or a lack of oxygen rather than water. However, roots go where water goes. It is important to place water where the roots are located and most absorptive. The preferred watering locations are the trunk flare and the dripline of the tree. It is recommended to water heavily at the root flare of younger trees, especially during droughts.

Sustainable tree growth relies on a better understanding of soil and water relationships. In order to determine the water needs of trees and plants in the landscape, the arborist should have knowledge of the tree needs and the resources available as well as the soil qualities affecting availability. Sustainability is about scarcity and when water becomes a limiting factor, the arborist must be able to identify strategies to ration consumption and use these resources to its fullest potential.

CHAPTER 4

URBAN SITES AND TREE PRESERVATION

LEARNING OBJECTIVES

- Describe the importance of a site analysis to improve planning and sustainability.
- Identify the factors to consider when conducting a site analysis.
- Recognize the impacts of construction damage on tree health and stability.
- Identify the common injuries associated with urban trees.
- Identify solutions and remedies for pre- and post-construction activities.
- Describe the tree preservation process and the process of creating a preservation plan.

SITE PLANNING AND ANALYSIS

Successful tree communities are the result of proper site analysis and planning. The process of creating a sustainable planting design requires a process of investigation and deliberation for "right tree, right place." An analysis of site conditions combined with a clearer understanding of client needs will provide the foundation for sustainable tree planting. The arborist or landscape horticulturalist may be familiar with the site or planting area; however, well-supported knowledge of the environment, hydrology, and soil properties are keys to establishment and long-term growth.

A thorough site analysis should include the following, in varying degrees of detail as determined during the initial site review:

- site plan
- climate characteristics
- edaphic conditions
- hydrologic factors

A site plan or map provides locations of all structures and physical features such as hard surfaces, buildings, fences, existing trees, and other infrastructure. Utility service providers should be contacted locating both aboveground and belowground utilities. Another useful feature of site plans includes investigating drainage patterns as well. This helps determine water movement, or stormwater flow, drainage, and other hydrologic conditions which may affect tree health.

TABLE 4.1 Site Planning Considerations

Consideration	Factors
Site Plan	•Physical features •Utilities •Structures
Climate Characteristics	•Sun exposure •Wind conditions •Hardiness
Edaphic Conditions	•Chemical and physical conditions •Texture and type •Profile
Hyrdological Conditions	•Water Holding Capacity •Drainage and Movement •Availability

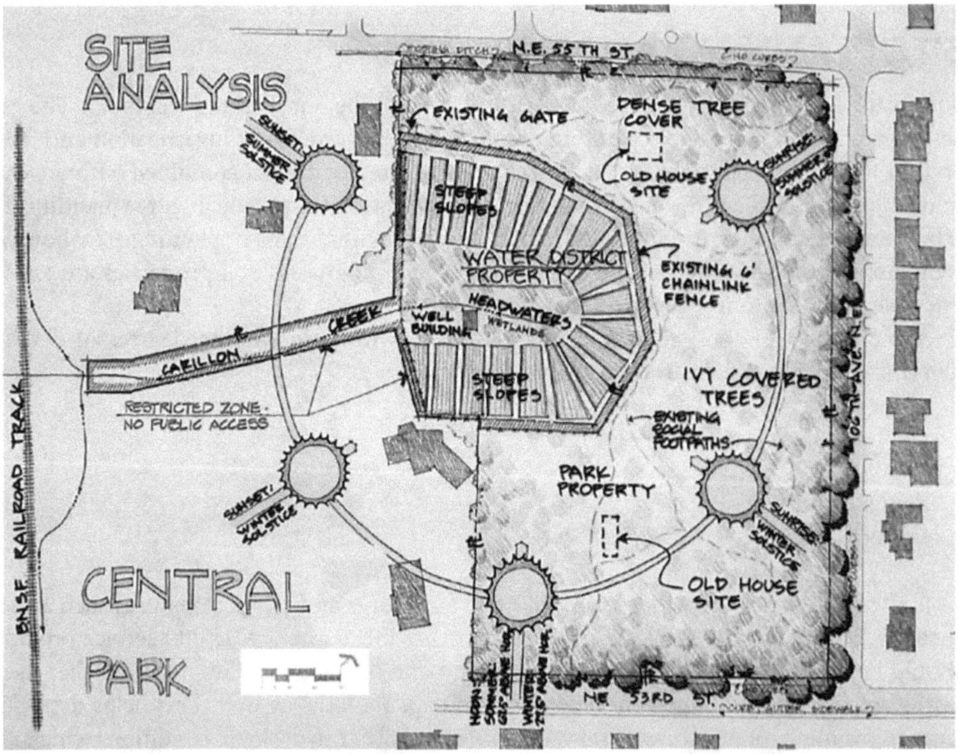

FIGURE 4.1 A site analysis will help plan for successful tree installations.

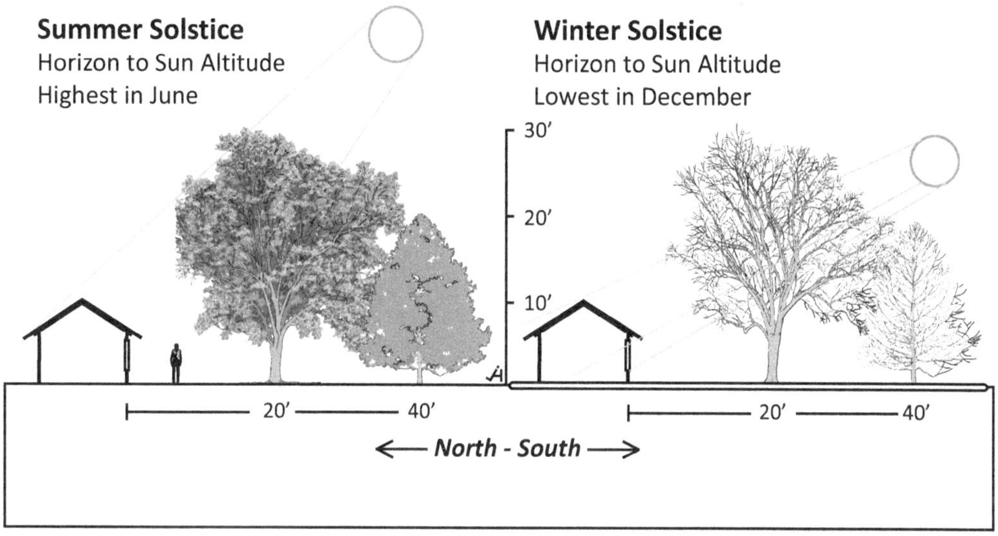

FIGURE 4.2 Deciduous trees properly placed provide shade in the summer and allow sun through in the winter.

The depth and detail of the site analysis will depend on complexity of the project, client expectations, and the difficulty of the site itself. Large-scale municipal tree plantings may require more depth in relation to site functions such as line of sight, or clearance interactions with infrastructure, including streetlights and traffic signals. Residential tree plantings may focus site analysis on moderating climate, such as winter winds and optimizing solar gain in the winter. Additionally, rights of way and easements should be included in the planning process as well.

When working on urban planting sites, special attention should be given to vehicular and pedestrian flow on public sites. Planning for future maintenance requires attention to mature size, including both height and spread. Trees often grow into undesired spaces creating structural conflicts with above and belowground infrastructure. If utilities are present, engage the utility service provider in the process. Most have utility foresters that can help with the process. Dense tree crowns can conceal signs or block vision for cars and pedestrians, creating dangerous situations. This will require collaboration with urban planners and the city forester to provide traffic patterns and limitations.

TREE PRESERVATION

Growth and development are constantly occurring right alongside the urban forest and preserving tree canopy cover compatible with land development is challenging. However, it is important to protect desirable trees regardless of the development, whether a large municipal expansion or a residential building site. Mature trees contribute significantly to the aesthetics and ecology of developed landscapes and these contributions are not easily reproduced with newly planted trees. Tree preservation is considered an advanced urban forestry topic because it synthesizes numerous subjects, including tree biology, soil science, and applied arboriculture. Trees are dynamic living organisms that have basic needs to survive and grow. These are air, water, space, and nutrition. Each are equally important for a long-lived tree and necessary for the arborist to know how to respond to damage as well as proactively keep trees safe from damage and injury.

FIGURE 4.3 Construction damage belowground destroys roots.

One of the most difficult tests for arborists is the fact that most trees will sometimes take years to express symptoms from construction damage. The severity of impact, type of damage, developmental age of the tree, and tree health going into the construction exposure play a role in response. It is challenging, if not often impossible, to repair damaged trees and reduce stress from construction impacts. These facts make avoidance obvious during all phases of the construction process. Successful preservation requires protection protocols be in place prior to construction activity and arborists should be a part of construction phases from the earliest stages of planning to the final walk-through of the project.

An arborist will receive numerous complaints and inquiries associated with construction damage. One of the most common questions is "Why is my tree starting to die just after I built my house?" Typically, trees don't die immediately from construction damage on or around trees. However, if severe damage to the roots or the crown occurs, immediate death will occur. It often takes several months and even years for symptoms to develop. Connecting today's dying tree to the damage of a few years ago is difficult for homeowners to understand and sometimes perplexing for tree experts to recognize and diagnose. The obvious question is "What kills trees in formerly tree-filled lots turned into subdivisions or commercial complexes?" The word "formerly" could be a clue. After selectively clearing all but the best trees or those trees considered to "fit" on the building site, landscape equipment such as bulldozers, skid steer loaders, and excavators begin to grade the site into a level, compacted plane. Building infrastructure, such as foundations, utilities, walks, and driveways, are carefully placed on the site according to engineering plans. As the building process occurs, several things may happen, none of which is good for established trees. A key perspective to keep in mind when dealing with construction and development is that not all impacts can be eliminated; development will happen. The objective is to minimize the impacts to any tree, being a team player with those involved in the development process.

DAMAGE AND THE IMPACT

Soil Disturbances

Soil and root system disturbance, the belowground destruction to trees, is the most significant and most difficult disturbance to reveal and remedy. Even minor changes to the soil grade or soil profile can produce negative responses from trees. When removing or adding soil around the roots of trees, extreme caution is advised. Undisturbed soils, such as those found in a forested area, are composed of layers, or horizons. The top layer occupies the upper 6 inches to 12 inches and is rich in nutrients and organic matter. This is the soil component usually missing in new site construction because of the grading and leveling process. The topsoil is typically removed, stockpiled, or used elsewhere in indiscriminate areas around the site.

Typically, the development process begins with landscaping on the site. This is the initial step to create a suitable surface to accommodate buildings, roads, and other infrastructure or create drainage flow, retention ponds, or other site features. Excavation and grading create a "hard pan" in the process and removes the valuable topsoil, leaving the remaining soil profile for development. "Hardpan" is a highly dense and solid layer of soil near the ground surface, which is usually comprised of clay or silty-clay soils. A hardpan soil surface or rigorously compressed soil layer is a man-made soil dysfunction which creates challenging growing impediments for trees and other plants because of the compacted, nutrient-deficient soils lacking in pore space.

FIGURE 4.4 Trenching near the trunk and equipment on the root zones results in serious tree damage.

A compacted surface is required to build structures. This is achieved by grading and compacting the soil. Large equipment is used to achieve the bulk density necessary to support the home or structure. Soil conditions required to support a house and the necessary infrastructure are vastly different from, and are in sharp contrast to, the soils of a formerly wooded lot or agricultural field. Undisturbed soils, such as those in a forest, or soils used in cropping are pliable, allowing roots to expand easily. The pore space of undisturbed soils is filled with life supporting oxygen and water. Conversely, those soils engineered to support buildings are nearly void of any pore space.

Many new construction projects occur on former agricultural fields and other undeveloped sites. While soil is frequently referred to as a "fertile substrate," not all soils are suitable for growing crops or trees. Ideal soils for food crops are balanced appropriately with mineral components, air, and water. The balanced contributions of these components enable water retention, efficient drainage, oxygen availability within the root zone, nutrient delivery to facilitate plant growth, and physical support for plants. This is the complete growing environment for trees as well. Basically, most plant life requires suitable soil media for sustainable growth, however, when human impacts enter the natural system, there are often consequences.

FIGURE 4.5 Arborists have different soil needs than engineers, who need compacted soils for building.

SOIL GRADE CHANGES

One of the most common causes of decline and death in trees on the construction site is cutting or filling around existing trees. In urban areas, typically, root expansion is at least spreading to the dripline of the crown or slightly greater. This is considered the critical root zone area that should be free from any disturbances.

The addition of new soil or "filling" graded areas on the site raises the level of soil and covers the original soil surface. This process of adding soil to the existing root zone can cause harm to the tree root system by disallowing the penetration of oxygen, preventing gas exchange between soil and air. Many variables determine to what extent soil can be added without damage. It has been found that fills of up to 6 inches can be added with minimal

FIGURE 4.6 Raising the grade around trees can be lethal by reducing gas exchange from the roots to the air above.

FIGURE 4.7 lowering the grade causes physical damage to roots.

impact to vigorous trees. This is provided the added soils have a very low bulk density and allow adequate gas exchange between the roots and surface. However, in many cases, as little as 2 to 3 inches in fill around trees can cause significant harm. Soil levels elevated from construction needs create aeration, water movement, and root growth challenges.

The best recommendation from an arborist is to advise against any fills. The reduced gas exchange and altered root environment creates poor growing conditions and potentially a stability issue as well. Studies have shown that fill soil over the root system can cause regression in the root system. In a short time, roots begin to decay and die back, causing overall decline in the tree and significantly increasing the risk of tree failure. Some case studies have shown that vigorously growing trees with fill added on the existing root/soil system will expand with new roots into the new soil layer. Then, the former root system dies leaving a shallow rooted tree prone to windthrow. Typically, this type of root damage doesn't reveal itself for several years. Symptoms may be observed, such as early fall color or declining and dying branches, in the upper and outermost parts of the tree canopy because of the compromised root system.

Removal of existing soil or lowering the existing grade is called *cutting*. This process is used when lowering the soil level is necessary for construction to proceed according to plans. The obvious concerns with this action around trees are physical damage to the root system, including the root plate and often the root flare. This is one of the most serious and acute forms of damage. Any root damage or removal below the dripline of the tree can seriously impair the tree's health and stability. Severing buttress roots at or very near the tree causes serious defects impairing stability and health. The small diameter roots near the tree protection zone (TPZ) does not necessarily increase failure potential but will have a negative impact on health. There is elevated risk concerns if most roots on one or more sides are severed. Also, the proximity of the root cutting to the trunk is critical.

Research indicates:

- When 1/3 or more of a tree buttress roots are missing, stability can be significantly reduced.
- Severe loss of stability is common when cuts are made at a distance 1 to 1.5 times trunk diameter.
- A tree is adequately anchored when less than 40% of the roots within the CRZ are damaged, decayed, or dead.

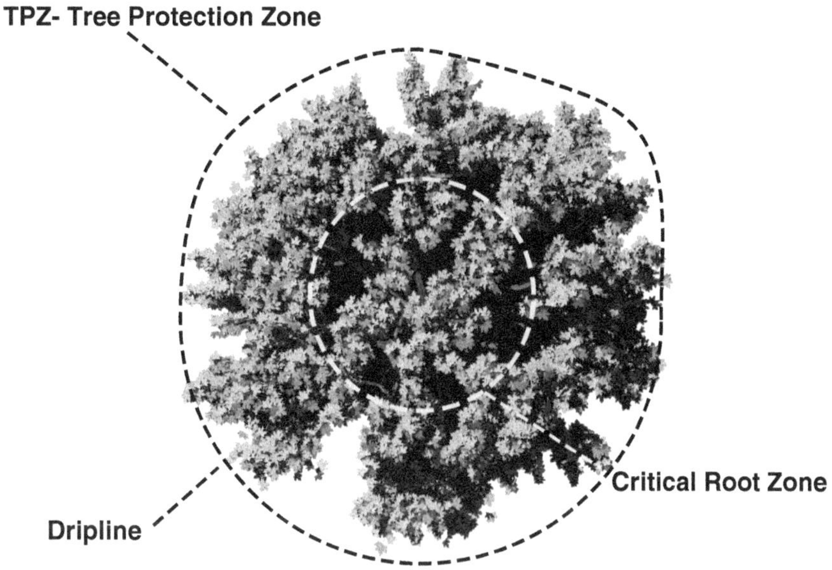

FIGURE 4.8 The TPZ protects roots from damage.

TABLE 4.2 Grading Around Trees

Filling Soil Within the Tree Protection Zone	Cutting Soil Within the Tree Protection Zone
Installation of aeration systems which include a piping arrangement below the ground surface to allow air movement.	Preconstruction root pruning near the TPZ.
Grade adjustments gradually from outside the TPZ gradually up to the trunk flare.	Prevent cuts within 5–6 times the diameter of the tree.
Retaining wall installation outside the TPZ.	Foundation wall installation outside the TPZ.
Grade elevations > 4–6 inches within the TPZ may require relocation or removal.	Grade reductions of > 3–4 inches within the TPZ with large roots damaged, consider removal.

Note. Solutions and remedies are relative to the type of activity within the tree protection zone.

Other forms of mitigation would include root pruning to improve recovery from excavation damage to exposed roots or during installation of infrastructure such as new curbs, walks, and driveways. Often roots are harshly severed during the excavation process, leaving ragged edges which do not regenerate well. If roots are exposed and torn, proper pruning cuts, like those made in the crown can apply. It is recommended to cut back damaged roots to a larger root so growth of the remaining roots is directed parallel or away from the competing structure. Proactive pruning and reparation pruning both help to stimulate regrowth in healthy trees. In some situations, it may be possible to preemptively inspect roots using forced air or water to wash the soil away to prune beyond the prospective constructive activity. Hand digging may be the only option to examine the root configuration to determine the best location for cuts.

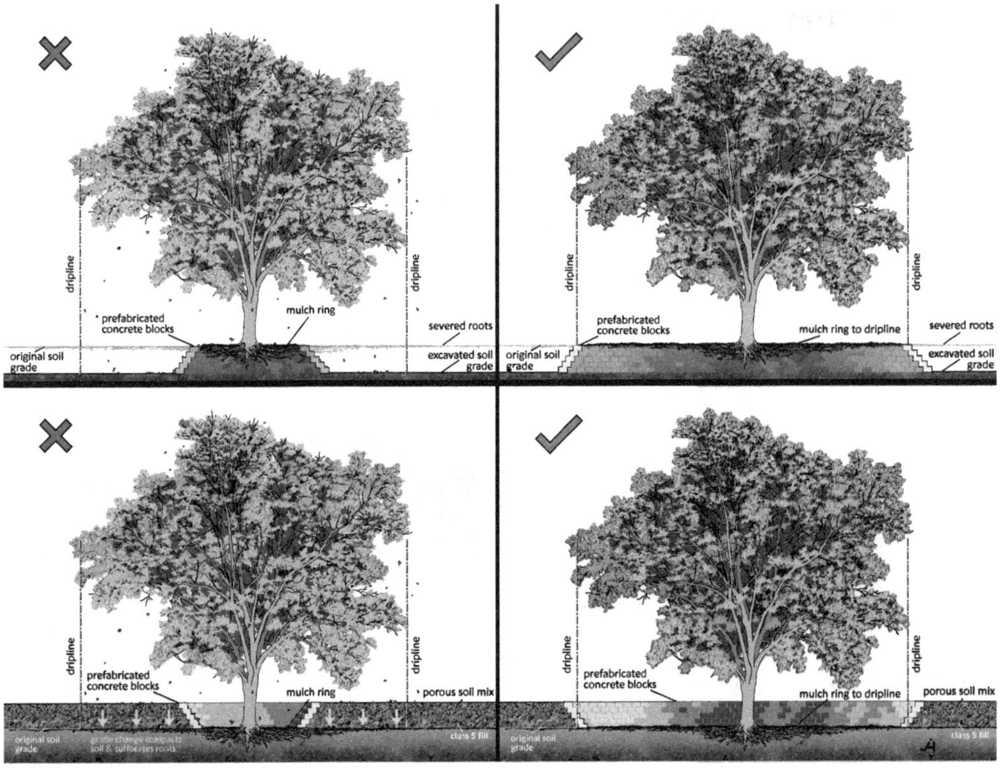

FIGURE 4.9 Retaining walls and tree wells should extend to the dripline of the tree to improve survival.

Soil Compression and Compaction

Compaction is one of the major health threats to trees and is caused by excessive traffic or activity on the soil. Compression occurs when excessive weight compresses soil particles tightly together, reducing the pore space between them that is essential for retaining vital air and water. This process leads to compacted soils, characterized by a lack of the necessary air and water spaces required for root penetration and the uptake of water in the soil, as well as for root respiration.

In addition to directly impeding root growth, soil compaction also limits tree performance in a physiological manner. As soil aggregates are broken down by compression from above, pore space (the area for water and air in the soil) is reduced. This is harmful to roots and soil organisms, primarily because it reduces oxygen and carbon dioxide circulation, which is exactly what aeration is within the soil. The results are anaerobic conditions, characterized by low oxygen levels in the soil, resulting in root decline and death.

Poor aeration reduces nutrient uptake and loss of important mycorrhizal fungi along with other soil organisms. In cases where the soils are already identified as deficient, compacted growing environments can exacerbate these deficiencies and their associated symptoms. Additional issues arise by reducing water movement through the soil leading to both water deficits and waterlogging. Increased surface runoff caused by poor infiltration into the hard, packed surface limits soil water recharge into the pore space and aquifers below. This situation then leads to the erosion of valuable soil and nutrient leaching.

There can be many contributors to compression of soil particles, including lawn maintenance equipment, vehicles parked on or driving on soil surfaces, flooding, and lack of drainage. Even constant foot traffic in the same areas can cause soil compaction. The pressure of

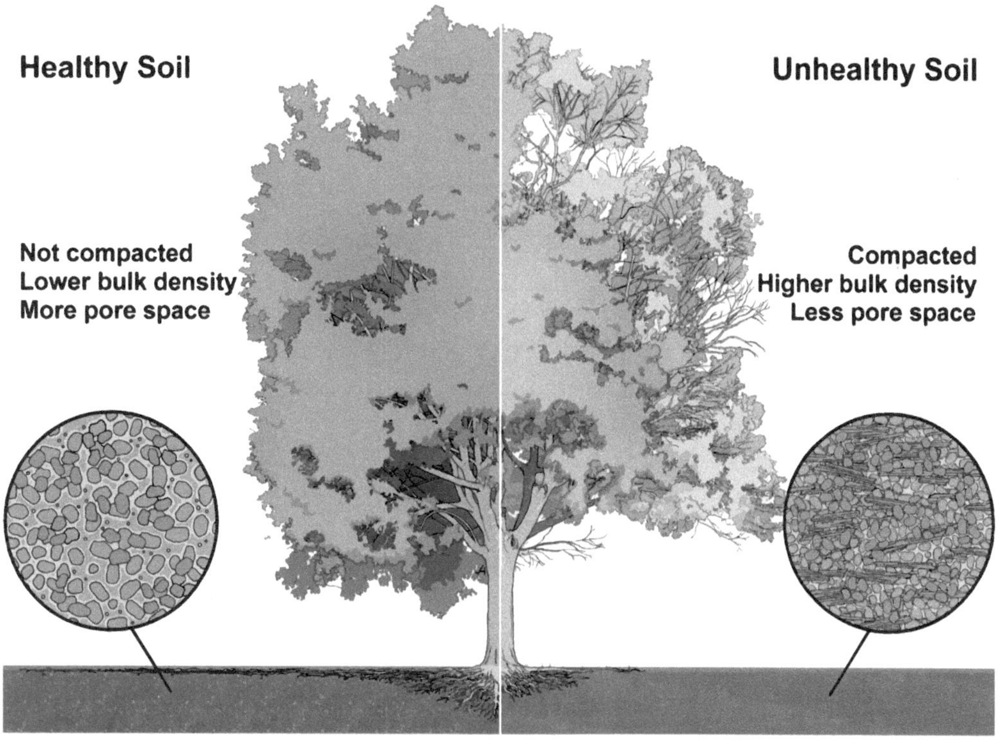

FIGURE 4.10 Comparison of health and unhealthy soil.

a compressing force spreads with the depth of pressure. This type of compaction is called surface compaction and is typically confined to the upper 6 inches of soil. Soil texture can help determine the extent soils are vulnerable to compression. Clay and loam soils, which contain finer particles, are more easily compacted because these smaller particles readily fill the pore spaces between the larger particles. This compaction is even more profound when these fine-textured soils are wet. On the other hand, larger particles, such as those found in sandy soils, are not as easily compacted due to their inability to fill small pore spaces.

The best remedy for compaction is prevention. Creating designated paths and implementing restrictions on activities such as parking on and driving around trees can help greatly. Proper planning and preservation strategies should outline this issue and provide proactive measures to reduce activities which lead to soil compression.

Recent research has provided support for methods of soil decompaction. In this research, two decompaction techniques, namely air spading and vertical mulching, were evaluated alone and in combination with the soil amendment biochar and/or a woodchip mulch. Effects on soil quality (bulk density, organic matter, vegetation ground cover, root dry mass, and earthworm counts) were monitored for over 5 years. A combined treatment of air spading, biochar, and a woodchip mulch layer proved optimal in improving the soil quality of a heavily compacted soil over the 5-year period. This treatment was, however, the most expensive and time-consuming treatment. A woodchip mulch was the most effective of the individual treatments and the most cost-effective. Air spading alone proved reasonably effective in improving soil quality over the 5-year study period. Effects of air spading could be improved by addition of a woodchip mulch. Vertical mulching alone or in combination with biochar had little influence on soil quality over 5 years. Results demonstrated that effective long-term soil decompaction measures exist for arborists to improve compacted soils and save trees.

Soil Compaction Modifications to Improve Structure and Air Infiltration
Mechanical tillage to loosen the upper soil layer with manual tools or small garden equipment.
Soil profile rebuilding by incorporating soil amendments with excavation equipment such as a mini excavator.
Forced air tillage and excavation within the tree root zone or planting area using highly compressed air. This creates a mixing action which modifies soil and allows for amendments.
Radial trenching by excavating radiating trenches of 4–18″ wide and 12–18″ deep from root flare to dripline using compressed air or water. Trenches are backfilled with replacement soil.
Soil fracturing injects compressed air into the profile to create horizontal cracks in the soil allowing for amendments to be injected and improving gas exchange.

Soil Contaminants

Soil on a construction site can be contaminated by masonry products, asphalt, paint, fuel, cement, oil, salt, and other materials. Any substance in the soil that exceeds naturally occurring levels and poses health risks is a soil contaminant. Contamination may occur across an entire site, or localized, such as concrete washouts and refueling areas. This damage can also occur in the landscape where there has been excessive use of pesticides and herbicides. Soil contamination is harmful to trees in several ways, including direct contact leading to death, changes in soil pH, and other detrimental effects. If there is uncertainty about the condition of the soil or suspect contaminants, obtain a soil test to identify any toxins.

Concrete products are one of the most common building materials found on every site. On sites with existing trees, precautions should be taken where concrete is being used because trees are sensitive to the washing out process from concrete trucks. The rinsate affects the pH of the soil dramatically. The ingredients in concrete products have a surprisingly high pH value of around 13; in other words, nearly at the top of the pH scale. This high pH stems from the cement used to bind the concrete together. Cement contains many alkalis, a highly basic category of minerals. In the case of fueling sites, petroleum product contamination harms soil microorganisms, reducing their number and activity. At high levels of contamination, seeds cannot germinate. In some cases, even mature plants cannot grow anymore. Products like diesel fuel and gasoline kills trees by poisoning the roots directly.

FIGURE 4.11 Fuel and other chemicals on the root zone will contaminate the soil and lead to damage.

Where and how much contamination are added to soils will largely determine how that contamination spreads throughout an area. The type of soil will also play a role in its distribution because of the faster infiltration rates of coarse-grained sandy soil types. Fine-grained clay soils or organic material in surface soils can hold contaminants tightly, which means the contaminants will accumulate if left undisturbed remaining at toxic levels longer.

Soil Remediation Strategies

A solution to the problem of soil contamination is called soil remediation. Soil remediation is a way of purifying and revitalizing the soil to improve healthy growth. In short, the goal of the process is to restore the soil to as near as possible to its natural, pollution-free state. Traditionally, there are three main soil remediation technologies: soil washing, bioremediation, and thermal desorption. However, only a few are feasible on most landscape sites. Soil washing is a process that uses surfactants and water to remove contaminants from the soil. The process involves either dissolving or suspending pollutants in the wash solution and separating the soil by particle size. Bioremediation involves the use of living microorganisms, such as bacteria and fungi and even trees, to break down organic pollutants in the soil.

Another more common solution to soil contamination is the application of activated charcoal. Activated charcoal is a porous, soft black substance made from heating carbon-based materials in spaces with a restricted amount of air. Most often this charcoal is produced from hardwoods and coconut shells. Charcoal adsorbs 100 to 200 times its own weight and has a strong attraction for organic chemicals such as pesticides. Organic chemicals will bind to the charcoal, which keeps them from contaminating soil and water resources.

Activated charcoal can be used to clean up and reduce effects of spills from organic pesticides, some petroleum products, and hydraulic fluids. This is especially beneficial for construction sites or instances of herbicide misapplication. Keep in mind that activated charcoal is only effective with organic chemicals, not inorganic chemicals (e.g., arsenates, lead, sulfur, borax, sodium chlorate). Also, it cannot be used for water soluble organic pesticides (e.g., MSMA, DSMA).

If the site history indicates no history of industrial use or other contamination, and soil test results are borderline, it may be possible to work with existing soil by following best management practices for tree planting modifications. Deep tilling, adding biochar, and adding organic matter will help dilute toxins. Although soil degradation and contamination are important concerns that should be addressed, they are not always a problem with urban sites. Site contamination is common on many sites; however, it is typically isolated in specific areas such as concrete washout areas and fuel containment areas. Often the issues are easily remediated with a few modifications.

MECHANICAL DAMAGE

Careless operation of equipment around trees can cause damage, compromising health and safety. Damage that creates conditions affecting health and safety can include aboveground and belowground parts of the tree by breaking branches, tearing the bark, wounding the trunk, or cutting trunk flares and roots.

FIGURE 4.12 Any damage to the trunk creates a wound which can affect health and create risk concerns.

Trunk and Crown Injury

The bark layers of a tree serve as a protective outer casing, providing the tree with armor against various forms of injury. The vascular system of a tree lies just beneath the bark, beginning in an area called the vascular cambium or cambium layer. This system includes the xylem and phloem, which are

FIGURE 4.13 Crown damage from construction equipment.

FIGURE 4.14 Root damage is often unnoticed until the tree begins to decline.

responsible for conducting water and nutrients throughout the tree. Any injury to the cambium layer on the trunk or branches creates a wound disrupting the flow within the conducting tissues. If the damage is severe enough, the tree can die. The affected area on the trunk or branches can take months or years to heal. In some tree species that don't compartmentalize well, the wound may never close fully. During this time, fungi and other wood-decaying organisms can enter the tree. Then decay can begin spreading to other parts of the tree, causing increased damage and destruction for the tree and increased risk for the owner or manager.

Construction damage causes health and survival issues for trees and creates risk issues for the people and property around them. Injury can include equipment colliding with the tree and knocking off bark, creating a large wound. Additionally, tree branches can "get in the way" of large equipment, leading to the harsh removal of these branches through the use of backhoes or other excavating equipment, often ripping them from their attachments on the tree. Tree damage aboveground and belowground creates weaknesses in the tree structure and reduces its ability to withstand storms and other weather-related events.

FIGURE 4.15 Rabbits and rodents will feed on the soft cambium of smooth-barked trees.

If the consulting arborist can determine from planning documents which trees are most vulnerable to physical damage due to the proximity to the construction activity, protective measures can be implemented. Trunk protection can be installed in the event the trunk could be brushed with errant equipment. Also, proactive pruning to improve clearance for equipment can reduce the harsh amputation of branches with construction equipment.

In addition to construction or human-induced mechanical damage, animals can cause damage to tree trunks. Some of the most common culprits include deer, elk, porcupines, beaver, mice, squirrels, rabbits, and birds. These animals can cause physical damage to trees by feeding on the inner bark and cambium layer just under the bark, resulting in trunk wounds. This damage will weaken the tree's ability to transport water and nutrients from the roots to the branches and leaves.

Mice and rabbits may girdle the trunks of small trees, especially in winter when less food is available, effectively destroying them. Deer and elk may devour the foliage on the lower branches of arborvitae, pines, and other evergreens. In fall, bucks rub their antlers on various trees to remove the dried velvet from their antlers and mark their territory.

FIGURE 4.16 Elk damage on aspen trees.

So, why does the tree seem to decline and sometimes die after construction activities, often delayed for years? The answer is a tree's reduced health and an inability to perform critical plant functions because of direct or indirect damage to the tree or its rooting environment. For example, as normal photosynthesis occurs, trees store carbohydrates for future needs and use some for immediate demands of plant processes. If a tree was healthy and growing preconstruction, functions such as processing food for energy storage were working to function properly. When that process or any other critical function is interrupted by root damage and soil changes, a reduction in photosynthesis and respiration creates an imbalance, using up carbohydrate reserves and less capable of replacing and restoring the carbohydrate balance. This overdraft in the *carbon currency account* weakens the tree and predisposes it to other issues such as insect and disease infestations. Trees unable to continue basic processes normally will result is gradual decline, dieback, and perhaps death. Protecting trees prior to any nearby activity prevents problems in the future. Preservation is critical to save desirable trees on the site.

TREE PRESERVATION PROCESS

One of the most challenging issues with any construction project is protecting and preserving existing trees on the site. There is a complicated interaction between tree biology and site construction which challenges preservation efforts.

Established and mature trees on a construction site can be preserved if provisions are made to ensure the tree trunk, limbs, and root system are not damaged or disturbed. Regardless of the reason, trees are preserved to provide ecosystem services on site. These include improving real estate values, environmental qualities, and aesthetics which are of prime importance to many property owners.

TABLE 4.3 Tree Preservation Actions

Action Item	Recommended Engagement
Selection	Choosing suitable species which are capable of surviving construction activities and are compatible with the site and its intended use.
Prevention	Place a focus on injury avoidance to reduce loss creating situationally appropriate boundaries.
Intervention	Requires involvement with the site, client, and planners, including mediation with protection challenges.

A tree preservation plan is a detailed proposal for achieving preservation goals on any site and guidance for those involved in the development process. However, the preemptive action often overlooked is creating a project design sensitive to trees on the site. If feasible, modifications to the layout or footprint of the structural elements to reduce impacts can create better results for trees at the conclusion of the project. The design process should include pre-planning consultation with a qualified arborist to determine how the biological and construction elements interact on the site and whether they can be compatible with ultimate building goals. This begins with a site analysis allowing the arborist, developer, and client to discuss client needs and construction necessities for a successful project. How arborists see the site and elements of the site is quite different from an engineering perspective. A final consideration should also include a tree appraisal for each tree designated for preservation. This allows for an actual monetary figure to be placed on the tree in the event of damage. Language should be added to preservation documents indicating liquidated damages will be enforced should injury occur to trees included in the preservation plan. Contractual penalties for violations should be articulated and imposed where appropriate.

TREE PRESERVATION PLAN COMPONENTS

- Preconstruction
  - conduct an inventory and assessment, recording species, age, health, and structural factors to determine suitability for the site
  - determine the protection zones
  - identify and complete work for pruning and removals
  - establish boundaries and physical barriers
  - implement preventative care with cultural practices such as mulching, irrigation, and supplemental nutrition
- Construction
  - monitor and inspect trees frequently
  - communicate with developers
  - enforce TPZ
- Postconstruction
  - implement tree health care
  - monitor trees for decline

PRECONSTRUCTION PHASE

The preconstruction phase begins with a site analysis and is necessary for tree preservation. This phase includes locating all structures and physical features that may affect the well-being of trees and other site amenities. An inventory should be conducted to determine quantity and quality of existing vegetation. It's important to recognize that not all trees can be saved, nor should they be. Accurate identification of sustainable species, compatible with the site and client needs is essential. The decision-making process should include an arborist or qualified consultant to determine which trees are providing functional benefits as well as which trees are compatible with the site design. Additionally, this preconstruction assessment provides an opportunity for implementing proactive tree work which is much more efficient and economical prior to construction activities. This may start with a tree risk assessment and health assessment. This will identify trees that require pruning or removal, presenting an unacceptable level of risk.

Root pruning prior to any site work can be helpful with preservation as well. Root pruning is the process of surgically removing roots before any excavation near a tree. This proactive process can help minimize damage to the tree's root system during construction or in preparation for transplanting. Also, root pruning may be necessary if mechanical damage has occurred by creating clean cuts back to suitable lateral roots. This is very similar to reduction cuts in branch pruning.

When pruning roots, roots less than one inch in diameter can easily be cut with less concern for health impacts and stability issues. However, cutting roots greater than three inches in diameter should be avoided. Roots of this size are very important for anchorage and tree health. If proactive pruning is found necessary, the further from the trunk that root cutting occurs, the better to reduce health concerns.

Depending on the size and species of trees in the construction zone, favorable trees interfering with necessary infrastructure could be relocated to more suitable areas on site. Transplanting desirable trees with a tree spade or other mechanical means is possible with assistance from tree moving professionals.

CRITICAL ROOT ZONE IDENTIFICATION

Damage prevention is best accomplished by identifying and enforcing the critical root zone (CRZ) and the tree protection zone (TPZ) for each tree or group of trees. These zones should be designated as off limits to any project activity. The roots and soil in these areas must be safeguarded to help ensure tree stability and health. If the protection zones are secure, there is less likelihood of serious, threatening injury to trees when prescribed construction limits are identified. The TPZ must be inclusive of enough area to retain sufficient root and crown area. It is important for the arborist to understand exactly how and where roots grow in relationship to the physical tree structure. It is generally accepted that tree roots can grow up to twice the distance away from the tree trunk, outside the dripline of the crown. However, the area under the dripline is considered the critical area of protection (CRZ). The exact size and outline of the zone is dependent on the species, shape, vigor, and project requirements. Rarely are these zones perfectly circled around the tree, rather they are consistent with crown shape and influenced by wind patterns. Often multiple trees in proximity may be included within the same boundaries of the TPZ.

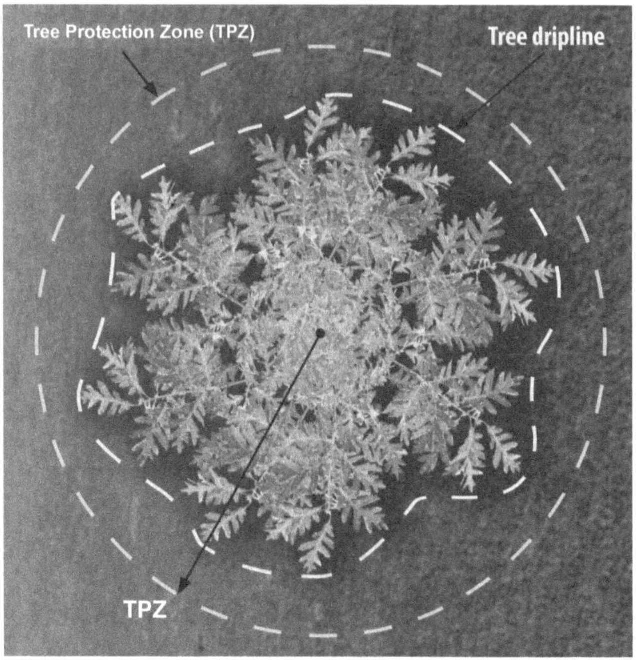

FIGURE 4.17 Plan view of tree protection zone.

TPZ DETERMINATION

There are various methodologies used for determining the TPZ. Regardless of the method, it is important to identify where the majority of the root system is located. Uniform distribution around the tree is typical, however, there are deviations. Impediments to root growth such as solid or impervious structures can alter the location of root growth. Also, environmental

TABLE 4.4 Tree Protection Zone Factors

Developmental Age	Relative Tolerance to Construction Activity	TPZ Factor
Young and newly established *(under 25 years)*	High Med Low	6 8 12
Medium-age *(25–50 years)*	High Med Low	8 12 15
Mature *(50–75+ years)*	High Med Low	12 15 16
Overmature	Low	24 or greater

Note. Based on ISA best management practices. Tolerance depends on tree species.

influences, such as prevailing winds, can create abnormalities in root configuration. Often roots create a more oval shape based on wind direction with the windward side extending farther past the dripline than the leeward side of the tree.

The most commonly applied methodologies used for establishing the TPZ includes using tree height, crown diameter, and trunk diameter as predictors of root spread. The application of tree height to predict root spread has been found to be less accurate in urban and landscape settings. Open grown trees have a wider root spread than forested trees in relationship to tree height. Sites that are subject to drought conditions have a wider root spread. Finally, there is a disparity in consistency among gymnosperms and angiosperms and even among species.

Crown diameter is convenient for visually estimating root spread, but the relationship between crown and roots is highly species and location dependent. This area, also called the dripline method, includes the soil and roots that lie directly below the outermost branches. To locate the dripline of the tree, imagine a ring just beneath the outermost branches encircling the entire tree as the protective boundary. This methodology can be safely used under less intense construction circumstances, designating the entire area within the dripline as the TPZ. However, research has shown no relationship between crown and root spread and these estimations can vary significantly. This means it may not include enough roots to successful protect the tree.

There are limits to which particular species tolerate construction activities, such as soil compression and root severance, as not all trees respond similarly to changes in their environment. For example, newly established and young trees may be more resilient with nearby construction zones. Conversely, older trees or those in a low energy state are more sensitive to disturbance. Additional protective measures should be taken for trees determined to be more sensitive due to their species characteristics and developmental age. There are several references available from research literature providing regionally specific recommendations on tree sensitivity to construction.

ROOT SPREAD PREDICTION

The predictor of root spread considered most accurate in urban locations is the trunk diameter methodology. This method is recommended by industry best practices and is consistently applied. To determine the TPZ with trunk diameter, measure the tree diameter in inches at 4½ feet on the trunk, also known as diameter breast height (DBH). Then, multiply that number by the appropriate factor based on sensitivity, developmental age, and vigor to get the distance in feet from the trunk where the protected area should be established.

For example, if the tree is 16 inches in diameter, multiply 16 by 18, which equals 288 inches (24 feet) to determine the distance from the trunk where protective fences or barriers should be established around the tree. The area inside this barrier is considered the TPZ.

FIGURE 4.18 Poorly installed tree protection is ineffective in reducing damage.

On many sites, especially in urban areas, the TPZ cannot be accomplished according to standards because of space limitations or access on site. The goal is to preserve the soil and root plate supporting the tree by minimizing disturbances within the protective area. Typically, research has demonstrated effective preservation can occur when no more than 25% of the roots in the TPZ is damaged. The proper protective distances for tree preservation should consider proximity to the activity, soil texture, developmental age, and health during the site survey and inventory. When engineers and developers inquire about what is necessary to help save a tree, the response should be, "I'll take all that I can get."

UTILITIES AND INFRASTRUCTURE CONFLICTS

In other construction activities, the installation of utilities, sidewalks, or curbs require cutting the roots, typically on one side of the tree. This scenario should apply a different protective methodology based on current research. This is a concern for stability, especially the potential for windthrow due to lack of supporting roots. Cutting large-diameter roots may make the roots more susceptible to root decay and failure. The maximum size root that can be cut and will not readily decay is not possible to determine. It is possible that cutting roots at a distance of 3 times the trunk diameter or less makes the roots more susceptible to decay than cutting those at a greater distance. Therefore, to be safe when cutting linear roots, cuts should be at the greatest distance possible from the trunk, preferably outside the dripline. Current recommendations indicate 5 times the distance from the trunk is acceptable. However, it is possible that 3 times the trunk diameter can still be acceptable, depending on tree characteristics and health.

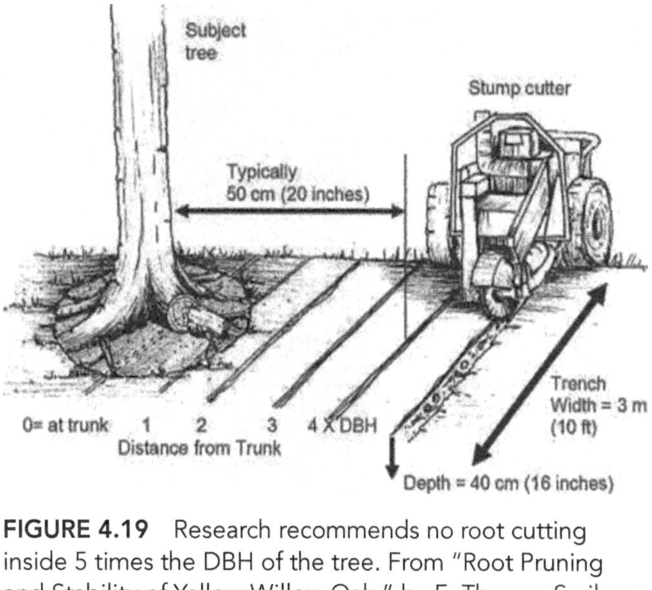

FIGURE 4.19 Research recommends no root cutting inside 5 times the DBH of the tree. From "Root Pruning and Stability of Yellow Willow Oak," by E. Thomas Smiley, 2008, Arboriculture & Urban Forestry, 34(2), 123–128.

Root severance facts:

- Linear trenching should not be closer to the trunk than a distance equal to or greater than 3 times the trunk diameter.
- Cutting roots at a distance greater than 6 times the trunk diameter minimizes the impact on tree health and stability (25% of root system lost).
- Additional cutting within the CRZ on two or more sides increases the likelihood of failure.

If it is not possible to meet the recommended allowances and belowground disturbances do occur, trees may be able to recover. Nevertheless, the key components to surviving construction impacts are the condition and health of the tree before the injury. A healthier tree always has a better chance to recover from damage.

Successfully preserved trees for the client requires that work begins in the planning phase and continues through construction to the post-construction period of any project. Good communication between the client, arborist, project manager and contractor are vital to saving trees.

Guidelines and protective measures that are typically applied include:

- Safeguard barriers: Install a physical and highly visible obstruction around the tree or trees to be preserved. Use durable fencing material, such as plastic construction fencing, snow fencing, or even chain link fencing. Place the barrier as determined in the preconstruction survey, identifying both the CRZ and the PRZ. Visual barriers are important; however, extra duty barriers may be required to physically restrain from construction activities.
- Signage: Place tree protection signs in protection areas and in visible places. All signage should be in both English and Spanish. Place signs approximately every 20 feet on

FIGURE 4.20 Barriers and boundaries must be established to protect trees from construction activities.

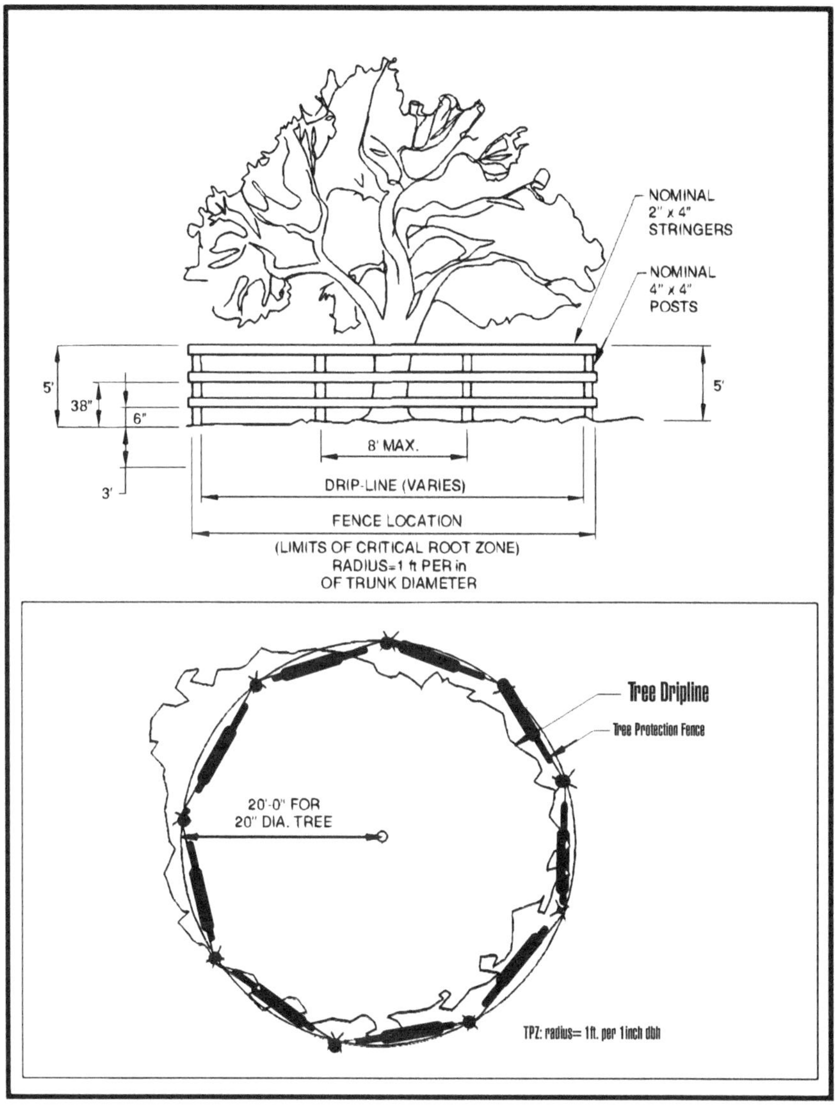

FIGURE 4.21 Sample TPZ specification.

protection fencing to insure visibility. Be sure to check that signs remain present and legible during the entirety of the project.

- Appraisal: An effective method of protection or perhaps a deterrent is a preconstruction appraisal of the trees on the site. This method requires a qualified arborist to conduct an appraisal to determine the replacement cost of each tree. These values are considered valid in the event construction activities cause irreparable damage to the tree. Often, these tree values are provided in the construction documents as a part of the contract, serving as liquidated damages for mitigation. Liquidated damage provisions are often used to identify the amount of damages that a contractor will owe the owner if there is preventable and unwarranted damage to the trees listed in the preservation plan.

CONSTRUCTION PHASE

During the actual development and building process various site activities should be monitored regularly for the benefit of the client and the trees. These include:

- Site access: Identify and prepare access areas for all construction activities. Limit vehicular traffic to designated paths, which minimizes loads on site soils. Prepare and maintain the access ways with at least 6 inches of hardwood mulch to minimize soil compression which leads to compaction. Limit crew and project parking to locations off site. Inspect the TPZ boundary apparatus to be certain it hasn't been removed, damaged, or relocated. Enforce the TPZ.
- Material storage: Do not allow storage of chemicals, materials, or supplies within the protective barriers. Do not allow fuel storage on the site and specify this in the preservation plan. Petroleum-based products can contaminate the soil and are highly toxic to trees. Require refueling, servicing, and maintenance of equipment and machinery preferable in an off-site area or predesignated area with safeguards for spillage and soil contamination.
- Debris and waste materials: Do not allow debris and waste from construction activities within protected areas. Washing down concrete or cement handling equipment should be prohibited on site. This compound is very caustic and can drastically alter the soil chemistry, damaging plants or directly killing them.
- Monitor and inspect: Review tree preservation documents and contractor performance in regular construction meetings. Enforcement, penalties, and mitigation regarding breach of the preservation policy should be discussed, agreed upon, and captured in all construction meeting notes. Place the details of the preservation policy on the project site for reference and reminders. Any tree damage or injuries should be reported to the arborist as soon as possible. Follow-up by the consulting arborist on the site is critical for success of the preservation plan. Random, frequent site visits are the most effective as part of the consulting arborist duties.

FIGURE 4.22 Debris and concrete washout can be lethal to trees.

FIGURE 4.23 The repetitive movement of trucks and equipment compresses soil particles, leading to compaction and root damage.

Some types of damage occur repeatedly on a site due to unintentional harm or just uninformed workers and their activities. Chronic damage that can occur includes vehicular travel and offloading materials on site, often in the same location. This can be prevented or minimized under the scrutiny of frequent visits. Some types of damage, like direct damage to a tree, can be assessed but are difficult to prevent. Many development activities and preservation efforts are mutually exclusive. Decisions must be made early in the planning process to maintain tree health and minimize injury.

POST-CONSTRUCTION PHASE

A comprehensive tree preservation plan doesn't end with construction. A key component of assessing development impacts on trees is the systematic evaluation of the site, site changes, and damages. Even with the best preservation plan in place, injury or damage can occur. Response to construction impacts will depend on the extent of site modifications and the preconstruction health of the tree. The sooner care begins after project completion, the better the chance the tree will survive the impacts. Post-construction care includes similar arboricultural activities as preconstruction preparation to maintain good plant health, primarily through providing good cultural practices. Minimize stress with supplemental irrigation during dry periods, maintain an adequate layer of mulch and monitor the plant for signs of stress or pest activity. Avoid any supplemental nutrition for at least one year after construction. Stress is an inciting factor for opportunistic insect and disease outbreaks and should be monitored closely. It is common practice to conduct post-construction site inspections for a period of up to 5 years to evaluate tree health.

The best strategy for building with trees is to have a qualified consulting arborist or horticulturalist involved from the start of the planning phase. The consulting arborist needs to be realistic and a team player with the developers and contractors. Ultimately, the development is going to occur, so it is essential to realistically minimize losses and collateral damage. Often, when a plan wasn't put in place, the tree professional is called too late when the damage has been done and rescue treatments for mitigation isn't feasible. Considering the important benefits and values that trees provide, it is wise for the contractor or property owner to employ the services of a tree expert and implement a well-supported preservation plan. It might seem easier and sometimes more affordable to just clear-cut the site and plant more trees, but mature trees provide considerable real estate and environmental value. Unnecessarily removing trees takes away the many environmental and economic advantages they provide. Large, mature trees provide many life-sustaining benefits, which humans need for enhanced beauty and quality of life. Preservation is equally as important as planting for the many benefits trees provide every day.

FIGURE 4.24 A consulting arborist involved with the planning phase is essential for preserving trees on site.

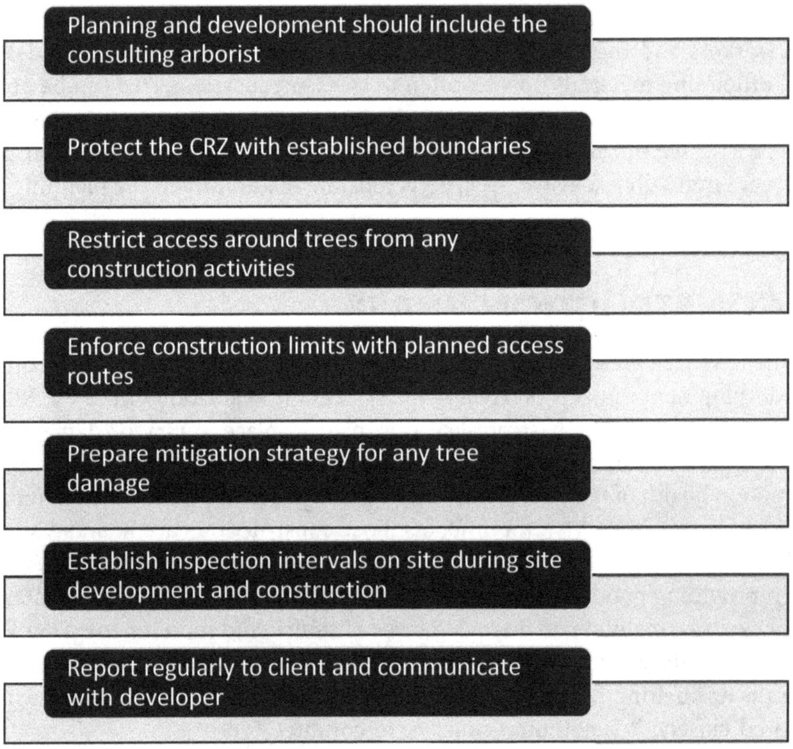

FIGURE 4.25 Preservation prompts.

CHAPTER

5

TREE SELECTION

LEARNING OBJECTIVES

- Recognize the importance of trees in the landscape including their functional and aesthetic value.
- Identify components of the decision-making process for tree selection.
- Analyze how site conditions affect tree selection and establishment.
- Discuss the importance of diversity and reducing monocultures in urban spaces.
- Recognize the impact of invasive species and how they affect the environment.
- Identify some of the common challenges and alternatives for urban tree planting locations.

TREE SELECTION

The built environment is a tough neighborhood for trees. It is an environment unlike any other due to its dynamic and often hostile nature. Let's begin with the belowground considerations. Typically, urban and suburban soils are fragmented, facing pressure from pedestrians and vehicles, impervious surfaces, underground utilities, and are constantly exposed to toxins and contaminants. These might include road salt, petroleum products, and others. The aboveground challenges include heat islands with temperature extremes, pollution, and all the factors associated with climate change. There are persistent extremes and challenges in this "unnatural" area as opposed to the more natural environment of the woodland setting. It's often considered a place where trees too often die young from lack of proper selection criteria, planting, and maintenance.

FIGURE 5.1 The urban planting area can be a hostile environment.

Our urban forests are disappearing at an alarming rate despite efforts to maintain and plant trees in and around cities. Recent research indicates that the average lifespan of an urban tree is less than 25 years; for suburban trees, it's less than 50 years.

FIGURE 5.2 Invasive pests like EAB can destroy large segments of the urban forest.

Construction impacts, poor soil structure, limited soil volumes, drought conditions, road salt, and temperature extremes all work against a tree's ability to thrive and survive in the urbanized growing location. It's no wonder why downtown, urban, and suburban ecosystems yield such short-lived trees.

The trees on every street, in every town, and spread throughout every city is a part of the overall urban forest. Often there is a misperception of only trees "downtown" or in the core city region are included in the urban forest, but wherever people and trees are together, it's measured as part of the urban forest. Trees provide economic, social, health, and environmental benefits for everyone. Trees could be considered the perfect biological machine with the ability to lower particulate pollution, reduce storm-water runoff, sequester carbon, decrease energy use in the summer, and improve air and water quality. Because more people are coming to realize this, society is placing greater focus on ways to reduce anthropogenic impacts on the environment, which means considering trees more closely. However, these functional benefits are only possible by creating longer lived trees through informed selection and realistic maintenance.

Successful tree selection requires the arborist or horticulturalist to think backwards—beginning with the end in mind—to get the right tree in the right place and installed in the right way. Let's identify some important factors of the decision-making process for tree selection.

DIVERSITY IN THE URBAN FOREST

Tree species richness is important for resiliency. Planning for increased diversity increases the resiliency of the urban forest and functional services. Tree inventories in most of our urban forests reveal a lack of diversity, with as few as two or three genera accounting for nearly half of the tree population. Poor tree diversity in the past has led to catastrophic results when there have been invasions by pests, such as Dutch elm disease. More recently, emerald ash borer and Asian long-horned beetles have destroyed thousands of urban trees. Due to these host-specific pests, entire populations of trees can be wiped out. For example, central Indiana is nearly 50% maple species and about 10% ash species. With 60% of the tree population composed of two genera, certain pest invasions could destroy 6 of every 10 trees Protecting the urban forest is possible with choosing trees with diversity in mind!

Climate resilient cities includes strategies such as reducing peak summer temps by increasing canopy cover with climate-ready species. However, it is important to recognize more trees isn't always the answer—it is a part of the solution. Just as important is the realistic maintenance inputs for better health and quality of trees. Actually, we need fewer trees to achieve the desired canopy cover for improved quality of life. The focus should be shifted to improved management techniques along with better selection criteria based on tree qualities and their functional ecology, especially in response to drought conditions.

There are a few popular tree genera that are particularly overplanted, as mentioned, primarily maple species in many communities. Identify those plentiful genera and species found in the inventory data and avoid them in planning future tree installations. The well-recognized guideline for diversity in the urban forest continues to be Santamour's Rule. The first published reference to the 10–20–30 rule was by the late Dr. Frank Santamour, research geneticist at the U.S. National Arboretum in his paper "Trees for urban planting: Diversity, uniformity, and common sense." The 10–20–30 rule is a guideline to reduce the risk of catastrophic tree loss due to pests. The rule suggests an urban tree population should include no more than 10% of any one species, 20% of any one genus, or 30% of any family, and has proved to be an effective formula to protect urban forests. Recent opinion is that the rule is inadequate primarily due to the globalization of our economy with international shipping and the exotic pests accompanying. Regardless, it is far preferable than the planting strategies found in most communities. If we consider the current issue with emerald ash borer (EAB) in North America, following the 10–20–30 rule means we would accept the loss of 20% of our urban canopy since both commonly planted ash species (*Fraxinus pennsylvanica* or *F. americana*) are highly susceptible to EAB.

However, don't think exclusively about native trees for the planting scheme but also consider monotypic tree selections as well. The fewer the species types a genus has, the less opportunity there is for pest issues. For example, Kentucky coffeetree (*Gymnocladus dioicus*) and katsura tree (*Cercidiphyllum japonicum*) have only one species for each and literally no pest issues. One of the most popular and over-planted tree species are the maples (*Acer*) which have over 125 species and over 200 known pests. Species-rich tree genera are prone to many pest problems. Species diversity is not as important as genus diversity; pests attack a genus, not just a species. Native isn't the answer; choose genera with the smallest distribution range and fewest species that seem to be resistant to catastrophic disease and insect issues such as emerald ash borer on ash trees and oak wilt or bacterial leaf scorch (BLS) on oak trees. Critical review of an inventory prior to planting is imperative for preparing a diverse, climate-ready urban forest system.

One of the inherent challenges in the 10–20–30 rule is implementation. Where is the tree species population in question? Are we talking about a city? A neighborhood? A block? If there are 10 trees on a block, should they all be different species? Some urban foresters have proposed alternatives to the 10–20–30 rule, such as the "look around rule." This guideline states that if you're choosing a tree, look around and if you already see that tree, select something else. The problem with diversity on a very small scale is we can end up with the "menagerie effect"—one of these, one of that, one of those—which often lacks aesthetic appeal and can be a maintenance challenge. Ultimately, this becomes a challenge for urban foresters and planners; working together, we incorporate diversity guidelines within established design principles.

NATIVE, INTRODUCED, AND INVASIVE SPECIES

The native tree for a region isn't always the most pragmatic or guaranteed choice for the best tree selection. Recently, people have increasingly focused on native plants as the answer to fewer pest issues and for long-term survival. However, native plants do not always guarantee sustainable landscapes nor resistance to insect or diseases. Native plants suffer equally from pests and environmentally induced decline, especially in the built environment. Besides, there isn't any tree native to the urban forest. In some urban areas, many native trees just don't perform well without significant maintenance inputs. Parking lots, lawn strips, and street tree planting vaults create harsh growing conditions for any trees. Typically, these sites have limited soil volumes and face exposures to pollution, deicing salts, and other contaminants. In these cases, some non-native, cultivated tree species may be better suited to the poor growing conditions, including compaction, minimal organic matter, soil texture, and tolerance to heat extremes. It is critical for planners and decision-makers to design for decline by providing planting locations with increased soil volumes.

Ecologists and environmentalists publicize the advantages of using native plants in residential and commercial landscape as well as urban plantings. Some of these endemic selections may perform quite well in less disturbed sites such as parks and green spaces. However, this narrow perspective is leading to problems. In an effort to "go native," inappropriate tree choices are based on inaccurate perceptions of the designation of a "native" tree. The term "native" can have varying definitions among concerned groups. Most definitions are based on geography and time scale, which are often determined by the point at which human intervention and impact occurred. The U.S. Forest Service in 2012 provided the definition, "A native plant is an endemic species that occurs naturally in a plant community, ecosystem, ecoregion, or biome habitat without direct or indirect human involvement." However, this definition may be considered too broad in many cases. Many organizations now refer to a native plant in the U.S. as having existed prior to European colonization. Most agree human influence at the landscape level was identified as a point in time at which many native species' home ranges began to shift. Since trees are such long-lived plants, there presents a need to begin planning for the shift of climatic characteristics within tree populations. Some of this conversation has centered on a shift in phenotype as opposed to a complete species shift. Many urban foresters and ecologists are looking toward southern genotypes of individual species as viable selections. In the face of climate change, a new definition of native may begin to emerge for many tree and shrub species. Recent research suggests that anywhere from 25% and 85% of species may already be changing their home ranges as our climate warms through acclimation.

REASON FOR PLANTING A TREE

The first step of tree planting is to determine why you are planting a tree. When it comes to tree selection, understanding *why* you are planting the tree will help decide *what* tree or trees to choose. This can help greatly with the installation's ability to provide the intended ecosystem services. In other words, knowing the desirable functions the tree is to provide such as flowers, shade, a screen, a barrier, or a windbreak, will help determine the best tree species for the role. Consider basic landscape purposes including beautification, screening views, providing a sound barrier, energy conservation, and wildlife habitat. Well-placed shade trees (i.e., overstory trees), can reduce cooling costs for homes or businesses and can increase pavement life. Also, properly planned ornamental trees (i.e., understory trees), can screen harsh views and reduce traffic noise.

BENEFITS TO PROMOTE AND PROBLEMS TO AVOID

FIGURE 5.3 Tree fruiting such as those from a Ginkgo tree can create safety issues.

Functional benefits (ecosystem services) are a primary consideration for urban foresters and city planners. However, the ornamental or aesthetic values of the tree shouldn't be discounted in tree selection either, especially for the private tree owner. It's obvious to look at tree characteristics in the selection process. During the selection process, consider seasonal leaf colors, tree form, winter interest, leaf texture, flowers, and fruits as important factors. Some tree species provide beautiful displays of color for specific time periods in the spring or summer. Other tree species develop fall foliage color displays, adding visual interest at the conclusion of the growing season. Tree foliage vary widely with textures and shapes and can also add visual interest with proper placement of coarse and fine-textured leaf species, providing illusions of depth in the planting. The contrasting textures of trees, shrubs, and other plants can be a design benefit all in its own if planned thoughtfully.

Trees bearing edible fruits and nuts provide excellent sources of food for many species of urban wildlife as well as people. However, some tree owners and managers consider some fruit- and nut-bearing trees to be "dirty" or a nuisance, especially around sidewalks, patios, and streets. These same trees may pose risk concerns in some situations. In the urban landscape, it's essential to consider fruits that are attractive to insects and other pests to minimize the risk of infestation. For example, fruit from the female Ginkgo tree creates an undesirable aroma, but also, it can create a slick surface on sidewalks. Likewise, fruit trees are welcome in the right place; however, their flowers and fruit can also create conditions of concern. Fruit trees in residential and commercial areas can attract rats, raccoons, and other urban animals, which may spread disease and damage homes. The flowering process attracts bees and other pollinators, which can be beneficial to our flora and fauna, but dangerous for those highly sensitive to the impact of a sting. Trees that produce copious volumes of pollen can cause health problems for people with allergies.

BEAUTY STILL MATTERS

Finally, be sure the tree offers the ornamental characteristics desired by the client, such as overall appearance, flowering, fall color, fruit, winter interest with its bark, or evergreen foliage. Trees are important for their functional qualities, but they should complement the landscape and provide visual beauty as well. A major focus of early research on the

FIGURE 5.4 Seasonal interest, including flowering and bark characteristics, should be taken into consideration when selecting trees.

social aspects of urban forestry was on how people perceive and value the beauty of trees in cities and towns. Since then, researchers have documented that the urban forest can generate utilitarian tree benefits. While the functional utility of trees is critical for our quality of life, there is ample research to show that beauty still matters as a reason for planting trees. It isn't necessary or prudent to promote one at the expense of the other; urban forest advocates should highlight how these two kinds of values, functional and hedonic, reinforce and support each other in enhancing the quality of urban life.

MATCHING THE TREE WITH THE SITE

Careful plant selection involves several important steps for developing a sustainable landscape. When planners, ecologists, and landscape professionals discuss sustainability, they are referring to trees and landscapes responsive to the environment, regenerative, and can actively contribute to the development of healthy communities. Proper selection of trees can yield longer-lived trees, which can provide maximum benefits with fewer maintenance inputs.

SITE COMPATIBILITY

Sustainability doesn't solely revolve around plant survival with minimal maintenance; it encompasses broader considerations. One of the most important considerations of selection is ensuring the tree is matched appropriately to the site. It is implied that the urban environment is different than most locations in a plant's native range. The urban environment is a unique ecosystem, unlike any other due to the extreme environmental pressures it exerts. These pressures should prompt landscapers and arborists to explore a wider range of plants with the ability to survive in tough conditions. Horticulturalists and nursery growers continue to cultivate native species, as well as discover and introduce new plant species, to broaden the plant palette and increase available choices in the face of changing climate impacts.

CLIMATE IMPACTS

Key factors influencing tree selection include not only shifting climate impacts but also water requirements. Drought often follows the path of shifting conditions to warmer and often more extreme seasonal variation. The adaptability of our urban trees will be more important in the coming decades. As mentioned earlier, due to research in

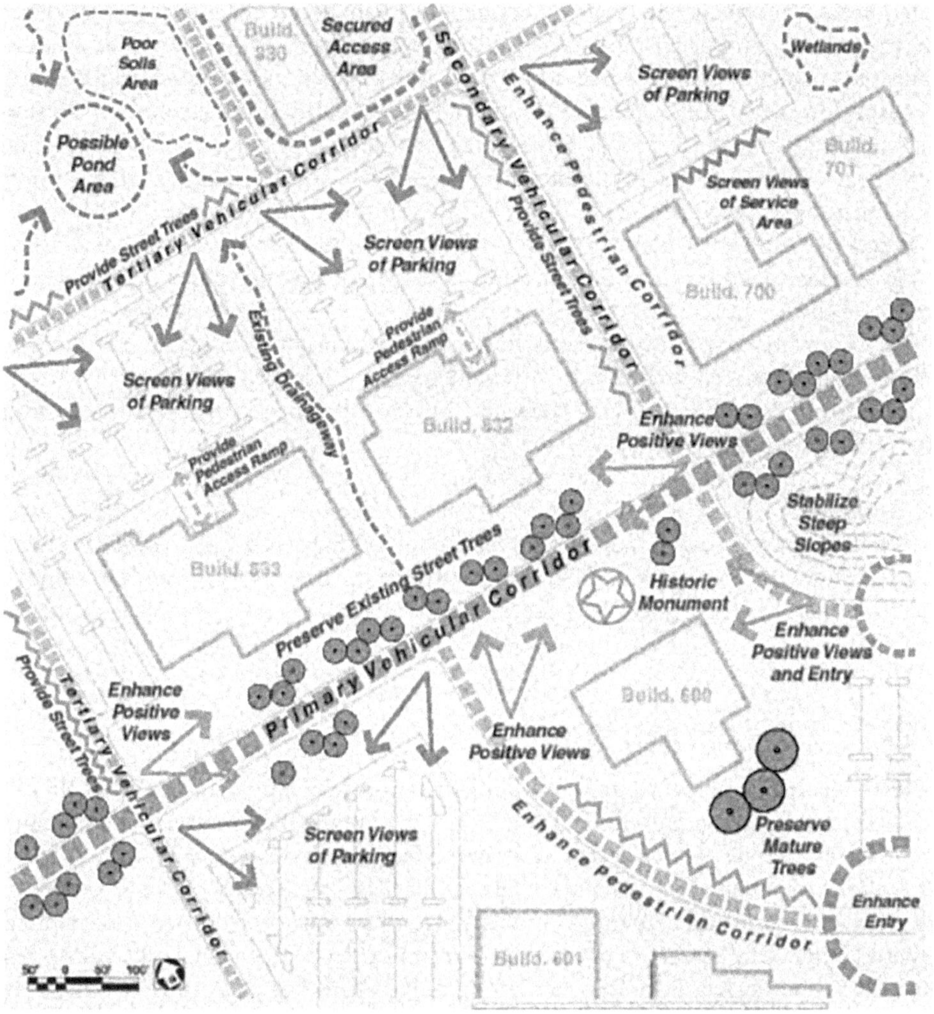

FIGURE 5.5 A site analysis presents site features and design features.

breeding and selection, well-adapted trees that are nonnative aren't considered at all. If it is determined that native tree selections will not perform well in the new location, nonnative species should be considered.

Unfortunately, simply considering moving warm weather trees from the South to the North won't work and is more complicated. Due to the polar region warming significantly more than other areas, the polar jet stream (aka polar vortex) is becoming disrupted, which moves extreme cold air into the United States. The result of this phenomenon are periods of much colder temperatures than average lows. Moving plants from the South to the North will result in killing these marginally cold hardy plants any time the polar jet stream becomes disrupted.

FIGURE 5.6 Dangerous pests like Spongy Moth can destroy entire oak stands.

The occurrence of climate changes is evident from an increase in global average temperature, changes in the rainfall pattern, and extreme climatic events. These seasonal and long-term changes affect our trees and the population dynamics of insect pests. Tree-damaging pests are noticeably increasing in urban areas due to elevated temperatures. The temporal dynamics and temporal asynchrony in our urban forest create increases in pest development with increasing damage to our overall canopy. Additionally, changing climate will increase the amount of abiotic (nonliving) stress to trees in the urban environment, thus increasing the amount and severity of pest outbreaks in the urban space. Studies indicate that trees grown under elevated CO_2 levels alter the nutritional value of plants, leading to changes in insect abundance and an increase in the consumption rate of insect pests. The rise in temperature plays a pivotal role in insect populations indirectly as well. Research has found than an increase in impervious surfaces around trees resulted in a significant increase in pests. Since much of the urban landscape is surrounded by parking lots, sidewalks, and driveways, trees in these locations are at an increased risk of pest problems.

Climate resilient cities includes strategies such as reducing peak summer temperatures by increasing canopy cover with climate-ready species. However, more trees aren't always the answer; they can be a part of the overall solution. Just as importantly, it is about maintenance inputs as well as better health and quality of trees. Fewer trees can achieve the desired canopy cover and ecosystem services if there is an increased focus on improved management strategies.

WEATHER CONDITIONS

The next consideration in the selection process is compatibility with local weather conditions, specifically, temperatures. Plants differ considerably in their response and tolerance to temperature extremes, as well as to inconsistent precipitation. First, temperature range must satisfy the development requirements at all seasons. Also, temperatures must not exceed the individual minimum and maximum limits during any stage of development. The parameters of these temperature extremes, specifically minimum temperatures, usually lead to winter injury. Each species has an average optimum temperature, but different functions of plant processes have optimum ranges as well. Trees and shrubs require a range of temperatures to complete their reproductive and vegetative phases, encompassing flowering, seed production, fruiting, leaf development, and dormancy.

Sun exposure is also a factor in selection. Typically, large canopy trees prefer full sun, allowing the tree to receive day-long exposure to the light energy from the sun. Many small, ornamental trees prefer at least partial shade since they are typically naturalized as understory trees, meaning that some protection from the summer heat or direct sun is important. Refuge from the sun's radiant energy in the afternoon helps reduce water loss and prevent leaf scorch. If full shade is required by the tree, protection from direct exposure to the sun most of the day is needed for best tree performance.

HARDINESS

Each tree has a recommended hardiness zone in which it will survive. The distribution and adaptation of trees to their environment are limited by temperatures in a couple of ways. The USDA Hardiness Zone Map indicates the zone for your area; it is the standard by which arborists, horticulturalists, and growers can determine which plants are most likely to thrive

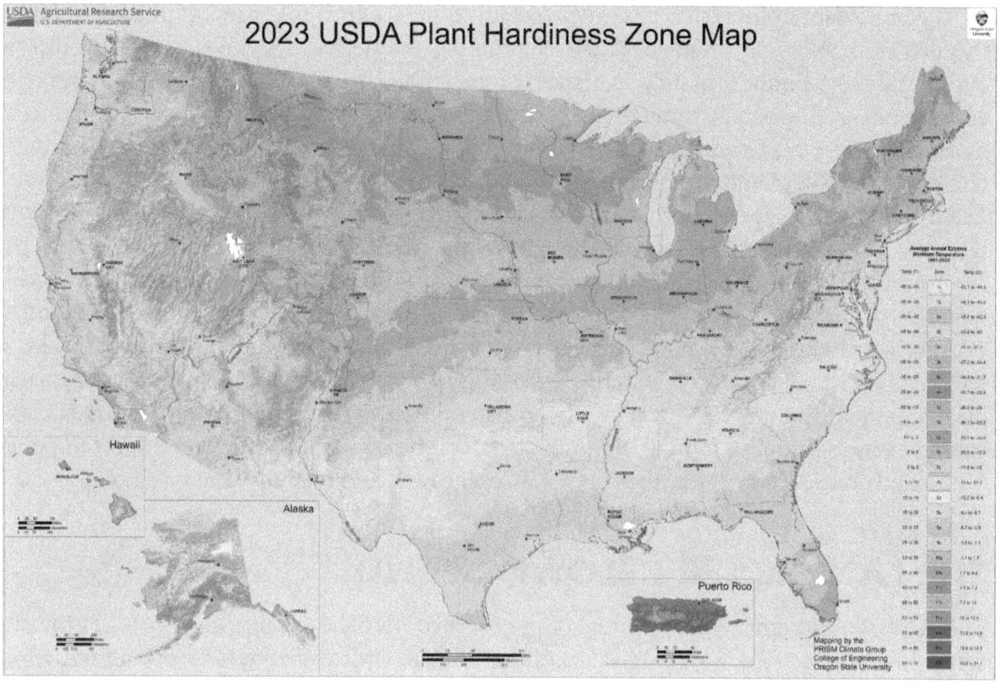

FIGURE 5.7 USDA hardiness zone map.

at a location. The map is based on the average annual minimum winter temperature divided into 10°F zones. Some trees can push the boundaries of their zones if placed in microclimates. Microclimates are localized areas where weather conditions may vary from the norm. A sheltered planting site with structures, or even other vegetation, may support trees not normally adapted to a region with perceived warmer conditions. However, the opposite can occur where a north-facing slope may be significantly cooler or windier than surrounding areas, limiting the survival potential of normally well-adapted plants. Be aware of microclimates on the site or property, as they can be either be a potential benefit or a detriment for trees.

Typically, in temperate regions, damage to trees from poorly timed cold snaps or extreme lows are of prime concern. The cause of this winter injury or damage isn't always hardiness of a tree or plant or necessarily low temperatures, or whether they were in zone 5 or zone 6. The problem is the sudden shift from warm temperatures to cold temperatures during a short time during the pre-dormancy or post-dormancy phase. If plants are not properly hardened off as they go into dormancy, plant cells are typically damaged by intracellular freezing. The term "hardening off" refers to the process necessary for a plant to become acclimated to its environment. In the fall, trees begin to harden off or acclimate themselves to winter conditions beginning with shorter days, followed by lower temperatures. A steady decrease in temperature until maximum dormancy is reached provides the plant protection from cold temperatures in the winter. Deciduous trees and shrubs undergo many physiological changes in preparation for the cold months. Acclimation for evergreens (gymnosperms) is similar. Solutes are produced to guard against freezing by lowering the freezing point of water, but evergreens don't fully reach a dormant state in the same manner as deciduous trees (angiosperms). They continue processes such as photosynthesis and respiration, as well as experience water loss through transpiration, during the winter months.

The important point is that it is not direct low temperatures that damage plants but rather the plant's inability to manage widely fluctuating temperature swings, especially during endodormancy. Endodormancy is a state characterized by the physiological inhibition of coming out of dormancy, induced by cold temperatures and/or short photoperiods. Cold damage can also occur during the later stages of dormancy. After the chilling requirement is completed, the plants are no longer in endodormancy. They are now in ecodormancy. Ecodormancy is a type of dormancy that occurs due to adverse environmental conditions, typically low temperatures. During this phase, the plant is actually ready to grow, but the environmental conditions, such as low temperatures, inhibit spring expansion. This is different from endodormancy, where something inside the plants is inhibiting development.

Another essential fact is that cold hardiness maps only help to predict winter survival and are certainly no guarantee! The hardiness of a tree or any plant is dynamic and dependent on many factors. They do not take into consideration other predisposing factors such as summer's heat, drought, rainfall, humidity, and soil types—all of which are important environmental considerations for plant survival and growth and play a role in hardiness.

THE LANDSCAPE BELOWGROUND

The belowground environment is one of the most overlooked factors when considering site compatibility questions. Some tree species can tolerate wide differences of conditions, while some trees have more specific requirements. Often, the best choices for sustainability are those trees that have the gained a reputation, frequently through anecdotal information from local arborists and horticulturalists, for their ability to overcome tough urban sites. Also, look to nature for insights into what trees prefer in their natural environment. Identify the trees growing in defined habitats such as riparian or flood plain areas and those trees growing in more upland, dry areas for hints on what tree does best for the intended planting location.

Trees capable of adaptation to their environment may naturally perform better in urban areas that present similar conditions. The planning process for tree establishment requires an understanding of the hydroperiod for the site as well. A hydroperiod can be defined as the number of days per year that an area of land is wet or the length of time that there is standing water at a location. This includes planting areas which may be waterlogged for an extended time. For example, riparian or river bottom species accustomed to prolonged wetness, which have higher moisture requirements can't be expected to thrive in dry urban conditions without supplemental irrigation. The opposite applies as well. Another consideration in urban forests related to natural environments is root development. Riparian trees, which experience continual and frequent flooding, have adapted with shallow root systems, allowing better survival during prolonged flooded soil conditions. These types of root systems may be more adaptable to compacted soils, which are frequent in urban conditions. Consequently, these tree types planted in urban areas can be expected to generate shallow root systems, causing potential infrastructure damage to walks and curbs. Conversely, the

FIGURE 5.8 Tree and infrastructure conflicts such as sidewalk panel lifting create risk issues.

FIGURE 5.9 Small tree planting pits; reduced soil volume typically means a short-lived unhealthy tree.

FIGURE 5.10 Larger tree planting pits; increased soil volumes create larger spaces for tree roots and a reservoir for water.

deeper-rooted system of the upland, mesic species, such as some oaks and beech species, are naturally deeper rooted due to the drier soils requiring deeper penetrating roots to seek necessary nourishment. In nature, trees that grow in clay soils with calcareous parent materials may also thrive in urban areas with similar soil conditions. Naturally occurring trees growing regionally in these dry, droughty conditions will likely cope more successfully with the same conditions in the urban forest.

One of the most important questions to answer includes: Is there enough space, or soil volume to support long-term growth? This is particularly important to consider in urban tree vaults and lawn strips. Large trees need room for roots to stabilize the tree as well as provide water and nutrition.

A tree with an anticipated canopy growth of 30 feet in diameter and a 12 inch-diameter trunk prefers a minimum of 1,000 cubic feet of soil. This is the equivalent of just over 30 full-sized pickup truck beds. When it comes to good tree growing conditions, the smaller the soil volume, the less of a reservoir available to the tree for water and root expansion. Typically, lawn strips should be at least 5–6 feet wide and 25–50 feet long to accommodate large-canopy trees and reduce sidewalk, pavement, and curb damage. Smaller, understory trees may survive with less soil area and not create infrastructure issues. In most cases, urban foresters must work with what is available. Often in harsher urban conditions, urban forestry managers, planners, and landscape architects are forced to design for decline. In the context of urban forests and trees in the built environment, it refers to the idea of creating tree installations that can adapt to or survive with a situation of decline, which means a reduction, deterioration, or loss of something. This is especially true when it comes to limited soil volumes. Sometimes they are included in the planning and development process. Basically, when planning for city trees, take all the soil space possible while still being compatible with infrastructure goals. When it comes to tree planting space, bigger is always better.

Awareness of *site history* benefits tree life predictions as well. Research of the site's past may reveal soils are disturbed and fragmented or compacted from previous construction. This information can provide preemptive opportunities for pre-planting mitigation to improve conditions. Site interventions, such as aeration, improve air and water exchange between the belowground and aboveground environments. Soil tests can provide information on potential contaminants such as deicing salts, past industrial site toxins, or other chemicals affecting soil health and plant survival when planting in urban areas.

FIGURE 5.11 Riparian trees have shallow root systems, incompatible with most street and sidewalks.

SOIL FEATURES

Soil conditions which include poor soil texture, high bulk density, and pH that is too high or low are often limiting factors for success, and selection and planting methods should be adjusted to fit soil conditions. Heavy clay soils and compacted soils found in many landscapes require careful thought during the selection and installation process. The water-holding capacity of the soil and infiltration rates are very important. Compacted clay soils tend to have poor drainage and most trees do not survive prolonged wet conditions. Additional information can be found in Chapter 3 Basic Needs.

Only species and cultivars tolerant of wet sites can survive in these conditions. Check out the trees growing naturally in wet areas and this will help you identify the right trees for those areas. When selecting trees, particularly those with inherently aggressive root systems like many riparian species, it is important to exercise caution

FIGURE 5.12 Benefits of trees can be quantified to measure the functional benefits provided.

because they often develop many surface roots. These large, shallow roots can disrupt lawn mowing operations and damage curbs, sidewalks, pavement, and other nearby structures.

Check the planting site for compaction and infiltration rates. This is a proactive technique that can preemptively improve planting success by implementing a perc test. Dig a few holes at least 12 inches deep around the site in and around the prospective planting area. If the soil is very challenging to dig into with a shovel, it is likely quite compacted. If the soil is fairly easy to dig into with a shovel, it is probably not suffering from compression that leads to compaction. In times of drought conditions, most soil will be difficult to dig into. This should be considered when assessing the site. High levels of compaction with increased bulk density is a prime determinant for drainage problems. Drainage can be determined by filling these holes with water and monitoring their leaching rates. Water remaining in the holes for a prolonged period will positively indicate poor drainage. Planting modification will be required to overcome this extended wetness in the root zone.

Tree species exhibit significant variation in their preferences for soil acidity. The *soil pH* value is a measure of soil acidity or alkalinity and should be tested in a soil lab. The pH scale ranges from 0 to 14, with 7 as neutral. Numbers less than 7 indicate acidity while numbers greater than 7 indicate alkalinity. The pH value of soil is one of several environmental conditions that affects the quality of plant growth. The soil pH value directly affects nutrient availability. It is important to have an idea of soil pH to avoid the inevitable nutrient availability issues that accompany trees planted in soils with inappropriate pH values. Site pH may vary too much to plant the same species across the entire site. Those planted in improper soil pH conditions often require continuous management with micronutrient applications, special nutritional programs, and soil amendments.

FREEDOM TO GROW

A crucial aboveground consideration for sustainable tree growth is providing a tree with enough room to grow naturally with minimal maintenance. Too often, trees are planted in spaces that do not permit mature, natural growth. Failure to recognize the anticipated size of overstory trees, and even understory trees, is a major reason why trees may require excessive, often damaging, and disfiguring pruning, ultimately leading to early removal. Spatial realism is recognizing that trees grow and expand beyond what we can visualize during the planning phase of selection. When planting a small tree, it is often difficult to imagine that in as few as 10 years it could be shading a large part of the yard. Many of our common landscape trees add over 2 to 3 feet of terminal growth annually! Closely planted trees can have health issues when they conflict with each other as well. Competition for light and space often

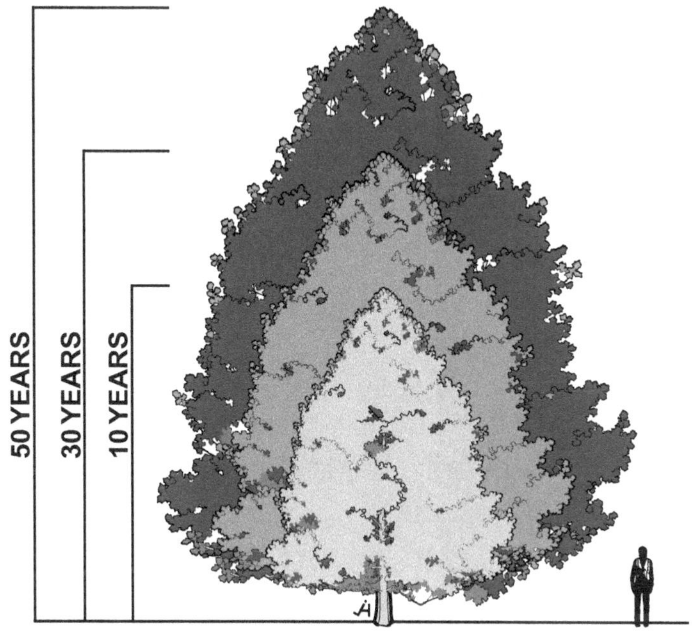

FIGURE 5.13 Sustainable tree selection requires understanding mature growth and size.

FIGURE 5.14 These trees will soon outgrow their design, competing with each other for space and resources, requiring removal.

leads to dieback for both competitors. Many homeowners desire the "instant landscape" which leads to overplanting. Overplanting is expensive, and money is needlessly wasted by installing too many trees, too close together. However, some plan for this situation with the intent to thin or cull out some trees as they mature.

Large shrubs and small, understory trees that grow larger, such as some species of magnolia and viburnum species, are viable competitors for landscape space. Consider mature size before planting to provide for informed spacing decisions. Inspect the area around the tree-planting site for potential conflicts with overhead and underground utilities. Responsible urban forestry and landscape design includes identifying trees which complement site features making them an integral part of the overall landscape. Site structures and trees can coexist in the same vicinity if planned appropriately. The goal of successful tree placement is to complement and not compete with buildings, street signs, traffic lights, and more. Unfortunately, many trees are planted and later needlessly removed or severely pruned just as they are reaching a functional size. Urban foresters are often faced with the problem of restricted spaces, and recognizing growth rates and mature sizes is imperative for sustainable plantings. Be sure the tree fits the location aboveground and belowground. Consider the mature size of the tree and if it will grow on the site without conflicting with site infrastructure, especially utilities.

FIGURE 5.15 Parking lot islands are considered heat islands stressing trees with limited water and soil rooting space.

HYDROLOGY AND WATER REQUIREMENTS

Another important indicator of sustainability is species selection based on their response to drought and limited *water resources.* The paved areas of parking lots and sidewalks provide little space for soil and supporting tree growth. The results are trees in a decline continuum that are only surviving and not thriving, which means reduced benefits from the tree.

Water is the most limiting factor on the site affecting tree establishment and survival. Tree species vary in their ability to manage abnormally dry conditions. Also, the requirements of tree species vary as well. Some tree species require greater amounts of water to survive such as many species of *Acer sp.* and *Platanus sp.*, in contrast to other tree species which are less demanding of this valued resource, including *Quercus sp., Beech sp.*, as well as other upland species. Selection can have a major impact on municipal budgets as well as any commercial or residential sites. Identify the maintenance requirements of the tree before installation to determine if it is acceptable and feasible. Basically, try to avoid trees on sites require constant care, such as pest management applications, pruning, and frequent irrigation. Right tree, right place considers all site influences.

PLANT SOURCING AND PROVENANCE

We're all familiar with the phrase "you get what you pay for" and does apply to many facets of our life, including tree selection. When it comes to selecting and planting trees, superior trees come from quality sources. Nursery stock should be inspected from top to bottom for functional roots, good branch structure and pest-free plants.

"In this world, you get what you pay for."

—Kurt Vonnegut, Author

The planning process involves sourcing trees, the process of procuring trees and plants in accordance with client needs and specifications. Reputable growers and nurseries typically provide well-managed crops which have been cultivated according to horticultural best practices and ANSI Nursery Standards. Responsible procurement includes knowing the provenance of your plants. If the trees are being sourced locally, then usually there aren't any concerns relative to local climate. This can be an issue if trees are purchased from a region with a distinctly different hardiness zone which may not survive the winters after planting. However, this information isn't always available or reliable, especially when it comes to the source of rootstock in grafted trees.

Successful selection requires knowledge of the genotype, provenance, and source. Functional ecology explains why and how plants are successful in their locations. Often, it comes down to species that can readily establish and possess strong reproductive capabilities. Ruderal species favor areas which have been disturbed. This is a tree layer primarily dominated by "weedy" or native tree species, sometimes accompanied by exotic/invasive tree species which is not a favorable situation for areas such as urban woodlots or neglected landscapes. Provenance plays a role in performance, with knowledge of their native range, genetic history, and cultivars as important determinants. This can be accomplished in many ways but requires diligent research and information which often isn't readily available.

FIGURE 5.16 Planning for planting involves careful selection, taking into account factors such as size and site.

The correct choice of tree species in relation to site characteristics and local climate is an essential prerequisite for sustainable plantings. Tree plantings intended for the future need to be capable of coping with adverse

FIGURE 5.17 Nursery tree with poor crown management including a codominant leader.

FIGURE 5.18 Improved crown development with structural pruning creating a single-dominant leader.

climatic conditions, in particular drought and flooding, in order to improve their longevity. Trees grown in local nurseries for several years have adapted to those local conditions. Trees produced in other regions with different climate conditions shipped to a new location may not be as adaptable to the new site location. Safer selection includes sourcing plants from similar climate conditions, especially considering factors such as average annual low temperatures and soil type, whenever possible.

Tree health during the selection process is important for the establishment of a quality, vital tree, surviving the planting process. Trees should exhibit vigorous growth during the most recent 2 to 4 years and have well-formed buds. Normal twig growth may vary from 6 to 12 inches annually and should be characteristic of the species. Examine internodal growth patterns to determine current health conditions. Trees should be vigorous specimens, free from insects and diseases. The bark on the trunk should be firm, with no indication of disease, wood borers, dieback, frost cracks, sun scald, or mechanical injury. Tree wrap should be removed so that the trunk can be inspected for hidden wounds. The trunk can be rewrapped after inspection to prevent wounds from shipping and transport. Trees with large trunk injuries should be avoided in the selection process at the nursery.

Crown uniformity is less important than trunk form and branch arrangement. However, a uniform crown represents a detail accomplished by attentive growers. Trees with irregular canopies, a dominant trunk, and well-arranged branches are far better than trees with uniform crowns and codominant leaders. The crown on well-structured trees will fill in as the tree develops. Choosing a nursery tree with good structure can postpone future pruning and maintenance. Trees with poor structure could require more pruning cuts, and a greater portion of the canopy will have to be removed to correct poor structure.

Branches should be well spaced, both vertically and around the circumference of the trunk, for strong attachments and unrestricted flow of water and nutrients. Branches should diverge from the trunk with the proper aspect ratio of 2:1 for good stability.

Better quality shade trees have one dominant central leader or stem, presenting a *good leader system*. Avoid any tree with codominant stems or leaders. Codominant stems are two vertical trunks or stems that are nearly equal in diameter and are competing for the same role, resulting in a weaker structure. Trees with codominant stems often have included bark (bark imbedded between codominant stems) and present an increased likelihood of failure and should be avoided. Multiple leaders represent weakness and can cause the tree to split apart as it grows and experiences loading forces such as wind. Some smaller ornamental trees such as crape myrtle, birch, serviceberry, and others may be grown with multiple stems and do not have to be corrected.

GROWING METHOD AND TYPES

Quality nursery stock should be chosen from a reputable grower. The commercially available types include container grown, field grown (balled and burlapped or B&B), and bare root (BR).

Container-grown plants are traditionally developed in smooth-walled plastic containers or pots. However, many growers have turned to alternatives, such as fabric containers, root-pruning pots, and wooden boxes for larger trees. Each type has its advantages and limitations. It is important to note that containerized plants are installed, whereas field-grown trees are transplanted. There are specialized containers incorporating root pruning to minimize the formation of stem-girdling roots, which are typical in smooth-walled pots. Nearly all container-grown trees are lightweight and easier to transport, handle, and install, unless they are the larger box containers designed for moving and growing larger trees. Also, the absence of a digging process allows availability throughout the year. The major drawback to container-grown trees lies primarily in the container itself, which typically results in smaller tree sizes and the potential for girdling/circling roots when compared to B&B trees. The often-well-developed root system is restricted within the container and the lateral roots will form a circle as they are deflected from the sides, primarily on smooth-walled pots.

Stem-girdling roots (SGR) are formed when new roots grow perpendicular to a cut root, or when the tree is growing in a container for too long. A quick field test can be performed to check root quality, especially in containerized or container-grown trees. Gently push the trunk back and forth. The trunk on a good quality tree will bend along its length and will be firm in the soil or medium. The trunk on a tree with a defective root system will often pivot at its base and will lean over before it bends, revealing a gap between the soil stem flare,

FIGURE 5.19 Smooth-walled containers create circling roots leading to stem-girdling roots.

FIGURE 5.20 Stem circling roots will grow with the tree, eventually girdling the tree.

often caused by circling roots. While passing this test doesn't guarantee defect-free trees, it is a good way to determine if there are concerns. Also, another quick test on smaller trees is to gently raise the tree out of the container to inspect the roots. Root-bound plants often lack soil media, with a very thick root mat present at the perimeter of the soil mass. Research has demonstrated the type of container strongly influences the conformation of the root system in the tested species. In these studies, plants grown in traditional, smooth-sided containers showed the highest percentage of dysfunctional roots compared to plants grown in containers with air-pruning or mechanical impediments.

Alternatives to the smooth-walled plastic containers for growing trees, include fabric packages and bags, just to list a couple. Fabric bags are especially popular for evergreen species in the nursery. These growing systems allow the young tree to grow with fewer circling roots, encouraging lateral growth. When planted, the roots can spread into the new surrounding soil more rapidly to establish a stronger network of roots.

When installing "containerized" plants or container-grown trees, root ball damage is minimized as the plant is transferred to the soil. They are typically available for year-round installation. Another advantage is that the entire root system goes with the tree into its new planted location. However, without treatment, a dysfunctional root system will rear its ugly head later as the tree develops. Root pruning or root control bags are like containers in that they are grown in the medium and bag for a period of time and harvested when the desired size is achieved. Planting procedures resemble those for container-grown plants and require the complete removal of containment materials. Minimum root ball diameter and stem caliper apply to container-grown trees as well. Generally, 2-inch trees should have 20-gallon containers, with container size progressively increasing as the caliper of the tree grows.

Field-grown trees are often harvested and sold as B&B plants and are produced in the nursery's native soil with no holding material. The roots / soil balls are harvested with special mechanized equipment or hand-dug and wrapped in burlap and wire baskets.

FIGURE 5.21 Fabric bags encourage better root systems.

Field-grown trees are typically harvested during the dormant season, either in the fall between when leaves drop and the soil freezes, or in the spring after the soil thaws and before budbreak. Waiting until plants are fully dormant reduces demands on the plant for water and nutrients and limits stress. During the harvesting process, as much as 80% of the existing roots are severed and left behind in the field. Often growers practice root pruning as a way to encourage more root growth within the area that will be harvested. For example, the grower will slice through lateral roots around a tree 5 years before harvest, and again 2 years before harvest, leaving the tree in the soil. New roots will branch from where the root was severed, increasing the number of fine roots within the harvested root zone.

Transplant shock is a term that refers to several stresses occurring in recently transplanted trees and shrubs. It involves the failure of the plant to adequately establish new roots, consequently the tree begins to decline shortly after installation. Newly transplanted trees do not have extensive root systems and are frequently stressed by lack of sufficient water. Research has shown that root-pruned stock has the highest survivability rate when transplanted into "moisture-limited" environments.

In *balled-and-burlapped production,* the root ball is lifted with a spade or similar machinery and placed in a wire basket that is lined with biodegradable burlap. The balled-and-burlapped (B&B) techniques allow larger trees to be harvested more easily than other nursery methods. Balled-and-potted, also known as field-potted (FP), stock is dug from the soil it was grown in and placed into a container instead of being held with the wire and burlap technique. While both of these harvesting techniques need to be watered regularly to prevent the root ball from drying out, they may not need to be watered as frequently as container-grown plants because the roots are held in natural soil. However, natural soil does weigh more than the materials used to wrap bare root plants or the soilless media used in container production. This added weight increases the cost of transporting B&B trees. Be sure the trees are consistent with nursery standards relative to ball size and caliper. For example, a general ratio for trees would be about 12 inches of soil ball width for every inch of trunk caliper. Undersized soil balls can be problematic in establishment and sustainability with weakened and inadequate root systems.

FIGURE 5.22 Balled and burlapped trees are larger in size and weight.

FIGURE 5.23 More than 80% of harvested B and B tree roots are left behind in the field.

B&B plants can be quite large and provide the largest cultivated trees available for purchase in most nurseries. Typically, these are available throughout the growing season. Usually, the best selection is in the spring and fall soon after the growers have harvested for that season. Of all the cultivation methods, balled and burlapped stock can be the most challenging from a physical standpoint, with the extreme weight of the soil ball posing a serious impediment to planting. A 3-inch caliper balled and burlapped tree can weigh anywhere between 250 and 400 pounds depending on the soil type.

Bare-root production includes field grown trees reaching the predetermined size for that species, much like B&B plants. However, in bare-root production, the soil is removed from the plant's roots before transportation to the end user. Bare-root (BR) trees are rarely marketed for commercial landscape design and construction. BR trees are usually purchased by nurseries to finish growing in the field or in containers. Bare-root deciduous trees are dug from field soil and receive no media covering on the roots. They are usually available only when dormant and in a limited size range. They are typically harvested as soon as possible before packaging and shipping or are stored dormant in coolers over the winter until delivered to nurseries in early spring. The greater risk for this growing type is damage from the root desiccation. Research has shown that bare-root plants have a higher mortality rate in the landscape

FIGURE 5.24 Balled-and-burlapped trees often have a larger caliper than other types of growing methods.

compared to container- and B&B-produced trees. Special handling and maintenance are required to keep these tree roots viable up to the transplanting phase. Because there isn't any weight from soil or bulk from containers, shipping bare-root plants is less expensive than other production methods and is easier on the tree planter. The post-planting procedures are like any other media. However, support systems are a must to keep trees standing vertical. Large bare-root tree transplanting is becoming increasingly popular and new planting techniques show promise for transplanting larger trees with no heavy soil to move. This protocol requires the use of air excavation equipment and is discovering increased application with moving larger, established trees.

PLANTS ESCAPING THE LANDSCAPE

Unfortunately, some of landscape species found in nursery and garden centers can escape into unintended areas, such as natural areas and low-maintenance spaces like roadsides and rights of way, creating ecological complications. Natural resource managers and urban foresters are having more difficulty with invasive exotic plants that have escaped from managed landscapes, displacing native plants, and disrupting natural ecosystems. The use of these invasive species is not advised, except in very limited situations.

An invasive plant species is typically non-native to the region it is introduced to and by introducing it into the landscape, creates environmental harm. Harmful characteristics include several traits favorable to trees, but when it becomes overly aggressive in nature, it's a problem. These traits include prolific seed production, which allows the tree or plant to spread rapidly into unwelcomed areas such as natural woodlands. The seeds of invasive plants are often readily distributed by birds, wind, or unknowingly by humans, allowing these plants to be moved anywhere. Some invasive trees have aggressive root systems that spread long distances and often grow so densely that they smother the root systems of surrounding vegetation. Additionally, some plant species produce chemicals in their leaves or root systems, which inhibits the growth of other plants around them.

The term "escape" is often used to describe the migration of a species into other neighboring areas through movement by birds, squirrels, wind, or other mechanisms that can disperse seeds. There is evidence around the country of several frequently used landscape plant species invading these unintentional areas as unwelcome visitors. Invasive plant species spread quickly and can displace native plants, prevent native plant growth, and create monocultures within favorable conditions such as a woodlot or open grass area. When a nonnative plant enters a new environment, it must adapt to survive. Invasive species may exploit a resource native species cannot use, which allows them to establish in their new environment. Some species can alter the environment in a manner that makes it more favorable for them, but less favorable for natives, which is called ecological facilitation. This is commonly the case with plants that can create toxic exudates (allelopathic chemicals) through their roots, leaves, and fruit which are toxic to other plants. These chemicals harm other plants, which allows the invasive plant to increase its range, as its chemicals wipe out native competitors.

It is important to recognize just because a tree or plant is available in the wholesale nursery or garden center, doesn't mean it is an appropriate selection. Often, trees and other plants for resale are determined by consumer demand with little regard to their invasive impact. In most states there are invasive species laws that prohibit selling, trading, moving, and transporting invasive species or those rapidly becoming a nuisance. Consult local laws to ensure a prohibited tree isn't being selected and planted.

TREE SOURCING AND AVAILABILITY

Availability from the nursery or grower is a factor in selection as well. It isn't helpful to propose a tree selection if it cannot be obtained within reasonable efforts. Availability can be a seasonal challenge based on harvest times or it can be influenced by supply and demand. The nursery stock should have limited pest issues, including diseases and insects.

There are several choices of nursery stock available from the nursery grower. Choosing the best nursery production method for each situation is essential. Each method has certain benefits, and a single type is not always best for every planting situation. Often nursery stock type is completely dependent on the grower and their preference. Investigate nursery growers and suppliers considered for sourcing trees. Check affiliations with nursery associations and look for references from horticulturalists, designers, urban foresters, and arborists. To learn about nursery stock standards, production methods, and other nursery-related plant materials, refer to the American Nursery Stock Standard (Z-60.1).

Listed are the recommended *minimum* rootball sizes, per trunk caliper, that can promote tree recovery following proper transplantation.

Trunk caliper	Rootball diameter
½"	12"
¾"	14"
1"	16"
1½"	20"
2"	24"
2½"	28"
3"	32"
3½"	38"
4"	42"
4½"	48"
5"	54"

FIGURE 5.25 B and B nursery stock root ball size recommendations.

MINIMIZE THE INPUTS, MAXIMIZE THE BENEFITS

Selecting the proper tree requires the arborist or horticulturalist to research the right tree for the right place, considering all the site and tree factors that can affect sustainability. Tree selection should be based on performance, not pedigree. This process also requires being realistic about the maintenance resources available for the tree after installation. A comprehensive, long-term plant health care program is necessary to help urban trees live longer with the ability to provide those important ecosystem services.

The primary purpose of any planting is not just to plant trees for the future, but also to plant trees with a future.

CHAPTER 6

THE TREE INSTALLATION AND ESTABLISHMENT PROCESS

LEARNING OBJECTIVES

- Describe the techniques and procedures used to plant and transplant trees.
- Identify the primary types of nursery stock, the exclusive installation techniques, and the benefits and limitations associated with each.
- Recognize the considerations identified during a site analysis which may influence installation protocols.
- Recognize the influence of rooting space on sustainability and tree longevity.
- Define the maintenance requirements of newly transplanted trees and the establishment procedures for early care.
- Describe the techniques used for tree support systems and the protocols for maintaining them.
- Explain the benefits and potential problems of staking or guying newly planted trees.

FIGURE 6.1 Right tree, right place provides a sustainable, long-lived tree.

As a society, we have been planting trees since as early as 2000 BC by the Egyptians for various reasons using varying methods. It is a seemingly simple task often understated, yet trees are living shorter lives due to poor planting practices. Research developments and discoveries continually improve tree installation and establishment protocols which help expand the lifespan of trees in the urban forest.

A properly planted and maintained tree will grow faster during the establishment phase and live longer by following best management practices and current research. It is more than simply digging a hole and burying the tree. The professional plantsman is responsible for designing a suitable growing environment which provides optimal growth and development, above ground and belowground. Survival is not the only goal for a newly installed tree, it is also to thrive and develop into a natural resource providing important functional benefits. This chapter will focus on some of the details with establishing commonly grown nursery stock and the essentials of early tree establishment.

TIMING FOR ESTABLISHMENT

Many times, we've heard the adage, "You can plant a tree anytime the ground isn't frozen." This anecdote may be true, however, there are optimal times to plant the tree, especially when considering the initial establishment period. Conceptually, trees can be planted almost any time of the year. Seasonal transplanting is important to consider primarily for environmental influences such as soil and air temperatures, moisture, and daylength. The other factor relates to physiological stages of the tree relative to dormancy, bud break, and flowering sequences.

In more temperate areas there is less biological activity in the fall and early spring, due to the changing stages of dormancy, which can increase planting success without major maintenance inputs. For example, there is increased precipitation typically in the fall and spring, a major factor in establishment for minimal inputs. Tree planting is only a success if establishment, which is basically expanded root growth after the planting process, is accomplished. There aren't many research-based guidelines providing preference to spring or fall planting efforts overall. However, studies indicate there may be optimal times for various species. Generally speaking, faster growing trees will perform better as a fall transplant, while slower growing trees will perform better in the spring. Unfortunately, there are too many variables related to climate, transplanting size, and nursery stock type to be consistent with any particular season of the year.

SPRING PLANTING PROCESS

Spring is a popular time for planting, especially in the temperate zones, due in large part to warmer weather and consumer demand. Post-planting survival is also likely to be greatest and post-planting care least demanding when semidormant trees are planted in early spring. Summer installations can be effective with the understanding additional care will be required. More frequent supplemental irrigation and monitoring is mandatory due to elevated temperatures increasing water demand from a smaller root system and full leaf set. Evapotranspiration rates often exceed the availability of water and the capability of roots to meet the demand of water loss through the leaves. This is the most stressful planting season for just about any type of nursery stock.

FALL PLANTING PROCESS

Trees transplanted in the fall may have additional time for late season regenerative root growth helping prepare them with additional roots and resources before the summer weather demands. Soil temperatures should be above 40°F for several weeks after planting to initiate new root expansion. Once soil temperatures drop below this threshold, biological activity is reduced. The longer the warmth persists into early winter, the greater the advantage, as it leads to expanded root growth during the establishment phase. Late fall planting increases the risk of desiccation, which is the drying of leaves or needles, as well as cold injury, and typically is not advised. This is especially true for gymnosperms, as they are more prone to water loss through their needles. It is advised to avoid planting evergreens in the fall and in the dormant months due to the considerable risk of water loss through the needles. Ensuring an ample water supply before the ground freezes is critical, and even providing supplemental irrigation when the ground thaws will help reduce tree loss.

Regardless of the calendar, neither spring nor fall have a consistent advantage for establishment. It is more about local conditions favoring transplanting requirements, primarily adequate moisture and realistic maintenance expectations after the trees have gone in the ground. Anecdotal evidence suggests that some trees do transplant better in the spring than fall and others transplant better in the fall than spring. However, there are many variables which dictate those parameters.

ABOVEGROUND SITE ANALYSIS

Several environmental factors, such as light intensity, wind, pollution, and existing trees, should be considered in the aboveground evaluation. Also, urban infrastructure, including utility lines aboveground and belowground, street and security lights, buildings, signs, and walkways, are considerations. Local ordinance and regulations play a role in the process. Many public rights of way require permitting for planting.

Adaptation to different regions of the country is mostly determined by their hardiness zones. The hardiness zone map, developed by the United States Department of Agriculture, specifies the average lowest winter temperature expected for regions in North America. When choosing trees for a planting site, be sure that their origin and growing location are compatible with the site.

FIGURE 6.2 Urban trees experience increased stress with limited planting spaces and heat.

SUN EXPOSURE

Trees suited for full sun to partial sun to partial shade will adapt to a site receiving 3 to 6 hours of direct sun. Trees that require direct sun protection are adapted to sites receiving less than 3 hours of direct sun. These species tend to prefer morning sun and afternoon shade. Most shade trees or overstory trees grow best in full sun. Sunlight reflected from glass or a wall on buildings can increase the heat load on a tree planted near a building, increasing transpiration rates and inducing marginal scorching. Drought-tolerant trees growing in full sun are best suited for this intense urban condition. In addition, providing a large area of soil for roots to develop often helps trees withstand reflected light because the trees have access to more soil from which to absorb water. However, the trees will likely require supplemental irrigation as well.

WIND EXPOSURE

Wind tunnels in "urban canyons" are a consideration in the core urban or downtown areas. The wind tunnel effect happens when wind encounters a tall rectangular building. As the wind hits the building, it changes direction. Usually, it will go in the direction of least resistance, typically downward.

Trees and people experience these accelerated wind speeds felt flowing between tall buildings. The closer proximity creates a smaller space for wind to travel, leading to a drop in air pressure, and causing the wind to move faster and circle between the two buildings. Wind increases the amount of water lost from a tree as it moves across the leaf surface drawing moisture from the green, leafy areas.

FIGURE 6.3 Wind tunnels found in downtown urban areas increase wind speeds and increase water loss through the leaves.

POLLUTANTS

Pollutants aboveground and belowground are a consideration as well on tree health and survivability. Air pollutants have a negative impact on plant growth, primarily due to interfering with resource accumulation. Many air pollutants, such as O_3 (ozone) and NO (nitric oxide), affect the metabolic function of the leaves and interfere with net carbon fixation by the leaves. Air pollutants deposited on the soil, such as heavy metals, salts, and industrial chemicals, affect the functioning of roots and interfere with nutrient absorption. When air pollution stress co-occurs with other stresses (e.g., water stress), the outcome on growth will depend on the vitality of the tree and its interaction with internal systems and external point source pollutants. Often, the results begin with a chronic exposure, leading to a decline spiral towards mortality.

PLACEMENT AND LOCATION

There can be many potential conflicts with trees and infrastructure in the landscape, along streets and rights of way. This can even include nearby trees. When placing new trees, it is essential to consider the interaction with adjacent trees in terms of available space and biological needs, such as water and nutrients. Trees planted between existing established trees may grow slowly or not at all due to intense root competition, shade, and resources available. Young trees do tend to develop large crowns over time and require full or at least partial sun (oaks, beech, etc.). They often bend toward the sunlight and develop a one-sided canopy when planted under a canopy of established trees.

FIGURE 6.4 Trees under power lines are subject to harsh pruning to maintain proper clearance for safety and reliability.

FREEDOM TO GROW

Spatial reality addresses issues around the planting area within the anticipated growth zone and the space above the tree, an area often overlooked. For example trees are far too often planted inadvertently too close to utility lines creating a serious safety issue for people and a bleak future for the tree.

FIGURE 6.5 Select cultivars with specialized growth habits compatible with urban infrastructure.

Trees are often located in parking lot islands that share spaces with overhead security lights. Eventually the tree canopy will grow into the structure, blocking the desired light. This requires regular pruning to clear the light, which results in a deformed canopy. Good planning involves placing trees and security lights away from each other, positioning lights about 12 feet from the ground so that the tree canopy can grow over the light, or selecting trees that remain small at maturity so that they remain under the light.

Building conflicts, which may involve garages, homes, offices, and other buildings, are not only a tree maintenance issue but can also cause damage to the structures themselves. Trees are most functional and stable in the ground when they develop a uniform root system with lateral roots distributed evenly around the tree. If a tree is close to a structure, the root system created will develop unevenly. Unbalanced root systems can reduce stability, leaving trees subject to windthrow. A tree with a narrow crown or fastigiate form may be a better selection when working within 10 feet of a building. Also, consider greater distances for reducing structural damage. If shade is desired, consider planting several small-stature trees to create a closed crown arrangement.

Street and traffic signs with trees nearby are frequently in conflict with each other due to poor planning and selection. Plant large trees near low signs and small trees near taller signs. Large-maturing trees could be in the way of a low sign for several years after planting, but if the tree is grown with a single trunk, lower limbs can be reduced and eventually removed so the sign remains visible. Additionally, consider line of sight when planting near intersections, driveways, and other vehicular and pedestrian entryways. Be sure to plant trees that follow local guidelines for spacing and height requirements when near public streets and walkways. Refer to local ordinances for direction.

BELOWGROUND ANALYSIS

As mentioned in the urban sites chapter, there are tree planting challenges belowground and aboveground which imperil tree growth in urban and suburban zones. Trees need freedom to grow naturally aboveground and especially belowground to develop a sustainable root system to help ensure longevity. Soil attributes affecting tree selection are pH, drainage, depth, salinity, and impediments to root growth, such as common hardscape. Many plantings fail because these factors are improperly evaluated or ignored completely with the mistaken belief there won't be a problem. A thorough site analysis will identify soil conditions and provisions necessary to improve it. Quality soil and volume is a valuable commodity for tree planting. Preconstruction planning gives the consulting arborist the opportunity to work with developers to compete for improved conditions and space.

SOIL ROOTING SPACE

The most limiting factor of urban tree root development is rooting restrictions. Tree selection should consider projected crown size and DBH with the soil volume available for root

FIGURE 6.6 Tree planting vaults have limited soil and water supplies for health tree growth.

FIGURE 6.7 Increased planting areas increases soil volume for rooting and available water.

growth. This strategy helps keep trees healthier and more stable. It also prevents damage to surrounding sidewalks, curbs, and pavement. Soil under pavement is typically poorly aerated and heavily compacted, a situation considered unfriendly for roots, unless the soil is coarse sand and well drained. More aggressive root systems of some tree species will follow the moisture from condensation that can gather beneath concrete pavements which often results in damaged surfaces.

The challenge of inadequate rooting space creates a major conflict for trees in paved areas. The traditional method of creating a tree vault in the walkway pavement is insufficient, regularly leading to a short lifespan and certain. There has been much research on making informed decisions on the amount of suitable soil space trees need to maintain vitality and achieve the desired crown development. Researchers have created methods to determine the recommended soil volume necessary for a tree to develop a healthy root system with its projected crown area.

The *mature crown method* involves calculating soil volume by estimating the projected area of the live crown and considering anticipated water loss through evapotranspiration. The minimum recommendation is 2 cubic feet of soil per square foot of crown projection for the selected species. Crown projection is the anticipated area under the drip line of the tree at a realistically expected maturity. The average crown spread is the average horizontal width of the crown, taken from dripline to dripline as one moves around the crown. It is important to consider functional soil volume to be approximately 3 feet for tree root area. This formula can apply to fastigiate and columnar habit trees as well. The belowground mass will eventually be quite comparable to any other crown form and likewise should have comparable soil volumes.

The *growth level method* found in Table 6.1 provides suggestions on soil volume based on the anticipated or projected growth of a tree in the urban environment. These simple suggestions are based upon the research of tree performance under varying conditions. There are limitations, however, to any methodology.

Climate conditions are an important factor when considering soil volumes. It is about understanding and accounting for ET rates. Urban vs. suburban can be quite different in water loss rates. Similarly, the same tree species in a hot, arid climate such as Texas would require more water and more soil volume than in Indiana. Basically, areas with hotter and drier climates require more soil volume (a bigger canteen).

TABLE 6.1 Growth Level Method

Growth Level	Suggested Soil Volume
Understory (< 25′ height)	500–600 cf of soil
Intermediate (25–40′ height)	1000–1200 cf of soil
Canopy (> 50′ height)	1600–2000 cf of soil or more

WATER AVAILABILITY

The key consideration for soil volume is water availability and the water holding capacity (WHC) of the soil. These soil characteristics should be identified for long-term maintenance predictions. WHC is the ability of a certain soil texture to physically hold water against the force of gravity. These soil particles hold the water molecules by the force of adhesion. As an example, a sandy soil has much less water holding capacity than a silt loam soil. Due to the size of the soil particles, the adhesive properties are much different between a sand particle and a clay or silt particle. It is important to know the water holding capacity of the soil to determine the water storage capacity of the planting area, and to determine how much supplemental irrigation should be applied. Basically, when it comes to negotiating soil volume with any development plans, take all the space available to get the best results for the tree.

SOIL CHARACTERISTICS

Soil chemistry should be acknowledged prior to selection. Many common species have a preferred range of soil pH in which they perform best. Soil pH governs the availability of nutrients to plants and affects the activity of soil microorganisms. These pH values may vary even on the same site or location, making it impractical to select the same species across the entire planting design. Most trees in the temperate zones can grow in soils with a pH between 5 and 7.5 levels. If the soil acidity level is near or less than 5, select trees known to be accepting of acidic soils. If the soil pH level is greater than 7, select trees tolerant of alkaline soil types.

FIGURE 6.8 Percolation tests can determine drainage rates. Poor drainage can be lethal to trees.

Most soils encountered in urban and suburban landscapes are often compacted and poorly drained. Construction and building activities compress soils to unacceptable growing conditions if precautions are not taken. Be sure to test for compaction and drainage before planting. This can be accomplished with a percolation test. This is performed by digging several holes or an individual hole at least 12–18 inches deep around the site. If the soil is very difficult to dig with a shovel, it is likely compacted. If the soil is fairly easy to penetrate with a shovel, it is more acceptable. Take into account soil moisture when digging since dry soils are much more difficult to dig. Drainage can be determined by filling these holes with water and measuring infiltration rates. If the planting pit is empty within a few hours, drainage is adequate. If the soil is very compacted and hard all the way down to the bottom of the planting hole, then wet-site-tolerant trees are most appropriate. However, expect many roots to develop at the surface. Soil probes or Dutch augers are handy and less invasive tools to investigate soil texture and compaction. These instruments can quickly determine soil moisture, compaction, and texture.

FIGURE 6.9 Soil probes are valuable tools used to investigate soil conditions.

UTILITY CONFLICTS

Last and certainly not least is checking for underground utilities. Minimum distance clearances are required by utility service providers and unacceptable proximity will result in severe pruning, disfigurement, and damage to the tree as it develops. Contact the local utility service provider for recommendations and rules regarding any tree planting with adjacent utilities. Many states and utility companies provide a free service for locating and marking underground utilities. This is very helpful for avoiding contact with buried wires and pipes when digging a planting hole. Always call 811 before you dig!

TREE INSTALLATION

There is no better way to promote a stable, sustainable tree planting than by following proper planting procedures up to date with research and best management practices. Trees have basic needs which are critical to long-term survival. The balanced tripod includes air, water, and suitable soil. An imbalance or lack of any of these components leads to mortality. The installation of trees requires special attention to creating the planting pit to accommodate the tree in its individual location. As simple as it may seem, there are important aspects to consider when developing the planting hole or pit to encourage rapid establishment and survival.

Research and practice have discovered planting hole depth is a factor in not only survival but also stability. Planting depth is about establishing the correct placement of the root flare. Excessive planting depth has been the culprit of many tree failures and premature tree death. Conversely, trees can be planted too shallow, where too much of the root ball is established above grade; this is typically found when planting new trees on berms and slopes. If the right tree is in the right place, and the arborist or horticulturalist can manage these important components, the tree can survive and thrive in the urban forest.

FIGURE 6.10 Excessive planting depth leads to failure and mortality.

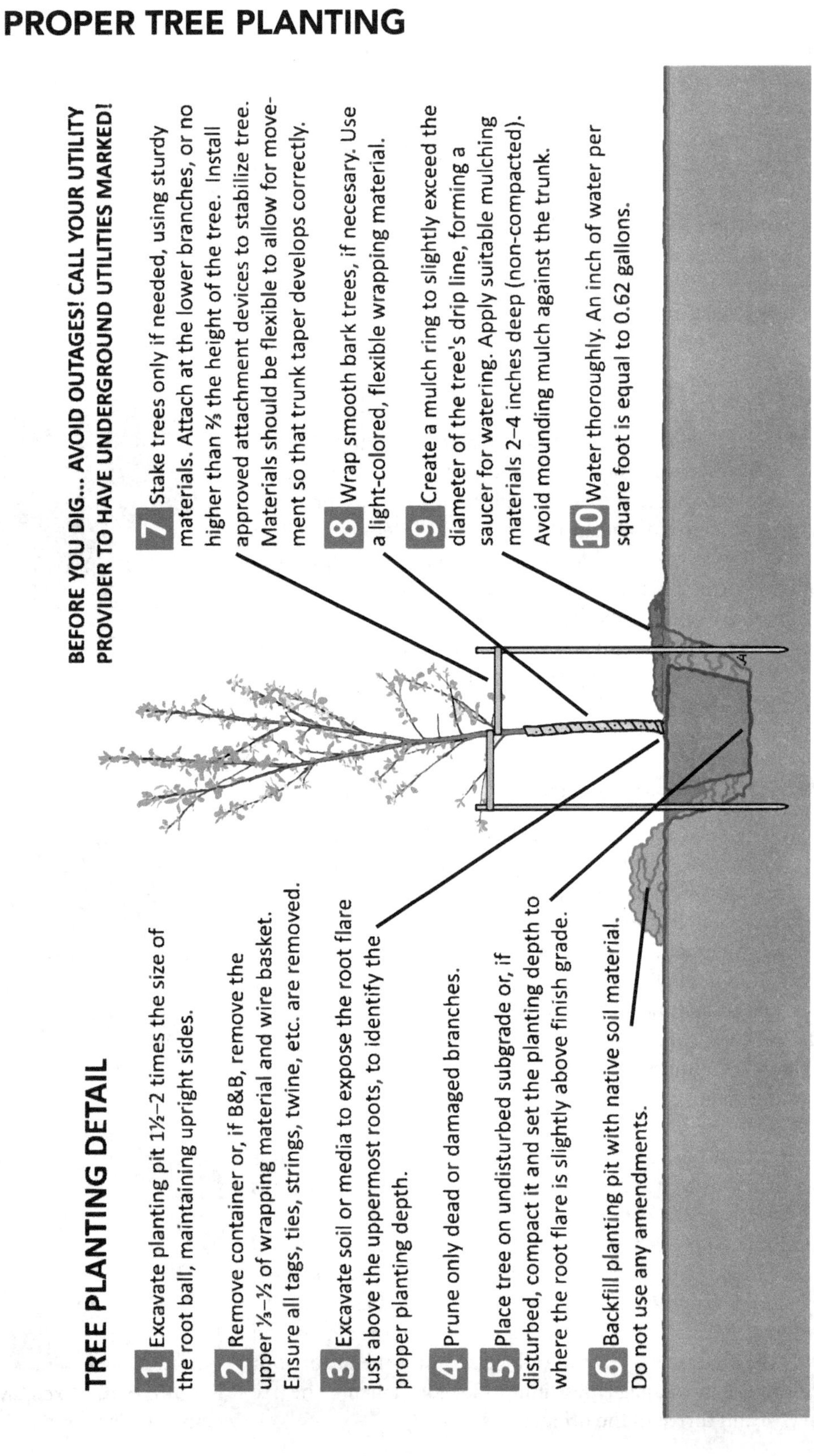

FIGURE 6.11 Accurate planting specifications are critical for developers and designers.

Planting Hole Preparation

Recent findings have shown that a shallower and wider planting pit is one of the better planting preparation techniques. The depth must be estimated with some accuracy by measuring the distance between the point where the topmost root emerges from the trunk and the bottom of the root ball. Then excavate the soil slightly less than this distance. No more than about 2 inches of the root ball (with root flare identified) should be above grade unless the site is found to be poorly drained. If the soil is poorly drained, modify planting protocols. If the hole was inadvertently dug too deep, add soil to the bottom of the hole and compress lightly. When considering width in the process, it's important to note this metric has been proven to be less critical as hole depth.

There are many recommendations for planting pit width and wider is often better. However available planting space is often limited due to walks and curbs in urban areas. Start the installation process with the hole 1 to 1.5 times the diameter of the root ball. Wider holes should be used for compacted soil and wet sites. This facilitates quicker root spread and prevents the slowing of lateral growth at the edge of the hole in compacted or clayey soils.

Digging in soils with high clay content may result in a planting hole with a glazed, smooth surface, especially when dug with a mechanical auger. This effectively creates a clay pot difficult for roots to penetrate. Rough up the sides to allow better root penetration into the new soil environment.

When planting pit size is limited or larger pit sizes are not possible, it is advisable to rototill a 12 to 24-inch ring of soil around the hole as deep as possible. If a tiller isn't available, use the shovel to cultivate the soil manually after planting. This will allow easier penetration into the once compacted soil by the developing roots into the surrounding soil space. If an air excavation tool is available, nearly any soil situation can be modified to improve rooting. The ability to hasten root growth translates into quicker establishment. Do not underestimate the positive effect these techniques have on tree establishment in unfavorable soils. The tree planter must make accommodations to meet the conditions and prepare the best possible tree and planting pit.

ROOT MANAGEMENT

It is important to manage the root system before actually planting the tree. This is the point where dysfunctional roots can be identified and corrected and hidden root flares can be located and exposed. The root flare is identified by finding the point where the topmost root emerges from the trunk and begins to flare outward. For B&B trees, this may require removing the rope that holds the burlap around the trunk but be careful not to completely remove the burlap or the wire basket (if there is one) just yet. Keep the soil of the root ball as intact as possible. The topmost roots and root flare in poorer quality root balls are often buried deeper down inside the root ball. It isn't uncommon to see the root flare buried 10 inches or more below the soil line of B&B transplants. If the topmost root cannot be easily located, continue to

FIGURE 6.12 B and B trees require the removal of transport materials and inspection to locate the root flare.

remove the excess soil to expose the root flare so that the distance between the bottom of the root ball to the point where the roots attach to the trunk (root flare) can be measured. This is essential to identify prior to establishing the final depth of the pit as well. If the depth was established prior to finding the proper root depth, it may be too deep requiring soil to be placed back in the hole. Caution should be taken to not mistake a grafting knot for the root flare or root collar. The root flare is often found 6 to 8 inches below where the scion and root stock are grafted together.

ROOT SYSTEM MANAGEMENT

Dysfunctional root systems often develop as a result their growing system, especially smooth-walled containers. Stem-girdling roots (SGR) are formed when new roots grow perpendicular to cut roots, or when the tree is growing in a container for too long. As the trunk increases in diameter, these roots may meet the trunk and begin to strangle it, hence the term stem-girdling roots. The trunk may become severely indented where the root was girdling it and in advanced stages, this can result in a fused connection between the root and the stem. This can cause trunk and root decay, which reduces the tree's ability to remain standing, especially in a storm. Circling roots do not always result in trunk girdling; however, trees may develop a severe lean after a windstorm due to circling roots. Often few, if any, branch roots develop on the outside of a circling root, providing less support on that side of the tree. Dysfunctional roots systems such as circling and kinked roots found in containers or even some field-grown trees, should receive planting pretreatment to prevent circling that leads to girdling root conditions.

Sever and remove any circling or kinked roots growing up above the topmost root. This will prevent these roots from strangling the trunk in the future. Kinked roots are roots that have been deflected from the bottom and sides of the container and turned back on themselves at almost 180 degrees. Kinked roots, unlike lateral roots, do not provide any mechanical support and hinder the movement of food and water.

Root-bound trees have many roots circling around the root ball mass, which creates a physical barrier in container-grown stock. This dysfunctional root system continues to grow in a circular pattern, preventing the tree from expanding roots into the landscape soil after planting.

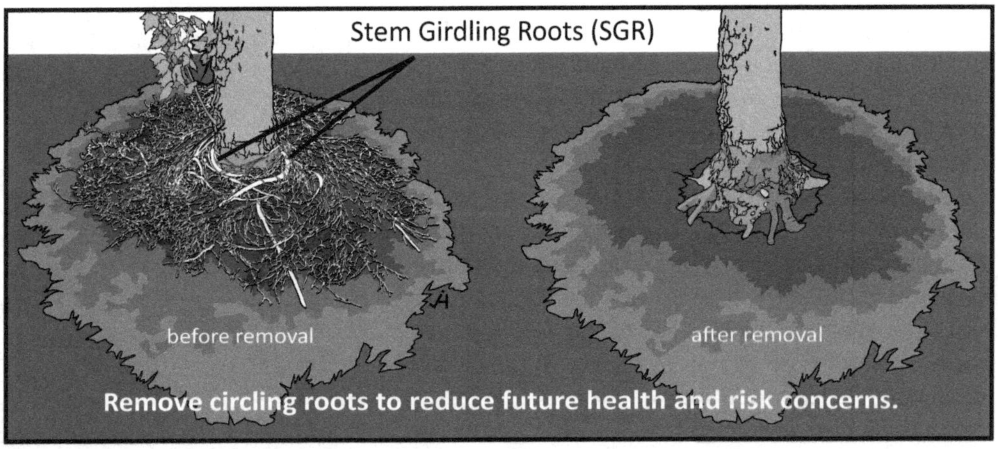

FIGURE 6.13 Circling Roots: dysfunctional roots should be corrected at planting time.

These roots should be cut when trees are being transplanted into larger containers and when they are being installed in the landscape. A lack of root treatment for circling roots can create future issues with health and stability. This can be accomplished by using a pruning saw dedicated to this process to box or shave the outermost roots before planting. This is done by sawing through the outer inch of the roots on all four sides and the bottom. This will remove the malformed roots and encourage lateral rooting into the native soil. Essentially, this is a form of root pruning that encourages new roots.

FIGURE 6.14 Container grown plants.

Also, inspect the bottom of the roots in containers. Often J-roots are found to be redirected up from the bottom of the container into the middle of the root ball, causing issues as well. Be sure to look for roots that circled when trees were in a smaller container, prior to upgrading to a larger container. These are difficult to locate and cut because they are hidden in the interior of the root mass. It helps tremendously with sourcing plants from a quality grower to avoid dysfunctional root systems. This is especially true if the grower uses root enhancement containers as opposed to the smooth-walled containers for trees.

TREE ORIENTATION

The positioning of the tree has been found to be an important consideration for improving establishment and reducing southwest injury on smooth bark trees. Research is limited but substantial evidence indicates trunk damage is more likely when trees are not planted with the same orientation as they were grown in the nursery. Some tree growers will apply a paint mark on the north side of the trunk as it was grown in the field or nursery. Sunscald or southwest injury appears as an elongated area of dead bark typically found on the south or southwest side of tree trunks. The area may be sunken with dried, cracked bark with dead wood beneath. This symptom is also called southwest winter injury because it is commonly found on the southwest side of the tree. It commonly affects recently planted trees, young trees, and trees with thin bark, and usually occurs during winter weather characterized by sudden temperature changes that affect the bark.

On a sunny, cold winter day, bark tissue on the south to southwest sides of the trunk are exposed to direct sunlight and as a result, the tissue temperature increases higher than the air temperature. The warmed bark responds by decreasing its ability to withstand freezing temperatures. When the sun goes down or behind a cloud, the temperature of the bark drops quickly to below freezing and the bark tissue then contracts quickly, resulting in cracking. This leaves a serious wound on the tree, creating future health and stability issues as the tree grows.

Achieving crown symmetry requires careful attention to orientation as well. Typically, some asymmetry occurs in the nursery due to consistent exposure to wind and sun from the same direction and competition during its formative years. In other words, one side of the tree is most likely less developed or shaped differently. Similarly, orient the tree so it faces the same direction as it did when growing in the field. This will promote consistent growth with more light and reduce bark injury from sun exposure.

PLACEMENT OF THE TREE IN THE PLANTING PIT

Much planning and thought goes into matching the site with a tree and into preparing the planting site. It is now time to install the tree into the planting pit. For B&B trees, it is important to prevent the root ball soil from breaking apart before the tree is positioned in the planting hole, so be careful with moving the tree. Lift the tree by carefully lifting the root ball, not by lifting the trunk. Place the tree in the planting pit with the root collar slightly above the established grade of the site. Typically, amendments are not recommended, but modifications are helpful. The soil removed from the hole makes the best backfill. Incidental modifications improve the backfill through the process of just excavating the planting pit. This provides aeration and loose soil backfill that is better for root development. Before final backfill, be sure the tree is straight and true in the upright position. Reposition the tree as needed to account for trunk anomalies and crown shape.

PLANTING DEPTH

Regardless of planting B&B, bare root or container-grown trees, root depth is critical during installation. The area where the topmost roots meet the trunk is called the root collar or root flare. Generally speaking, for young nursery-grown trees, the large woody roots should be no more than an inch below the soil surface and the root system should be well-distributed in all directions. When the root flare is deeply buried in the soil ball, decline symptoms will present. If some roots emerging from the base of the trunk can be seen at the soil surface, it is an indication that the roots are at the correct depth. The top of the soil ball may have to be carefully excavated to locate the flare and top of the roots. Movement of the lower portion of the trunk, forming a gap between the trunk and the surrounding soil, is a sign that the roots are too deep. Many nursery-produced trees are grafted, with the trunk and branches of one species grafted onto the root system of another. These trees will have a graft union that should be visible about 1 to 2 inches above the soil. This anomaly in the trunk can often be confused with a root flare when the tree has been grown to an excessive depth in the nursery.

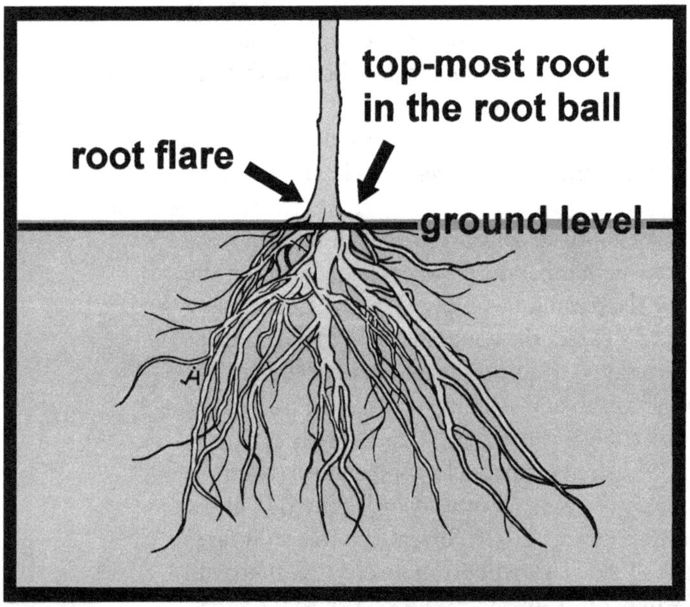

FIGURE 6.15 Find the root flare for proper planting depth.

Be certain that the root flare is visible above the soil line. Always handle the tree by the container or the root ball. Do not lift any tree by its trunk. This can sever the soil and root contact in the container or ball, creating stability and transplant concerns. If measured carefully, the root ball will not have to be removed from the planting hole to adjust for mistakes in hole depth. Larger trees may settle more into the hole, so position these an inch or two higher to accommodate for excessive depth. Most research arborists agree that it is better to plant the tree too high than too deep, but at grade is by far the most preferable except in special circumstances. The simple method to check depth is to lay a shovel or

FIGURE 6.16 Grafting knot and root flare.

FIGURE 6.17 The root flare and upper roots should be visible after removing soil.

long-handled rake across the top of the planting hole to double check root ball depth. If the tree is too deep in the hole, remove it from the hole and firmly pack soil in the bottom of the hole to raise the root ball. If it is only a little bit too deep, tip the ball to one side from the bottom, not using the trunk, and slide some soil under it; then tip it back the other way and slide some more soil under the ball. However, if there is a risk of the soil ball crumbling, do not remove the root ball. This essentially results in a bare-root tree, which creates many more challenges and changes in planting protocol. Once the tree is at the appropriate depth, place a small amount of soil backfill around the root ball to stabilize it.

FIGURE 6.18 Proper tree planting depth is determined based on the location of the root flare.

REMOVAL OF B&B TRANSPORT MATERIALS

While the tree is on the ground, within easy reach, remove any twine, tags, or other materials from the crown and trunk. If planting a balled and burlap tree, cut and remove the string on the outer portion of the burlapped ball. Be sure to remove anything around the trunk. Then, cut and remove the upper one-half of the wire basket and burlap after placing the tree in the planting pit. It is

FIGURE 6.19 Before placing the tree in the planting pit, remove tags and twine on the crown.

FIGURE 6.21 Remove only the upper third to half of the wire basket and burlap when transplanting trees.

FIGURE 6.20 Trees grown in fabric bags require the bag to be fully removed.

ill advised to remove all wire and burlap. The wire baskets are primarily used to help keep a root ball intact during shipping and handling and removing the upper 1/2 to 1/3 of the wire is consistent with best practices. Some tree planting methods include attempting to remove all the wire before backfilling. There is little research documenting the detrimental effects of wire baskets on trees, but there is data that indicates increased mortality on bare root trees, which can happen if the basket and burlap are removed. If the wire must be completely emoved, do so after the tree is positioned in the hole. Otherwise, there is a high risk of losing the soil/root interface. If the tree planter is comfortable with complete wire basket removal, it will allow for root inspection. Nonetheless, basket removal will likely require staking or guying the tree to stabilize it in the planting pit.

SOIL BACKFILL AND INSTALLATION

Gently backfill the hole with soil excavated from the prepared planting pit. The final grade should be level or slightly above grade. Gently slope soil up towards the trunk to cover the remaining root ball above grade. This is especially important if the planting pit was prepared for poorly drained soils. Attempt to pulverize large soil aggregates as much as possible. Finer soil backfill creates fewer air pockets and better soil/root contact. Do not compress the backfill, instead use a tamping tool such as a wooden stake or handle to settle the soil, remove air pockets in the backfill soil, and stabilize the root ball. After the planting hole is filled with soil, the root ball should remain 1 inch above the backfill soil for small trees and up to 3 inches above for larger trees. Do not excessively tamp the loosened soil, especially when the soil is wet. Do not pack backfill soil to eliminate air pockets because this will cause soil compaction, especially in clay soil. A preferred method is to water the backfill soil during supplemental watering, which will help eliminate many of the air pockets, allowing the soil to fill the voids. Add soil as needed to maintain the appropriate final grade.

POST-PLANTING PROTOCOLS

Mulch Benefits

The finishing touches of tree planting should include providing a 2 to 3-inch layer of mulch around the tree and planting pit. The benefits of appropriate mulch products include reducing soil temperature fluctuations, conserving moisture from evaporation, adding organic matter to the soil, reducing mechanical damage to the trunk, and the aesthetics of the planted tree. Generally, a 1- to 2-foot diameter circle of mulch per inch of tree trunk caliper will provide the best mulch area for newly planted trees. Maintain a layer of approximately 2 inches over the root ball. Deeper layers of mulch on the root ball can prevent irrigation and rain from reaching the roots and encourage surface roots. It is advised to keep grass as far away from the trunk as acceptable to aid tree establishment, prevent mechanical damage to the trunk from lawn equipment, and reduce soil compaction. Common mulch materials include pine needles, compost, bark, and wood chips. Peat moss should not be used since once dry it can be very difficult to wet, becoming hydrophobic and possibly preventing water movement into the soil profile. Inorganic materials such as rock and crushed stone can be used but it is not recommended. They provide no organic matter and often work their way into the soil. Stone may also elevate soil temperatures in the root zone by holding heat in the rooting area. The practice of placing plastic sheeting and geotextiles, such as weed barriers, is common in tree

FIGURE 6.22 Excessive mulch is damaging to trees.

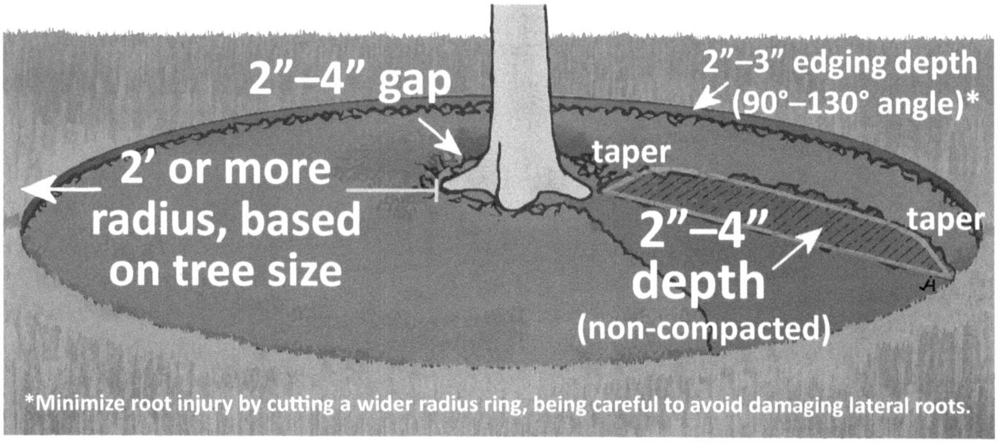

FIGURE 6.23 Expand mulch beds as needed to include the drip line.

planting. This may be effective for controlling weeds but is troublesome for trees. Avoid the use of these materials under mulch to allow improved water penetration into the root zone.

Too much mulch is possible and often creates issues for the tree. Do not pile mulch in a volcano-like manner against the trunk. This can encourage decay on the trunk and create a habitat for rodents that will feed on the soft cambium during the winter. Excessive mulch can keep roots too wet in poorly drained soils. Roots will often grow up into this excessive pile of mulch, creating shallow roots. Stem-girdling roots form from this situation on some trees and cause stress and decline. Some arborists prefer to create earth rims or small berms to hold water when irrigating from a high-volume delivery system such as a hose or water truck. The berm will ensure that water penetrates to where it is needed most, in the root ball. If soil is sandy or very well drained, a berm may not be needed. The berm should be removed once the tree is established. In temperate climates, by the end of the establishment period, a tree will have regenerated enough roots to stay alive without frequent supplemental irrigation, especially in areas where roots can expand uninhibited by competing structures. Supplemental watering should be sensitive to the demands of the tree. Frequency and amounts should be monitored to prevent overwatering and underwatering.

Finally, plan to expand mulch areas as the tree grows to keep the organic layer of mulch on the root zone and to the dripline of the crown. This allows for a healthier root system, protection from lawn equipment, compaction and improved water management.

ESTABLISHMENT AND IRRIGATION

The establishment period refers to the time it takes for a tree to regenerate enough roots to thrive and survive without the need for constant supplemental irrigation. During this period, shoots and trunk grow slower than they did while still in the nursery. When growth rates become consistent from one year to the next, the tree is considered established. In moist climates, by the end of the establishment period a tree will often have regenerated enough roots to stay alive without supplemental irrigation during normal rainfall periods, especially in a landscape where roots can expand uninhibited by conflicting belowground structures. In drier climates, irrigation systems may have to supplement rainfall to provide enough water for survival after establishment. Establishment rate is determined by several factors such as soil conditions, tree health at planting, and planting protocols.

During the establishment period, or anytime the tree experiences prolonged periods without normal rainfall, trees should be irrigated 1 to 2 times weekly with 3 to 5 gallons per inch of trunk caliper. Water should be applied only to the top of the root ball and around the edges of the excavated planting pit. Research shows that the frequency of irrigation has a greater impact than the volume. In other words, infrequent watering cannot be substituted by adding larger volumes of water less frequently. This practice requires monitoring closely for signs of moisture stress and examining leaves for wilt symptoms. If a minimum of 1 inch of rain per week is not available, supplemental irrigation is needed. If leaves are wilting or scorching, irrigate the trees slowly enough to allow the water to soak into the tree root system and planting pit. Check soil moisture with a probe or garden trowel for dampness and drainage. Get into the edges of the soil ball and planting pit to examine moisture

FIGURE 6.24 Irrigation is essential and various delivery systems are possible.

conditions. Overwatering can be just as harmful as underwatering and often present with similar symptoms. Just as importantly, a consistent watering regimen in the fall before the ground freezes is recommended for a healthy plant in the spring. Water percolation improves when the mulch layer is loosened. After mulch applications, using finely or "triple shredded" mulch, organic fibers often bond together forming a tight mat. This layer can become hydrophobic, causing water to run off and away from the root ball system. Coarse mulch sizes such as single grinds and bark will reduce the potential of a matting effect.

FIGURE 6.25 Check the soil for moisture conditions before watering.

FERTILIZER, SUPPLEMENTS, AND TREATMENTS

Typically, fertilization is not recommended for newly planted trees and will not significantly improve survival or growth in the first few years. Fertilizer may be beneficial later, after establishment, depending on the soil and conditions. Typically, it's advisable to apply a slow-release fertilizer in the second or third season after planting, and for larger tree transplants, this practice should be continued during subsequent years. A fertilizer N:P:K ratio of 3:1:1 or 3:1:2 is recommended. Fertilizing with nitrogen on a stressed tree, which includes a tree that has been recently installed, can intensify stress conditions. It is important for root growth to dominate during this time and fertilizers containing higher nitrogen ratios will stimulate top growth at the expense of roots.

More recently, the application of bio stimulants such as growth hormones, mycorrhizae, organic products, and other exogenous compounds have been found in tree establishment protocols. The application of these organic products at planting has shown limited or no benefit with aboveground and belowground applications. The use of mycorrhizal products has proven to be somewhat helpful in establishment when applied to very poor urban soils. However, it has also been demonstrated that even in poor urban conditions, mycorrhizal inoculum is present in adequate amounts. This indicates that additional applications aren't necessary and growth rates are unaffected when treated at planting with commercial inoculants. It is far better to improve soil conditions which promote the proliferation and effect of these fungal friends than add these supplements to poor soil. The quality and species of the inoculum is often a consideration in the efficacy as well. The application of biochar at planting has shown evidence of improving the establishment of new trees. Biochar can increase soil porosity, improving the water-holding capacity, making it easier for plants to take up water and nutrients, as well as providing an improved environment for microorganisms. Application rates are critical and can be detrimental if applied at too high a rate. Recommendations for woody plants indicates a 5% rate by soil volume (1:20 ratio) appears to provide the best performance. Follow label recommendations when using this product.

Tree growth regulators (TGR) are becoming more popular chemistries involved in tree installations for improving establishment rates. The active ingredient in many brands, paclobutrazol, which is used for shoot growth reduction in turf and trees, is used extensively. This compound has been shown to increase root growth in some species and improve growing conditions by reducing water needs through cellular modifications within the leaves.

Water loss through the leaves is a natural and vital process for trees. In fact, over 90% of water loss occurs through the leaves. However, it can be problematic for recently transplanted trees if there is inadequate moisture in the root zone. Applying antitranspirants to the foliage can reduce water loss through the leaves. This product forms a thin layer on the leaves that blocks stomatal openings. These products can be successfully used in summer plantings for angiosperms and in the winter for gymnosperms when applied in late fall plantings. They can be an issue with tree health by interrupting the normal flow of water vapor and CO_2 diffusion. In some cases, trees can "overheat" due to plugging the stomates and building up pressure within the leaves. However, a light application during challenging planting conditions will provide benefits with limited adversity, coupled with good planting practices.

TREE PRUNING

Pruning is often necessary when the arborist or landscaper assumes responsibility for the tree from the nursery. Often trees have crowns with codominant stems, multiple leaders, and clustered branches, which improve crown density and appearance but are not good for long-term stability.

The essential components of structural pruning:

- develop and maintain a dominant central leader;
- identify the lowest branch in what will become the permanent crown;
- prevent branches below the permanent crown (temporary branches) from growing larger than half the trunk diameter;
- space main branches along the central leader by removing or shortening others over time; and
- reduce vigorous upright branches and stems back to lateral branches or remove them.

FIGURE 6.26 Prune to create a dominant leader on newly-installed trees.

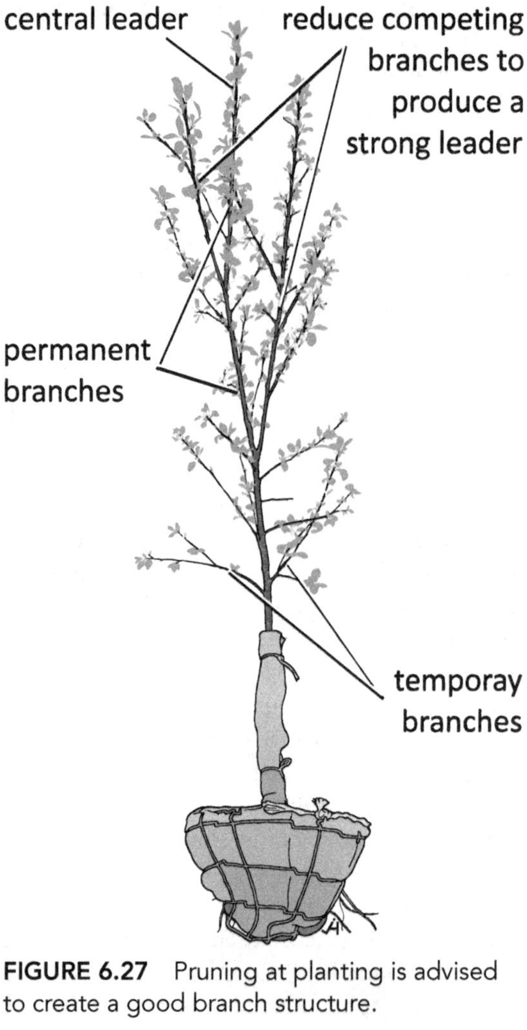

FIGURE 6.27 Pruning at planting is advised to create a good branch structure.

Structural pruning is necessary to develop a strong trunk and branch framework to support the tree crown. When the structure is strong, it is less likely the tree will fail. The pruning objectives should focus on leader development and creating a structure that represents the desired form at maturity. If there isn't any planned pruning for future crown development, pruning at planting is essential. Focus on branch structure, attachment, and spacing. Eliminate or suppress the upright growing stems to develop the leader and prevent a codominant structure.

Young, healthy trees tolerate greater amounts of live branch removal than mature trees or those under stress. If severe structural defects are present, it is prudent to remove more living tissue to correct the dysfunctional form, especially when dealing with codominant stems or multiple leaders with poor branch attachments, or to slow growth of temporary branches, allowing the leader to proliferate. Pruning a branch slows growth on that branch in proportion to the pruning dose and can direct future growth by invigorating parts of the tree that were not pruned.

Recently planted trees are not tall enough to isolate any permanent branches established. Planning for this later will be critical for establishing clearance and compatibility with nearby walks, roads, and structures. All branches requiring pruning should keep their diameters smaller relative to the trunk. Consider pruning non-beneficial tree parts, including broken branches, crossing, or rubbing branches, and upright sprouts. Additional guidance will be provided in the pruning chapter.

TRUNK PROTECTION

Young trees may need protection against animals, frost cracks, sunscald, and mowing equipment. Mice and rabbits frequently girdle small, newly established trees by chewing away their tender bark. Since the tissues that transport nutrients in the tree are located just under the bark layer, a girdled tree often dies in the spring when growth resumes. Weed whackers or string trimmers also are a common cause of girdling. The good way to prevent this is to install an expandable wrap that protects the bark, buffers temperatures and allows air flow. Most white plastic guards will protect from these threats and can be reused if it is not biodegradable. The use of trunk wraps is optional and depends on site conditions. If trunk wraps are applied to new trees, periodic inspection is required. It is advised to remove the guards in mid-late spring and place them on in the mid-late fall. The key factors are ensuring that the trunk wrap is loose-fitting and allows for adequate air flow to prevent excess moisture on the bark and the creation of a habitat for insects. Canker diseases and borers are commonly found under wraps that have been left for too long and have not been managed properly. Avoid tight-fitting paper or synthetic wraps.

The use of tree wraps is a debated point among researchers and practitioners. However, in temperate climates, it has long been the practice of wrapping or covering thin or smoothed bark trees to prevent injury, especially southwest injury. The serious damage that occurs often doesn't appear until several months later. Typically, a long, vertical crack appears accompanied by dead bark shedding along this cracked trunk area. If it is indeed winter damage, it usually is found on the south or southwest side of the trunk.

FIGURE 6.28 Tree trunk wraps protect trees from damage.

FIGURE 6.29 The use of paper wraps can increase pest and decay occurrences under the wrap.

FIGURE 6.30 Southwest injury is common on smooth bark trees.

FIGURE 6.31 Flexible, perforated plastic trunk wrap.

Bark cracking is a phenomenon occurring in many species of trees and can have many causes. This winter injury occurs during the colder months on the lower section of the trunk on the southwest side when there is a sudden temperature drop (e.g., when the sun goes behind a cloud during the winter). The freeze-thaw cycle happens very quickly when there is a change from very warm conditions to cold conditions, resulting in a crack. If there is a snowpack, the reflected rays of sunlight on the bark will actually increase the temperature of the bark. Usually southwest injury occurs on thin-barked trees, such as *Acer spp., Cercis spp., Malus spp.,* and others. A thin-barked tree lacks the larger amount of cork cells as in thicker-barked trees, thus the vascular tissue is located more closely to the bark surface. Thick bark trees tend to be more resistant to cracking due to a greater lag in the freeze/thaw cycle as well. Trees under stress due to environmental factors, herbicide injury, and/or insect and disease are more susceptible to cracking.

In some circumstances a healthy plant may heal, compartmentalizing the damaged area. However, the resulting injury will be present for the life of the tree. This is a wound which must be considered in risk management as well. It is a defect which requires monitoring for progress of the damage or the recovery process.

Trunk wraps have been applied to trees as a protective layer against these fluctuating temperatures for many decades. However, the wrapping may, or may not, be effective, according to research. Paper wrap has not been shown to be an improvement over an unwrapped trunk, but continues to be the accepted standard, and users consider it to be effective in preventing *sunscald*, which is the term for winter bark injury.

White, flexible, light-colored, plastic wraps appear to be the safest for providing protection for the trunk. These loosely fitting wraps allow air circulation to buffer temperature extremes and prevent excess moisture from accumulating between the wrap and the trunk tissue. Also, they will not encourage insect infestation or disease development. Wrapping also may protect your trees from physical injury from animals gnawing or rubbing against them. At the very least, protective trunk wraps do help prevent mechanical damage after a support system is removed and continues to guard against mechanical damage.

ROOT EXPOSURE

Roots are not exclusive to weather extremes. Generally, soil can effectively buffer the extremes in temperatures so damage is rare. However, it can occur in unique planting situations such as raised beds or poorly formed planting berms. Typically, tree roots have little tolerance to low temperatures, as opposed to the aboveground portion of trees. Although the crown of a tree may tolerate -30°F temperatures, the root killing temperature of exposed or elevated roots may occur even between 20–30°F. Additionally, root desiccation and heat injury can occur in summer weather. Trees planted on raised berms or above grade on compacted soils often leave roots zones partially exposed or barely covered, leaving them vulnerable to injury.

Winter burn is prevalent on evergreen foliage in temperate zones. This type of winter injury is desiccation or drying out of needles or broadleaved tree species. Damage can occur when there are extended cold periods while soil is frozen, thus moisture is unavailable to the roots. During bright, sunny days, trees are still losing water through their leaves, but are unable to replace it due to the frozen soil. Windy conditions can create even greater water loss through the exposed leaves. A combination of winter sun exposure and frozen root zones results in an overall water loss in the plant, often resulting in severed *necrosis* and death.

FIGURE 6.32 Nursery stock stored on site are vulnerable to weather conditions.

FIGURE 6.33 Shallow planting with exposed root balls can lead to drought stress and winter injury.

HIGH TEMPERATURE DAMAGE

Occasionally, high temperatures can create lethal conditions for trees aboveground and belowground. Trees have optimum growing conditions across the range of temperatures from 70–85°F. Higher temperatures can injure and kill trees by interrupting the normal plant processes. A thermal death threshold can be reached at approximately 115°F. The thermal death threshold varies depending upon the duration of hot temperatures, the highest temperature reached, tissue age, water content of tissue, and ability of the tree to adjust to temperature increases. Tree temperature usually runs just at or slightly above air temperature. Trees dissipate internal heat through transpiration (water loss from leaves). Trees can dissipate heat effectively if functioning normally with adequate soil moisture. Unfortunately, hot temperatures can greatly increase the air dryness, causing leaf stomata to close in some tree species in an effort to reduce water loss. However, this action limits transpirational cooling. When transpiration is limited by hot temperatures and the tree is surrounded by hot, impervious surfaces, such as parking lots, leaf temperatures may rise above the thermal death threshold. This usually presents itself in the form of marginal leaf necrosis, chlorosis, and dieback. Also, research has shown that photosynthesis declines because of drought. Trees will reduce growth and respiration to conserve energy. This is accomplished usually through a strategically prompted reduction of green tissue the tree system must support with water and carbohydrates. The tree may induce marginal scorching, common on species such as dogwood trees when planted in full sun, or by shedding leaves when under a drought induced deficit. Additional problems can result in loss of photosynthetic area and water uptake with fewer leaves and less transpirational pull. However, it is an amazing strategy for survival until supplemental watering or natural precipitation occurs.

Occasionally, soil temperatures may exceed tolerable ranges as well in harsh urban conditions, such as heat islands found in parking lots. Extreme heat may build up within planting areas, especially when bare soils exist without mulch or other ground cover. Planting pits in urban areas and parking lots often experience high heat levels which are transferred from nearby surfaces. This elevates the soil temperatures in uninsulated soils creating a lethal condition.

Research reveals for most trees, maximum root growth occurs at temperatures between 70–80°F with growth reduced as much as 90% at 95°F. Similar studies have found some trees will suffer lethal damage after only a few hours at 117°F. Because of the lack of available moisture, respiration, and other contributors, roots will recede, decline, and dieback completely.

TREE SUPPORT SYSTEMS

Support systems such as staking and guying are, in most cases, unnecessary and can even be detrimental. Movement caused by the wind is crucial in helping saplings develop into strong, structurally balanced trees with adequate caliper and anchor roots.

However, in some situations, staking, guying, or similar systems may be needed to hold trees upright until adequate root growth anchors them firmly in the soil and until trunk strength is developed. When determined to be necessary, the support system must be installed properly and removed at the appropriate time to prevent damage. More information is provided in the remedies and reparations chapter.

FIGURE 6.34 Guying and staking stabilizes the tree until the root system develops.

FIGURE 6.35 Close up of webbing material used for guying trees.

FIGURE 6.36 Staking is another option for tree support.

Bare-root trees and container-grown trees with small, lightweight root balls may require support until lateral anchor roots develop.
Large evergreen trees with high wind resistance may need to be supported to prevent tipping.
Open sites exposed to strong winds, such as new housing development sites, sites with sandy soils, or elevated settings present conditions where support of newly planted trees is essential.

Taller trees with undersized root balls may need to be staked as well. Without support, trees may become tilted and movement of the root ball in the planting hole may damage the fine, absorbing roots trying to establish themselves in the new soil.
Trees in areas with high rates of vandalism should be considered candidates for support systems. Staking or guying has helped deter theft and damage in many instances.

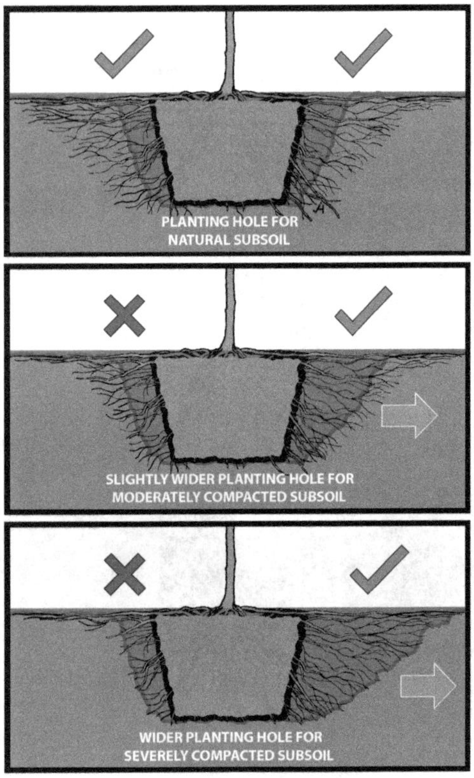

FIGURE 6.37 Shallow, wider planting pits are often necessary in poor, compacted soils.

SPECIAL PLANTING PROTOCOLS

When planting in severely compacted soil, breakdown and pulverize the substrate as best as possible to allow the roots to anchor into the surrounding soil. Roots developing in compacted soils are likely to remain very shallow, then begin to form circling roots if preparation is inadequate. It has been demonstrated that trees benefit from larger, wider planting holes in these situations. Avoid doing this beneath the canopy of existing trees. Significant root damage could occur, causing decline.

Radial trenches can help alleviate conditions associated with compacted soils and encourage rooting beyond the planting hole as well. Air excavation tools create trenches which can be filled with amended soil media or organic products, improving structure and air exchange. A coring machine can be used to enhance growth in compacted soil. It pulls soil cores out of the soil, creating holes for water and air penetration. The soil cores are left on the surface of the soil to decompose naturally. This is referred to as vertical mulching.

Modified original soil backfill or amended soils can be placed back into the trench. There is no evidence amended soil increases root growth more than backfilling with original soil. Although this may not provide all of the benefits of loosening the soil around the entire planting hole, it may be less expensive, and roots should be able to grow better in the loose, aerated soil in the trenches. This technique can also increase root growth on existing trees in many situations.

FIGURE 6.38 Radial trenching improves gas exchange in the soil.

PROLONGED WET SOILS OR POORLY DRAINED AREAS

Sites with poor drainage create many problems. It may be necessary to consider whole site drainage as opposed to individual tree corrections. Drainage for individual trees may be improved in several ways. A ring of perforated plastic drainage tile around the root ball in the bottom of the hole connected to a pipe discharging the water at a lower level can be effective, but labor intensive.

Another approach is to plant the top of the root ball partially above grade. No more than one-quarter of the root ball should be above grade, and the soil should be gradually sloped between the top of the root ball and the original grade. This method may utilize excess soil from the planting hole if of sufficient quality, or additional soil may have to be added. It is critical to add adequate soil in the bermed area to sufficiently protect the tree roots from desiccation and low temperature injury in the winter. Roots above grade are more vulnerable to lower temperatures freezing the roots, killing them. In other words, don't just place the ball and throw soil around it. This poor practice can be problematic in that the soil in the upper part of the root ball and backfill will dry out quickly, causing root death. Supplemental irrigation will be necessary until the roots can spread beyond the planting area, finding more resources for survival. In wet soils and areas with poor drainage, the best strategy is to select appropriate trees during the tree selection process. Choose trees tolerant to poor growing conditions.

FIGURE 6.39 Raised planting areas with berms can accommodate poor drainage, however this is not a good example of a raised planting.

LIMITED SOIL VOLUME

Urban tree plantings don't often offer the soil volumes needed for sustainable tree growth. There is a major disconnect between what arborists need and engineers need when it comes to soil. Because engineers need compacted soil to place sidewalks, streets, and buildings, tree root growth is usually confined and restricted in urban environments. Several innovative design methods are available to increase rooting space. One of these is to include the use of specialized support cells (Silva Cells). The Silva Cell is a modular framework for containing unlimited amounts of suitable soil beneath paving while supporting traffic loads and accommodating surrounding utilities. The cells or modules are filled with quality,

uncompacted soil to grow trees and manage the rate, quality, and volume of stormwater. This suspended system is becoming more prevalent in many urban green spaces.

HOW MUCH SOIL IS NEEDED?

The more intense the environment relative to heat and exposure, the more soil space is recommended for sustainable growth. The planting space is not just for root growth and stabilized tree structure. It is about the volume of water that can be held (think canteen!) in this space and can be made available for the tree. The shape of the planting area isn't as important and can be adapted to meet development demands.

Not all areas planned for trees can provide the kind of soil volume suggested for every tree. Budgets, site constraints, and more are all factors in the final decision. However, through proactive approaches like soil sharing, creating root corridors from tree pits to nearby larger soil volumes like lawns and parks, and engineered solutions such as suspended pavement, we can get a whole lot closer to getting our city trees what they need. When an urban forester is asked how much volume is needed, the answer should be as much as I can get!

ENGINEERED AND STRUCTURAL SOILS

Engineered soils are mixtures of gravel particles and soil designed to meet compaction requirements for street base and planting designs. The application of engineered soils facilitates deeper rooting to reduce infrastructure conflicts such as curb sidewalk panel heaving by tree roots. These special soil products have been developed to increase soil volume for trees planted in confined spaces, and research has shown that these designs improve growing conditions. Engineered soils like CU-Structural Soil can improve tree root development, leading to the growth of larger, healthier trees, compared to tree plantings in existing, disturbed, and compacted soil. There are companies specializing in engineered soils that provide custom blended mixes to meet project parameters such as proper plant pH, soil texture, cation exchange capacity (CEC), total suspended solids, phosphorus index, soluble salts, nutrient values, or other specifications.

Engineered soil is a type of soil manufactured by blending a combination of products for a specific purpose such as for trees to grow in an urban streetscape. Often these specialized soil mixtures are utilized when sidewalks and trees are installed together in the same area. Structural soil, on the other hand, is a type of engineered soil specifically designed to meet compaction requirements for street base and planting material. Structural soil is just one type of engineered soil, and there are many other options or recipes which may be better suited for the planting conditions.

Engineered soils have many benefits, especially in urban forestry. They provide a better growing environment for trees planted in confined spaces. Structural soils improve water storage, which helps to retain water on-site for infiltration and use by soil microorganisms and plant life. This can reduce surface runoff and erosion as well. Additionally, engineered soils containing compost in greenscapes can reduce the cost for stormwater infrastructure, shift maintenance burdens away from local governments, provide better control of nonpoint source pollution, and provide hydrologic control of small, frequently occurring storms.

> As tree and landscape professionals, we must focus on not just planting trees for the future but rather planting trees with a future.

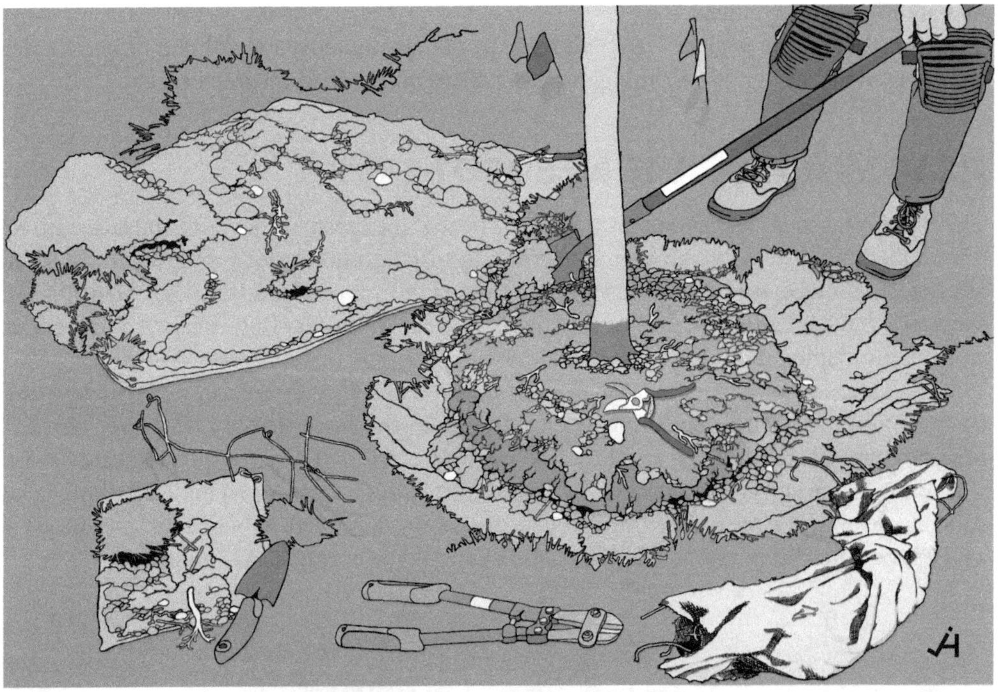

FIGURE 6.40 Adjust planting protocols based on soil conditions.

CHAPTER 7

TREE PRUNING PRACTICES

LEARNING OBJECTIVES

- Explain the various pruning objectives for trees in the urban forest.
- Recognize the biological and mechanical response to pruning.
- Identify the tree components involved in proper pruning and branch attachments.
- Describe the industry-accepted pruning practices and techniques, including removal, reduction, and heading cuts.
- Identify the common pruning tools and their applications in pruning trees.
- Demonstrate the structural pruning process on different developmental ages of trees.
- Recognize how pruning strategies are modified based on developmental age of the tree.
- Describe the components of a pruning specification and how to prepare them.

Proper pruning is one of the best things that can be done for a tree; improper pruning is one of the worst things that can be done to a tree.

—Dr. Alex L. Shigo

IT STARTS BEFORE PLANTING

Good pruning actually starts with planning—and choosing the right tree. The goal is to minimize inputs and maximize the benefits trees provide, and this starts with proper tree selection and placement. Planning for the right tree in the right place planted in the right way reduces the need for frequent pruning. Know the mature size of the tree for its location; determine if it will fit the intended space as it matures and grows before planting.

FIGURE 7.1 Objective-based pruning results in better tree health.

DESIGN WITH MAINTENANCE IN MIND

Reduced-maintenance plantings involve the smart selection of good quality plant material. Choosing trees is much like purchasing any product: you get what you pay for, and the tree you select can determine long-term maintenance obligations. These included supplemental watering and pest management. Start out right by purchasing from a reputable source, such as a dedicated nursery or garden center. Recognize how to select trees that are structurally sound and healthy.

SPATIAL REALITY

Also, spacing trees properly will reduce complications later as trees mature and develop. Spatial reality is the recognition of mature tree size in a three-dimensional perspective. Overcrowding trees and other landscape plants can create maintenance headaches and unnecessary costs, including excessive pruning and even the removal of plants to accommodate conflicting growth from nearby plants. Pay attention to the projected mature height and spread of the tree during the selection process. Right tree, right place can have a major impact on pruning and other maintenance requirements. More information on selection and planting can be found in those chapters of the book.

OBJECTIVES OF PRUNING; PRUNING WITH A PURPOSE

Trees are the largest, oldest living organism on the planet and can live long, beneficial lives if planted in the right place with their basic needs provisioned appropriately. Trees are

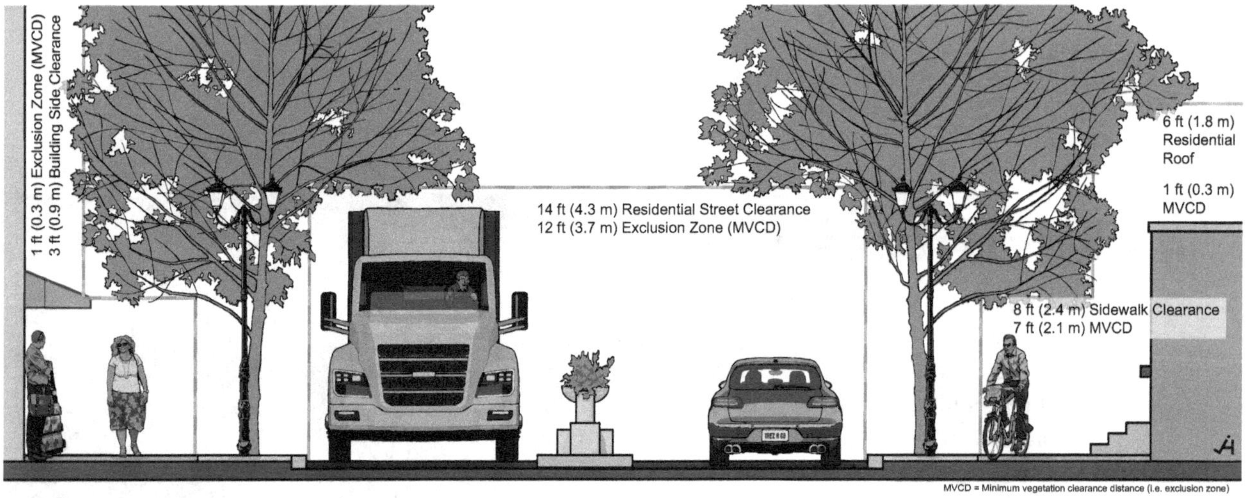

FIGURE 7.2 Pruning for clearance is essential to make trees urban compatible.

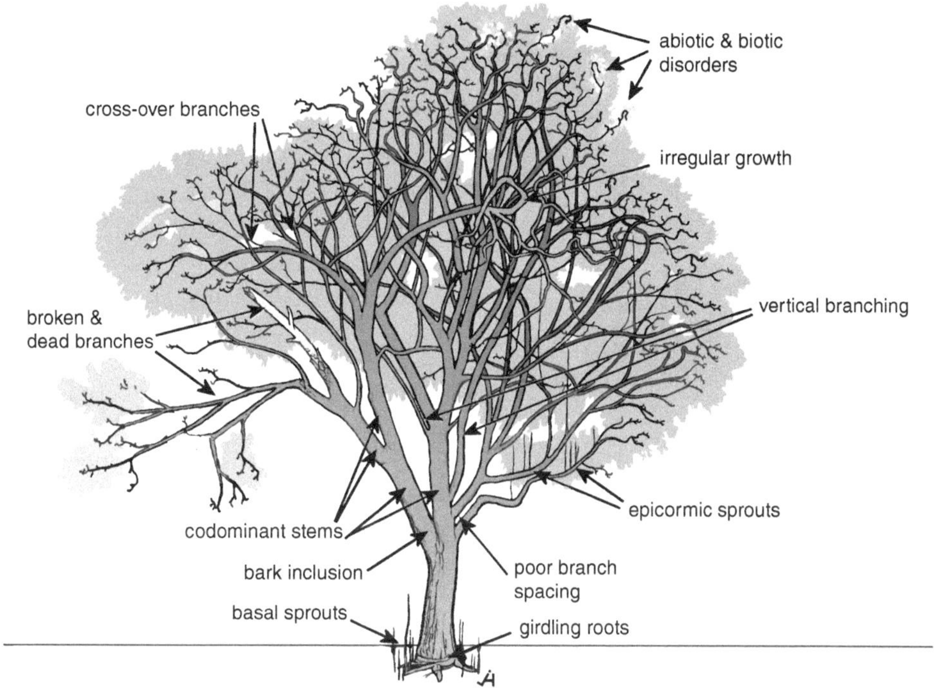

FIGURE 7.3 Pruning focus should include mitigating poor structure and non-beneficial plant parts.

often placed in less-than-favorable growing locations that don't allow natural root and crown development and often require pruning as an intervention to develop a stable and durable structure, improve clearance, and maintain an acceptable appearance.

Pruning has been called one of the best, worst maintenance practices performed on trees. The process wounds trees and has a major impact on plant processes including carbon balance, hormonal response, and resource allocation. Improper cutting on a tree causes severe damage or in some cases, even death. However, tree pruning is a necessary and common practice, making it important to understand both the proper strategies and techniques and the tree's response to the intervention.

Pruning objectives are prepared to mitigate the conditions of concern expressed by the tree owner and observations made by the arborist. No tree should be pruned without first establishing clearly defined objectives. These objectives answer the question of "why" we are here to prune.

CONDITIONS OF CONCERN

Let's begin with why we want to prune a tree in the first place. The most common reasons include appearance, clearance, and reducing risk. Typically, people prune to improve the aesthetics of the tree by reducing the length of fast-growing stems or unwanted growth.

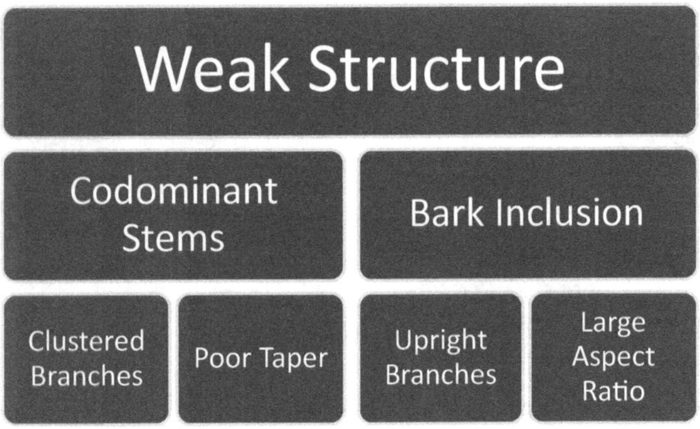

FIGURE 7.4 Structural concerns in trees.

However, too often trees are pruned only to maintain a desired shape or size to fit within the designated location in the landscape. This happens due to poor design or the wrong tree being selected for the intended space.

We often don't get what we want from pruning because it can be difficult to describe and specify the actions required. Objective-based pruning is considered a best management practice, leading to improved pruning results. Objectives clearly communicate expectations for the arborists performing the work and provides clarity on goals to meet client needs. Also, the biological needs of the tree should always be a consideration. For example, critical thinking is required for pruning doses balanced with meeting client needs. No arboricultural intervention should cause harm or damage such as excessive pruning.

REDUCE RISK OF FAILURE

One of the most important objectives of pruning is to reduce the risk of tree failure. This includes removing or reducing defective branches or pruning to improve a structurally weak crown. Risk reduction involves improving tree stability and creating a durable crown with good branch structure. The ideal time to initiate this type of pruning is when the tree is being planted, or when the tree is young or newly established in its location. A larger, mature tree often requires advanced equipment to remedy structural concerns involving risk and safety.

FIGURE 7.5 Conflicts are often the result of poor planning and design.

Trees form their branch structure naturally; however, in the nursery, trees are pruned with different goals in mind that may not be compatible with an urban setting. Sustainable, long-lasting trees are the result of structural pruning when the tree is young; however, a tree of any age can benefit from this type of pruning. Structural pruning emphasizes branch aspect ratio, strong branch attachments, branch orientation and spacing. The overarching goal for trees that grow into large shade trees is a strong, dominant leader system. Crown architecture is dependent on its natural form, which can be either excurrent or decurrent. Excurrent tree species, such as most conifers, will develop a central leader naturally. Decurrent species may not inherently develop a dominant leader and may require intervention to improve structural conditions. Some of these problems are caused by misguided pruning performed in the nursery. This is where the focus is on symmetry and lush crown development.

FIGURE 7.6 Decurrent trees.

FIGURE 7.7 Excurrent trees.

Sometimes, pruning is necessary to remove dead or dying branches or those affected by insect damage or disease. The objective of health maintenance is an important aim for the arborist. This type of pruning helps defend against the spread of some pests and prevents further damage. Damaged or rubbing branches can also be a part of this objective as well.

When trees are not selected or placed properly on the site, they often outgrow their intended space. Also, new structures placed near trees will require pruning as the tree continues to grow and expand. The objective of clearance pruning is reducing conflicts with buildings, streetlights, signs, or even other trees. Utility vegetation management is focused heavily on clearance pruning to maintain regulated distances between trees and utility lines.

SIZE MANAGEMENT

Trees with nearby conflicts will require size management. This objective requires the arborist to make the overall crown size smaller. This is often the case with utility conflicts and demands branches be removed or reduced to meet clearance requirements. Pruning for size management requires that the arborist pay careful attention to making appropriate reduction cuts. This is a pruning situation that can easily lead to topping, which is an unsustainable and unhealthy pruning practice. Topping is the reduction of tree size by heading back many or most

FIGURE 7.8 Trees too close to utilities require excessive pruning for clearance.

FIGURE 7.9 Topped tree with excessive foliage removed.

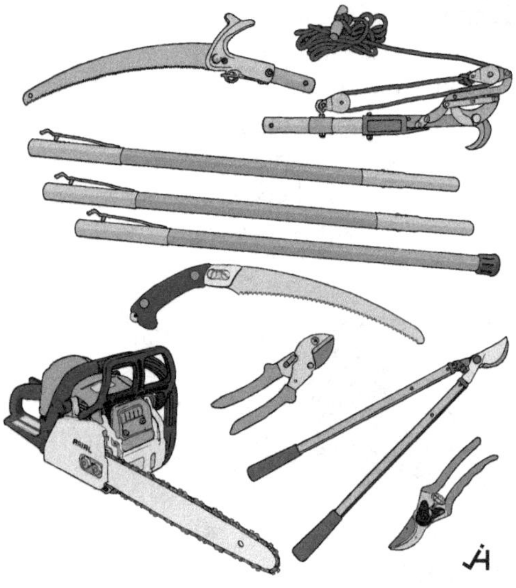

FIGURE 7.10 Common pruning tools.

large, live branches without regard for tree health or structural integrity. Topping manages height and spread but leaves behind woody stubs that cause decay, weak branch attachments, and increased likelihood of failure.

VISIBILITY

Pruning to improve or maintain views is another important objective for visibility and safety. Sightlines are a particularly important consideration with road junction layout and urban planning. Insufficient sight distance can be a contributing factor in intersection traffic or pedestrian accidents. Intersection sight distance is typically defined as the distance a motorist can see approaching vehicles before their line of sight is blocked by an obstruction near the intersection. The driver should have an unobstructed view of the intersection, without the interference of any hedges, trees, or other structures. This is an important consideration in tree selection and placement, which will also impact pruning intensity and frequency. Additionally, clients may request pruning in order to maintain the visibility of views such as a lake or landform in the vista. Maintaining these important vistas require input from the property owner and collective discussions with the arborist.

PRUNING EQUIPMENT

The basic hand tools used for pruning include hand pruners or secateurs, loppers, and pruning saws or hand saws. The tools used in tree pruning are determined by the pruning process, size of the tree and its components, and size of cuts to be executed. Be sure proper training and instruction on the use of each tool has been taken into consideration before operation, especially chainsaws.

HAND PRUNER; SMALL BRANCHES

Secateurs and hand pruners are used to cut small branches, typically less than 1/2 inch in diameter. There are two styles of pruning shears: anvil and bypass. Anvil style pruners have one sharpened blade that cuts into a thicker, flat anvil. Typically, the anvil style pruners and loppers are considered less preferable. This is because they create a crushing cut on the anvil side, as opposed to a clean cut achieved with bypass cuts. The reasoning behind this is the extended wound recovery time of the severed end, often caused by the ragged, crushed edge. The preferred type of hand pruners is the bypass pruner which works as a scissor, bypass action. They have one sharpened blade cutting past a curved, reinforced edge. When sharp, they make clean cuts and get into small branch angles easily.

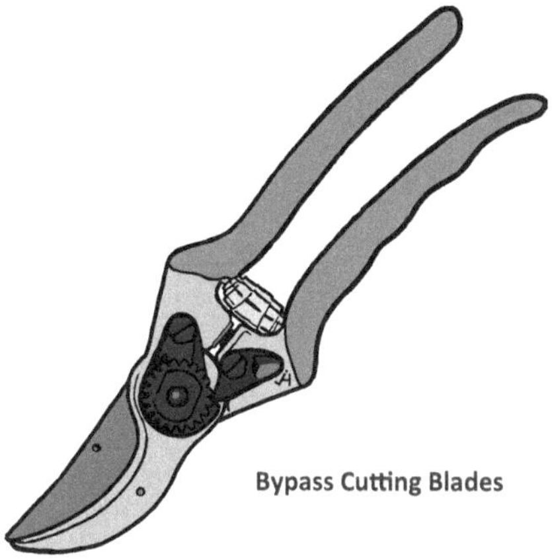

FIGURE 7.11 Bypass pruners leave a clean cut for better healing.

FIGURE 7.12 Anvil pruners smash the lower edge of the cut branch which slows healing.

LOPPER: MEDIUM BRANCHES

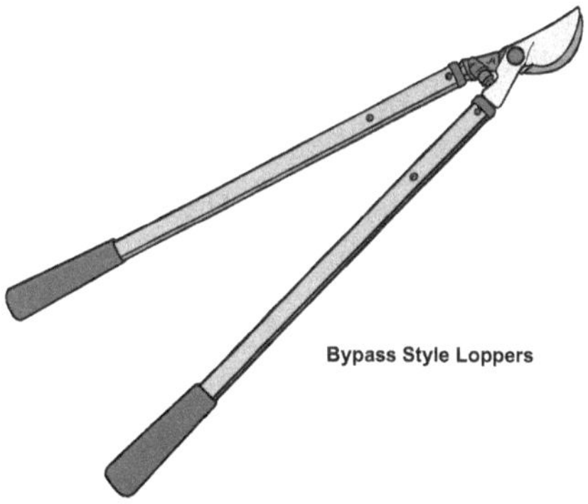

FIGURE 7.13 Loppers are useful for larger branches, cutting similar to secateurs.

The next tool for larger branches is the lopper. This long-handled pruning tool is applied to branches that are 1/2 inch to about 1 inch in diameter. Like secateurs, loppers come in anvil or bypass style, and some brands may offer the ratchet style. Loppers are available in various lengths—from about 12 to 40 inches with handles of various materials such as wood, metal, or fiberglass. Each one has advantages with length being the primary consideration. The longer the handle, the greater the leverage on the cutting end. As with hand pruners, bypass style cutting ends are the preferred style. Also, the manual effort required to execute the cut is the same as well. If excessive force is required, the tool is undersized. The next step up would be the pruning saw. There are mechanical pruners which include pneumatic, hydraulic, and electrically operated shears or loppers. Most are battery powered and cordless, which can be quite helpful for the professional arborist and horticulturalist.

The proper tool for the appropriate size cut should demand reasonable force. Regardless of the tool used, do not try to cut a stem that is too thick for the equipment. This is signaled rather easily if the effort required is more than comfortable or requires both hands to close the pruner. The result is often a poor cut with damaged tissue and equipment. This is a common mistake and once the bypass system is damaged, it rarely is capable of repair.

HAND SAW: LARGE, STRUCTURAL BRANCHES

When cutting branches and limbs larger than 1 inch in diameter, the pruning saw is the correct choice for executing the cut. A pruning saw is the preferred tool for most arborists due to its versatility and cut size capabilities. Virtually any branch size can be tackled by the hand

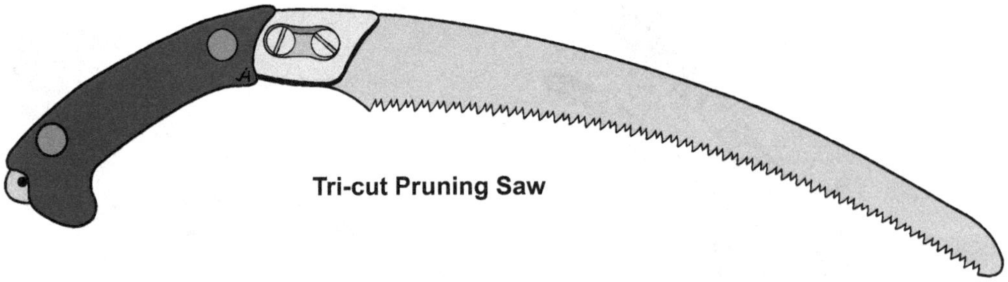

FIGURE 7.14 Pruning saws are specialized to cut only on the pull stroke and used on larger branches.

saw; however, a powered saw is more efficient for larger branch cuts. Also, a hand saw can get into tighter branch angles and make a much cleaner cut that is less likely to damage living tissue. There are several brands, lengths, and styles. There are fixed and folding styles as well. A fixed blade saw stores in a scabbard, while a folding blade folds with the handle covering the teeth. Arborists prefer fixed blades for their easy application and reduced need for maneuvering with blade opening and locking while in a tree. A scabbard straps to an arborist's lower leg or hangs on the climbing saddle where it's easy to reach.

A straight blade is available as well as curved blades; however, a curved blade makes a faster cut. A straight-blade saw will have less contact with branch wood than a curved-blade saw, which makes it easier to execute a careful linear cut. However, a curved-blade saw cuts fast because it's cutting on more sides of the wood—there are more contact points between the saw teeth and the wood. The difference in saw blade teeth is measured by the depth of the teeth, which is called the well. Larger teeth will clear more material, while finer teeth will clear less material. For arborists working on live branches, a fine-tooth saw allows for a more precise cut and wound edges. A blunt-tipped saw helps an arborist prevent damage outside of the cutting area.

Professional hand saws have blades cutting only on the pull stroke. This translates to more cutting power along with the advantage of the curved blade. The different style teeth on the pruning saw separates them from a wood saw. Pruning saw blades have teeth that are three-beveled, often called triple-edged or tri-cut, which allows for a fast, clean cut. There are several options available and variations in teeth size and blade kerf size as well. Smaller teeth and narrow kerf allow for cleaner cuts on smaller branches. Whether ground pruning or working aloft, these saws are extremely sharp and caution should be exercised during both use and storage.

POLE PRUNERS; FOR THE LONG REACH

Often extra reach is necessary when working aloft, or if working from the ground. The pole saw or pole pruner is a saw or lopper with an extended handle to allow for greater reach. Some models have both lopper and pruning heads on the same pole for versatility. Also, the pole pruner can eliminate the need for a ladder. The choice of whether to use the pruning head or saw head depends on the cut size and access angle. From the ground, it can be difficult to achieve the proper cutting angle creating a pruning wound that may not be allow for quick healing. If feasible, a pruning head would be the best choice, however size dictates the choice.

It is recommended to use nonconductive fiberglass materials. Typically, the poles come in sections and are quick locking, which allows for smooth sliding through your hands and more efficient cutting. Single poles and telescopic poles are also available for pole pruning equipment. The dielectric fiberglass poles made especially for tree clearance work do not conduct electricity and should be the only choice for a pole pruning selection. These extended pruning tools place the arborist closer to utilities so particular caution must be taken, while remaining vigilant of any utility line. Never prune within 10 feet of an electrical conductor, unless trained and accredited to do so.

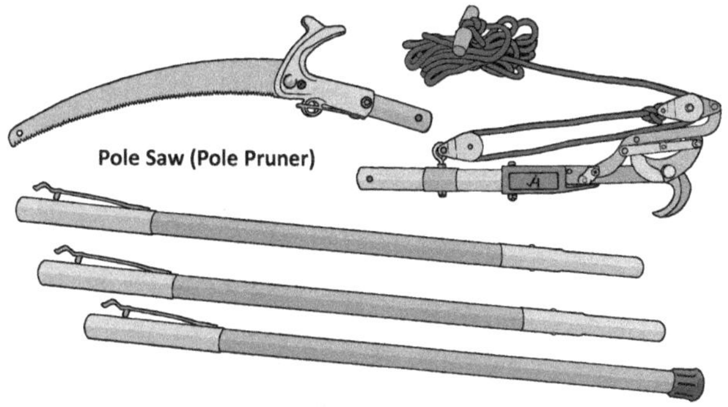

FIGURE 7.15 Pole extensions provide a longer reach for ground-based pruning.

CHAIN SAWS FOR LARGE LIMB REDUCTION AND REMOVAL

Finally, the power saw or chainsaw is necessary in tree pruning for many reasons. Chainsaws are highly efficient, albeit potentially dangerous tools. Most arborists or tree climbers use chainsaws quickly, carefully, and skillfully to prune larger tree branches and limbs. Chainsaws are invaluable when doing prolonged pruning of large-diameter limbs and ground clearance work, only after proper training.

The appropriate length of the guide bar is determined by the tree or branch size and to some extent by the user's level of expertise. A shorter guide bar weighs less and is easier to maneuver when doing limbing or pruning work. The long guide bar is used for larger tree sizes.

There are numerous chainsaw brands available, including gas powered and electric or battery-operated options. The selection is based on application and personal preference. As with any equipment, try it out, discuss chainsaw needs with professional dealers and consult with colleagues and peers on their experience with selecting equipment for the job. Remember, as with any tree or landscape work, use proper personal protective equipment (PPE), which includes eye protection, hearing protection, chaps, head protection, and steel-toed boots. Check with ANSI standards and OSHA rules for the appropriate protective gear based on the work being conducted.

Chainsaw size selection for the objectives depends on the following factors:

- Proficiency and experience

Choose a smaller chainsaw with a less powerful engine if new to chainsaw work. A small saw is more maneuverable than a large one and best for routine pruning. If the chainsaw is too heavy, fatigue becomes a factor presenting a safety risk.

- Tree or branch size

Choose a larger chainsaw with more power if felling larger trees. If the equipment is too small, this will subject the chainsaw to greater strain and unnecessary wear. Pruning with a chainsaw will require a smaller saw for safety, maneuverability, and executing good cuts.

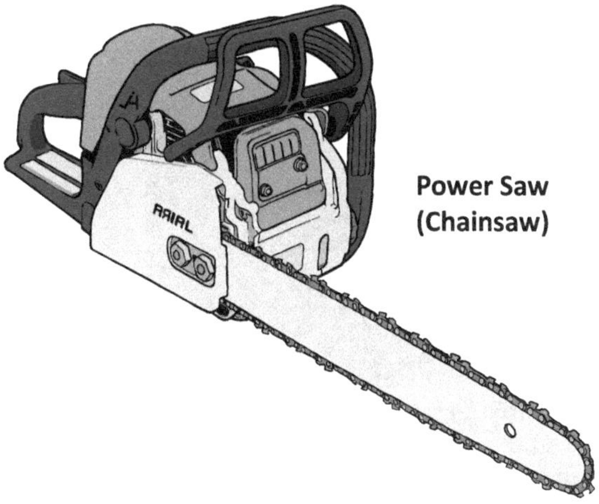

FIGURE 7.16 Power saws require proper training for safe use.

TREE COMPONENTS AND BRANCH ATTACHMENTS

Trees with a strong, stable structure are a priority in the urban forest. This begins with the arborist recognizing the significance of good branch attachments, strong trunks, stems, and well-supported crowns. Many planted trees develop defects or weaknesses if they are not structurally pruned at or in the first few years after planting. The larger the tree becomes, the larger the parts and loads on defects become, increasing the likelihood of failure. Large trees support more weight and are exposed to greater forces from wind than smaller trees because wind velocity increases with height above the ground.

The perceived strength of a tree's structure can be determined by examining the trunk and branches. Figure 7.17 below highlights important characteristics to keep in mind.

All of these characteristics contribute to the strength and stability of a tree under static and dynamic loading, such as its own weight and wind, for example. When any of these characteristics become conditions of concern that affect stability or increase the likelihood of failure, it may be necessary to intervene with structural pruning.

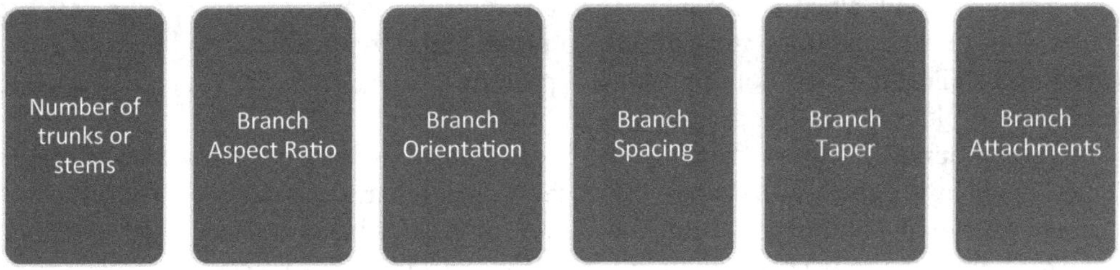

FIGURE 7.17 Tree characteristics affecting stability and strength.

FIGURE 7.18 Codominant stems should be reduced or removed to develop a dominant leader.

NUMBER OF TRUNKS

Trees with a dominant leader well up into the crown are stronger and have a durable form. Conversely, trees with two or more main stems with similar diameters (codominant) are weaker and less resilient, but it is also important to point out that trees with more than one main stem can be stable. The following sections point this out in more detail. Mainly, it is about the overall size of the tree, its parts, and the loads experienced by the tree. Basically, any tree can remain unharmed for its entire life, so long as loads do not exceed its capacity to carry that load.

BRANCH ASPECT RATIO

Branch aspect ratio is a term used in the industry to describe the size of a branch compared to its trunk or parent branch. It is important as a branch attachment feature because of its effect on union and attachment strength. The aspect ratio value is determined by estimating the ratio of the diameter of the branch to the diameter of the trunk or parent branch where it originates. For example, where a 6-inch branch is extending from a 12-inch trunk, the aspect ratio is 2:1 or 50%. Each branch borne from the previous branch should be roughly half the size of its parent branch.

Strong branch attachments are more likely to develop where:

- the size of the branch is less than one-half the diameter of the parent stem (branch aspect ratio less than 50%);
- a single branch, rather than several branches, arises at one location; and
- bark is not included within the attachment.

Strong branch attachments develop when the annual growth rings (wood) of the trunk and branch overlap, creating laminated, dowel-in-post-like structures. The result is the development of an enlarged area at the base of the branch is called a collar. The branch protection zone located between the branch bark ridge and branch collar limits the spread of decay from the branch into the trunk.

Branches with a small aspect ratio of 1:3 (about 30%) are more strongly secured to the trunk than those with a large ratio of 2:3 (about 70%). This is owed to the fact that the trunk and branch wood overlap and wood intermingles with each annual growth increment (ring) at the attachment. When branches are much smaller in diameter than the trunk (e.g., 40%), a natural protective area called a branch protection zone develops within the branch base extending inside the trunk. This important zone of cells resists the movement of wood decaying organisms into the trunk when the branch dies or is pruned away. The branch protection zone does not develop on branches with large aspect ratios (e.g., 70%) and allows more decay movement and less opportunity to compartmentalize.

BRANCH SPACING

The distance between branches is a feature not often given consideration in tree maintenance. Trees with main branches spaced apart are considered stronger than trees with close spacing between main branches. Important connective tissue forms properly when branches are spaced appropriately. Weaker connections result when a nearby branch prevents wood from developing an effective collar around the base of the branch. Branches spaced along and around the trunk are better connected to the tree than branches that are clustered together and originate from one position. Clustered branches and codominant stems with bark inclusions (included bark) are more prone to failure than trees with one trunk and a more dominant leader.

FIGURE 7.19 Poor branching habits should be avoided when sourcing trees.

Where several branches arise at one point on the trunk (multiple attachments), there is not enough

FIGURE 7.20 Multiple leaders are weak forms which increase likelihood of failure under loads.

space for trunk wood to wrap around the base of each branch in the normal fashion. The clustered branch attachment is therefore weaker, and there is a greater likelihood that one or more of the branches will fail. For small-statured trees, vertical branch spacing can be 6 to 8 inches apart. For medium-height trees (less than 40 feet at maturity), 1 or 2 feet between branches may be adequate; for large trees, 4 or more feet is considered ideal.

Another suggested consideration for spacing would be that the largest branches should be spaced apart at a distance equal to or greater than 5% of the ultimate height of the tree. Also, optimal structure has branches spiraled around the trunk so that no branch is directly above another branch. On excurrent trees like sweet gum (*Liquidambar styraciflua*) and most conifers, vertical spacing is less critical because the branches are usually only moderately vigorous and more horizontally oriented, which means they seldom compete with the leader. Little or no pruning is needed except when an occasional branch becomes overly vigorous and competes with the leader.

FIGURE 7.21 Pruning redirects growth.

BRANCH ORDER

Branches grow from the main stem(s) and development begins with buds found beneath the cambium layer of the stem. The availability of necessary resources, such as light, hormones and energy, causes these buds to grow at certain rates and shapes to develop into the tree height and form. The central leader grows from the apical meristem and branches begin their branching sequence from this origin.

Lateral branches grow from larger branches called parent branches. In pruning language, branch development involves branch orders. First (1st) order branches, regardless of size, grow directly from the trunk. These larger branches are often referred to as scaffold branches; however, definitions need to be considered when prescribing pruning plans.

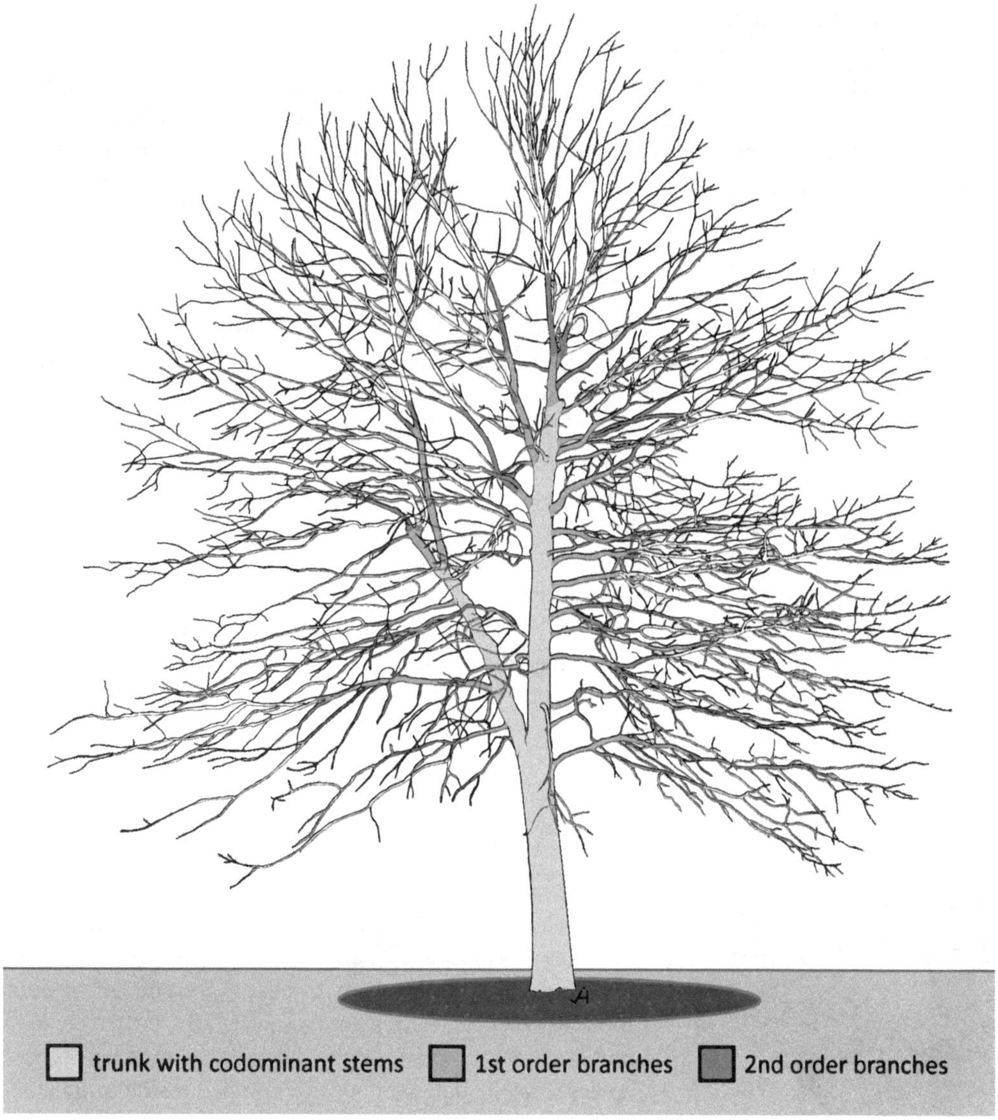

FIGURE 7.22 Branch order system

Branches are considered first order up to the point where they divide into equal diameter branches. This can occur close to the trunk or out at the edge of the crown. The two or more branches beyond the dividing point are considered second (2nd) order. Most pruning operations should be performed on the largest first order branches; smaller first order branches should be left alone. Best practice and research indicate first order branches on mature trees are best left on the trunk. This is due to the increased amount of heartwood, meaning less sapwood, which increases decay potential. Unless absolutely necessary to meet objectives, first order branches should not be removed.

Smaller lateral branches arising from first order branches are considered second order branches. Pruning often requires the arborist to reduce a first order branch back to a second order branch. This reduction pruning cut is described in the ANSI A300 standard as pruning back to a 1/3 lateral or larger than the reduction cut diameter. Next, second order branches can be reduced back to 3rd order branches or removed to meet objectives. Third order branches are lateral branches growing from second order branches. Additional details are found in Appendix 1.

BRANCH ORIENTATION

The arrangement of branches on a tree helps with identification. This begins by determining whether the tree has opposite or alternate branching, which refers to the orientation of the twigs with the branches. Opposite branching refers to a branching pattern where side branches and leaves grow from the stem directly across from each other. However, it can refer to its positional attitude relative to a flat or upright orientation.

Upright branches on some species grow more vigorously than do the more horizontally oriented branches. Also, vertical branches are more prone to forming bark inclusions in the union when they grow, which is more at an acute angle with the trunk, typically less than 45 degrees. Typically, ideal structure includes a more flattened branch orientation than upright. Branches with an upright orientation may become long and heavy enough to bow outward, away from the central leader, because they are not strong enough to support the increasing force on the branch. Such branches are more likely to fail during loads created by wind, rain, snow, or ice. This requires the arborist to choose branches to remain which have a flatter orientation when pruning.

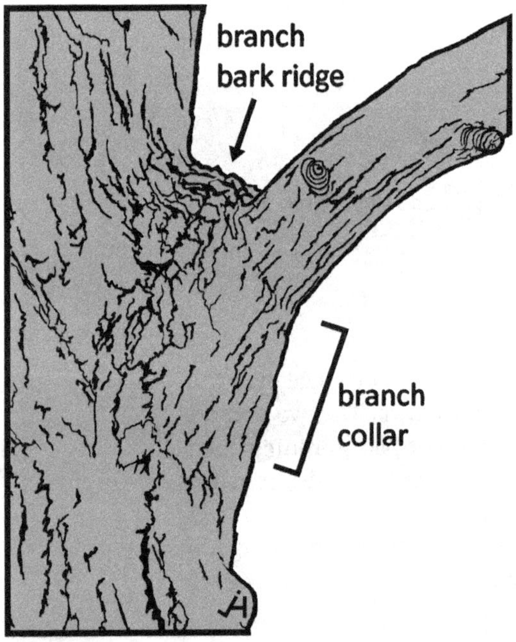

FIGURE 7.23 Branch collars are indicative of strong attachments with a favorable aspect ratio.

BRANCH ATTACHMENTS

Branch attachments are identified by two components, branch bark ridge (BBR) and a branch collar (BC). Every attachment has a branch bark ridge; however, strong branch unions develop when a branch collar is present. BC are common on branches with a small aspect ratio, but rarely form on branches with a large aspect ratio. If the branch emerges at an acute angle, bark can become trapped inside or included (i.e., squeezed or embedded) within the union instead of growing on top of it. Branch unions with included bark cannot form a strong branch connection. Inclusions in unions are among the most severe branch defects; the form of inclusion varies considerably from species to species or even from branch to branch within a tree. Inclusions on codominant, large, and long branches are of greatest concern because the mechanical stress from the heavy load being supported by the weak union is more likely to result in failure.

BRANCH AND TRUNK TAPER

A well-tapered branch or trunk decreases noticeably in diameter with length or height. Taper allows distribution of loads over the length of the branch or trunk.

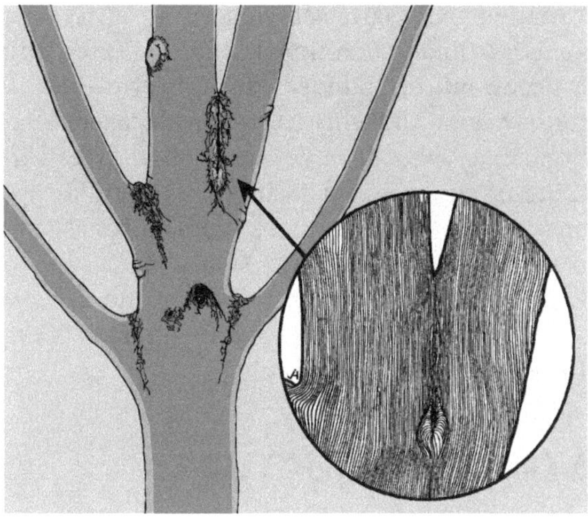

FIGURE 7.24 Included bark is often found in codominant stems.

How do you create a stronger, more stable tree?
Below are the essential steps for pruning to improve structure and restore a strong central leader system:

- Identify the most appropriate stem to become the dominant leader.
- Determine which stem or stems are competing with the leader.
- Remove or subordinate the competing stems.
- Identify the lowest branch that will eventually become part of the permanent crown.
- Prevent branches below the permanent crown (temporary branches) from growing larger than half the trunk diameter (aspect ratio).
- Space main branches along the central leader by removing or reducing others over time.
- Reduce or remove vigorous upright branches and stems back to lateral branches.

Long branches with poor taper are more likely to break than branches with noticeable taper because branches with good taper are stronger under dynamic loads such as wind. This is a well-supported concept in engineering which can be applied to arboriculture. Tapered cantilevered beams are found everywhere around us, including trees. A simple cantilever beam (branch) has no taper and is consistent from end to end in size, making it a weaker branch. Whereas a tapered cantilever beam (branch) adds additional mass at its connection point often called response growth, increasing strength. Providing a tapered shape to the cantilever beam (branch) improves stability by increased resistance to wind and ice loads. Good branch taper is created by maintaining foliage and smaller lateral branches along the length of the branch or trunk. The opposite of this would be lion's tailing, which removes many interior branches, pushing the center of mass of the branch towards the end of the tree, reducing taper, increasing end loads and thus, increasing the likelihood of failure.

FIGURE 7.25

STRUCTURAL PRUNING PROCESS

Tree structure refers to how branches and stems are arranged on a tree. Branch attachments, orientation, and arrangement influence stability and failure potential. There are many

FIGURE 7.26 Strong structure requires a dominant single leader. The figure on the left is a strong central leader and on the right multiple leaders representing a weak structure.

variations and variables, depending on species, genetics, and environment. Tree species with a decurrent, or rounded, open crown—such as most maple and elm trees—produce clustered branching, which can be weaker than trees with more of an excurrent form, or an upright, pyramid shape—such as sweet gum, pin oak, and most conifers. An arborist should be capable of identifying poor branch structure and demonstrate how to improve structure implementing the appropriate pruning cuts.

STRONG STRUCTURE

As discussed earlier, a strong central leader system is important for tree stability and for creating a resilient, sustainable tree. Many researchers have demonstrated this is one of the most important factors to help prevent structural failure under loads such as ice and wind. Trees work mechanically as a whole structure and all parts contribute to the desirable strong structure. It begins with the trunk at ground level and continues to the first order branch, at the lowest part of the crown, usually at the largest lower branch or group of branches. The continuation of a single trunk tapering into the crown is called the leader. A dominant leader may be absent in more open-grown and planted trees, where the trunk or leader divides into two (codominant leaders or stems) or more (multiple leaders) equally sized stems.

CODOMINANT STEMS

The shape of the codominant union at the point of bifurcation can be an indication of union strength. Research has demonstrated that U-shaped unions are considered stronger than V-shaped unions because of the intermingling of twisted, tortuous wood present in the U-shaped union. The U-shaped bifurcation is indicative of the presence of a branch bark ridge, which basically holds the stems together. V-shaped unions do not possess a branch bark ridge and are usually replaced with included bark and decay, creating a predisposition to stem failure. As the tree matures, parts get larger and loads increase, which elevates the likelihood of failure. The stems remain similar in size, without a single, dominant leader. This results in a structure proven to fail under stress conditions, such as high winds and storms.

FIGURE 7.27 Codominant stem failure is predictable and preventable.

Many times, trees with codominant leaders are an existing issue for the arborist. However, early intervention with structural pruning in the tree's life can remedy this situation before it becomes an increased risk. In other cases, an individual may purchase and plant a tree without knowing exactly how to select trees with good branch habits. Often, just as the tree develops into

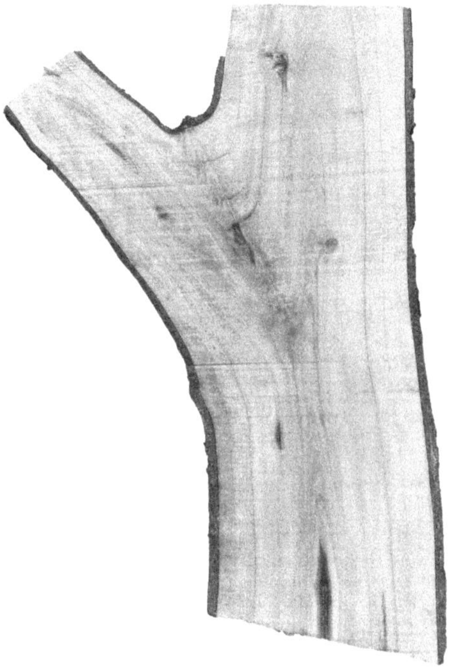

FIGURE 7.28 Branch attachment with good aspect ratio and branch bark ridge formation.

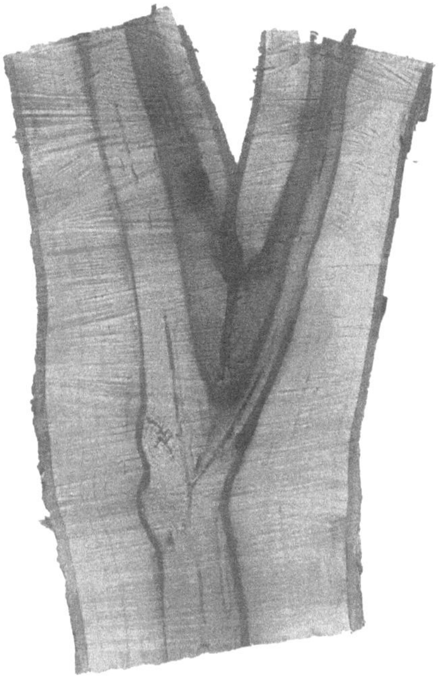

FIGURE 7.29 Branch attachment with poor aspect ratio and v-shaped formation with developing decay.

a significant element in the landscape that provides important services such as shade, along come the strong winds, and the tree splits. This is frustrating and costly, leaving the tree owner perplexed and frustrated. This is where a clear understanding of branch aspect ratio and established single leaders is an important key in pruning to reducing failures. It is critical to be able to identify poor unions and branch attachments. This is especially true in more advanced situations of codominance. There will be included bark in the center of the union, which will appear as bark pushing out on both sides of the stem union. Research has shown this advanced issue is 30% more likely to fail under excessive loads. Basically, consider choosing the leader which has the best branching, such as spacing and aspect ratio on smaller branches above the union. It isn't always the most vertical, but it will eventually correct itself.

AT WHAT DEVELOPMENTAL AGE SHOULD TREES BE PRUNED?

Pruning to improve structure is appropriate for trees of any age. However, older, mature trees may require distinctive pruning considerations with a more measured approach. On older trees, much of the pruning should be performed on the sunlit exterior branches at the edge of the crown. The crown periphery is the pruning zone. Important interior branches are shaded and contribute to branch diameter growth or taper and should remain on the tree. Large branches or groups of branches in the same location on the trunk or leader, called branch clusters, can contribute to weakness and should be corrected.

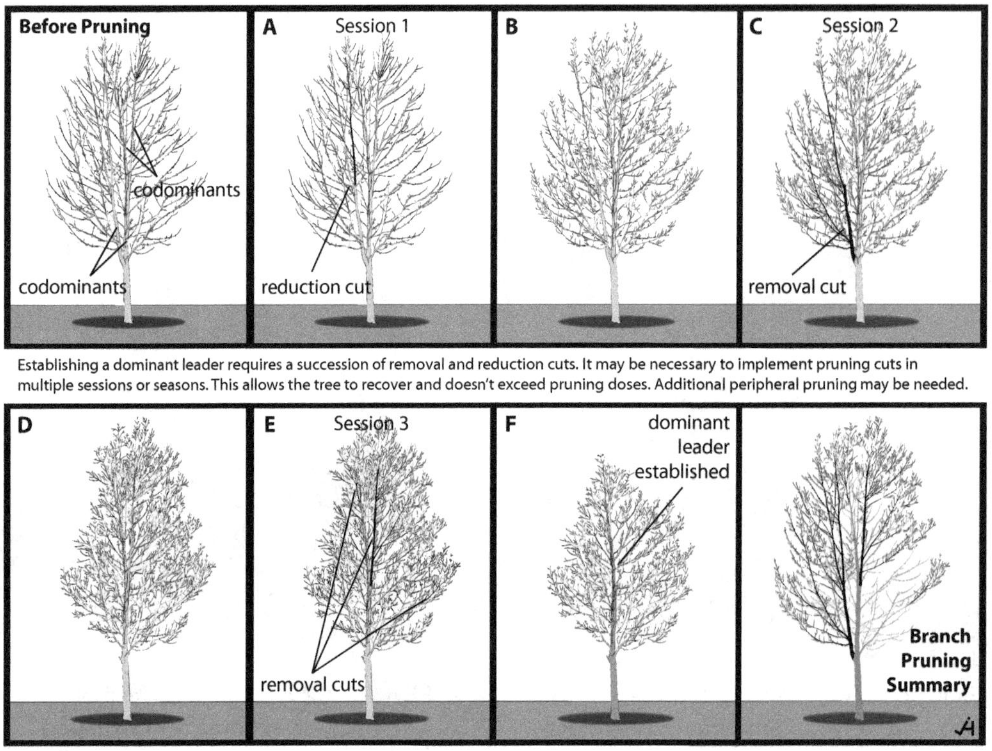

FIGURE 7.30 Establish a dominant leader for a stable structure.

Structural pruning should begin at planting and continue as necessary, through the first years of establishment and even longer, if needed to reach objectives. The strategy differs greatly from newly planted trees to retrenchment age-trees. Basically, as trees age, pruning needs should be minimal. However, neglected trees may require additional pruning for risk and compatibility issues.

Pruning to improve structure is vital for cultivating long-lived trees. If a pruning strategy isn't in place, poor structure is left to develop, requiring attention later for reducing the likelihood of failure. Another consideration is that the longer pruning is delayed, the more expensive it will be to correct from the tree owner's perspective. Additionally, it is more problematic for the tree with a deferred maintenance approach, as larger wounds are created during pruning, which can lead to increased decay and slowed healing.

PRUNING NEWLY PLANTED AND EARLY ESTABLISHED TREES

Let's look at pruning after planting to improve poor structure. Even if the stems and branches are small, structural pruning is an important part of tree planting. If the tree is young and has been established early, pruning to select the dominant leader is relatively simple. Select a branch that is in the center-most part of the crown, free of any wounds, relatively vertical, and most importantly, has strong, well-spaced lateral branches. Then remove the competing stem back to the trunk.

> Pruning cuts made when branches are small avoids having to make large cuts years later. Smaller cuts (2-4 inches in diameter and smaller) result in faster healing and less decay than larger cuts.

There may be a slight void in the crown after pruning but the tree will fill in the crown over a period of a few years as it develops, with resources for growth now directed toward the new leader.

TEMPORARY AND PERMANENT BRANCHES

Young trees will go through a branch development process which will require structural pruning over a period of time. It may be helpful to think of the young, newly developing crown as two parts, temporary branches and permanent branches. Branches originating below the lowest permanent branch (e.g., below 15 feet over a street or 8 feet over a walkway) are considered temporary on the tree. Those above that height become the permanent branches in the crown. Temporary branches are important for the tree for providing food and energy and good taper to the trunk. As the tree grows taller, it will develop new branches in the upper crown, allowing the lower temporary branches to be removed, pushing the desired crown height upward.

Since most young and recently planted trees are not yet be tall enough to have many, if any, permanent branches, treat the branches as temporary by pruning them to keep their diameters small relative to the trunk diameter. Each year, remove the branches with the largest diameters. Be mindful of the pruning dose relative to establishment time and vitality of the tree. If the tree is not well established or stressed, remove less foliage or establishment may be slowed. Once the tree grows tall enough to select permanent branches, gradually remove the temporary branches over time. Formative pruning of young, newly planted trees is also very economical in comparison to pruning larger trees.

PRUNING MEDIUM-AGED TREES

The larger, medium-aged trees found in our urban forest and landscapes can be structurally pruned to improve stability. Issues such as overextended branches, or branches competing with the leader, can be reduced or removed to reduce likelihood of failure. Long branches extending outside the main body of the crown are called overextended branches. They may have an increased likelihood of failure if the branch attachments cannot support the length and load of the branch. Check the taper of this first order branch to determine if there is enough fiber to support the weight. The advantage of reduction cuts over removal cuts for those larger (4 inches or greater) branches is that no large opening (i.e., pruning wound) is created on the trunk. A combination of small branch removal and subordination is a strategy creating a more attractive result while accomplishing desired objectives. Aggressive structural pruning, even on trees that are 30 to 40 years old, may be applied and can reduce one codominant stem redirecting growth into the non-pruned side, forming a more resilient structure.

The process of restoring or developing a central leader becomes a little more challenging for medium-aged trees. For practical purposes, medium-aged trees include those that are 15–50 years old. Most trees of this age and in good health are young and strong enough to provide the resources for sealing the pruning wound and reallocating food resources to growing the central leader. Be sure to investigate the overall architecture, then consider the branching structure, searching for codominant stems and poor branch aspect ratios.

The age and vigor of the tree, as well as the size of the stems, will determine the best approach and amount of pruning required to improve branch structure. In correcting the

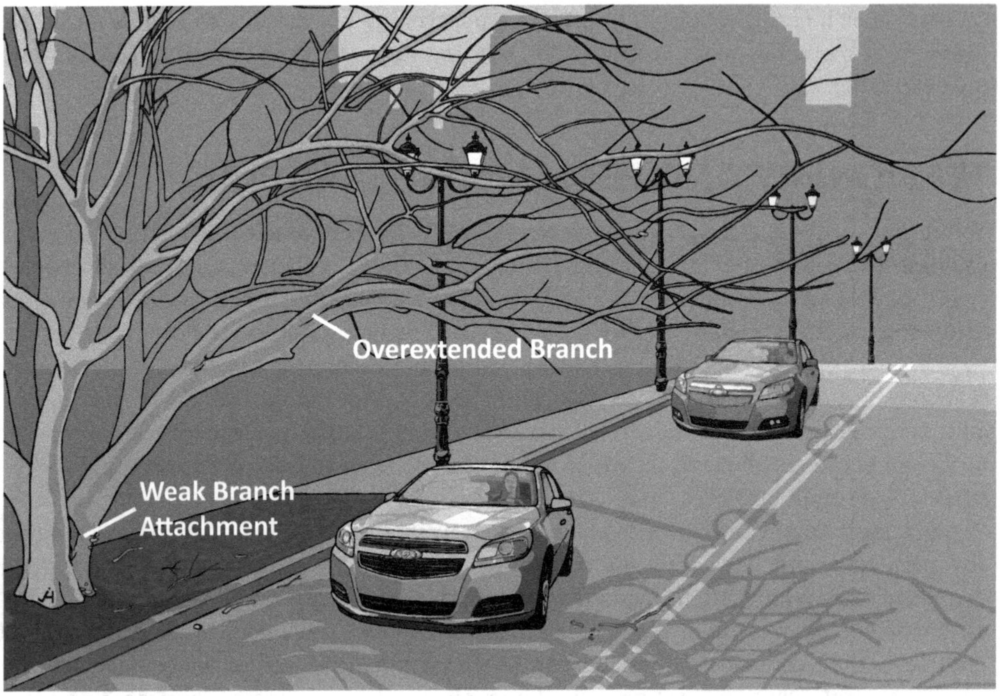

FIGURE 7.31 Overextended branches require a strong branch attachment.

codominant issue, the leader doesn't need to be perfectly straight. It should be dominant or considerably larger in diameter than all the other branches. Choose the one with the best branches. One of the rules for maintaining durable crown structure is that the leader or parent stem should be twice the diameter of its attached branches.

Pruning goals at this stage of development are to improve clearance, if needed, by removing the lower branches back to the trunk where practical to provide adequate clearance, to continue pruning to one dominant trunk where possible, and to balance the crown for stability and aesthetics. When selecting and pruning branches and stems, pruning cut size is a primary factor to consider. What can the tree withstand as far as removing living tissue without health consequences? After all, pruning is removing the tree's capacity to create food and energy, which is needed to help with recovery from pruning.

PRUNING DOSE FOR MEDIUM-AGED TREES

When deciding how much live tissue can be removed or reduced through pruning, consider the size of the branch or stem being cut. The critical factor is the amount of sapwood as compared to the heartwood present. Sapwood is more resistant to decay compared to nonliving heartwood. In application, the more sapwood present, the faster the pruning wound will seal. Increased heartwood or dead wood on larger branches take longer to close, often resulting in increased decay. Typically, on most species, larger pruning wounds, such as those exceeding 4 inches, can be difficult to close, and decay may develop in the wounded area, leading to further tree health and safety concerns.

> Tree pruning response is determined by developmental age, vigor and genetic capacity to heal.

PRUNING MATURE TREES

Pruning objectives switch from directing growth on young trees to managing loads on large and mature trees.

Mature trees, not yet entering retrenchment phase, are characteristically different than younger trees. Attributes of older trees include dead branches and missing bark; perhaps limb, trunk, and root injury and decay; cracks and cavities; and declining branches. They likely have substantial amounts of nonliving heartwood and a small ratio of sapwood to heartwood. The vitality of old trees can vary dependent upon growing conditions. Some mature oak trees may be 75 to 100 years old. Older trees or those having reached their developmental maturity can still experience benefits from structural pruning. The use of pruning strategies to redirect growth from the lower and middle crown to the center and top part of the crown creates a more stable, sustainable structure. This is accomplished through the careful subordination of branches and minimal pruning cuts.

Pruning on large trees focuses primarily on minimizing tree failure. Secondary objectives can include clearance and appearance. Cuts required to minimize the likelihood of failure or ensure compatibility with nearby structures may involve reducing mass where needed, especially toward the ends of the largest and longest branches. Implement cuts on branches with decay or vulnerable attachments to the trunk. If pruning was not executed earlier in the life of the tree, clearance for pedestrians and vehicles may be needed. However, large removal cuts on the trunk can be troublesome for older trees when it comes to wound closure. Larger cuts on branches with increased heartwood are more prone to decay.

In large part, these trees require no more than periodic (every year or two) dead branch removal and a small amount of pruning at the ends of the largest live branches to reduce the load. The arborist should remove no more live branches than needed to accomplish objectives.

OBJECTIVES FOR VETERAN AND MATURE TREES

The primary objective when pruning veteran and mature trees is to minimize failures and reduce excessive loads. Typically, it's too late for structural pruning to manipulate branch aspect ratios and branch spacing on mature trees. Many of these older, veteran trees have grown old because they lack serious defects, so significant pruning usually isn't necessary.

There are several possible objectives when pruning mature trees, but the primary objective is to identify, analyze, and evaluate the likelihood of whole-tree or branch failure. If tree risk assessments indicate a need for mitigation, options will most likely include pruning and perhaps support systems such as props. Essentially, this means managing the weight distribution (load) in the crown. When the objective is to reduce the likelihood of whole-tree failure then the tallest, widest, or leaning part of the crown is reduced in length. Where only one or two branches are causing the concern, those branches are reduced in length; pruning on other parts may be needed to improve crown balance or appearance. Pay particular attention to treating large branches with poor taper, especially when most of the foliage is concentrated toward the branch's end. These are among the most likely to fail. The condition becomes more acute when these branches have received large pruning cuts in the past or when cavities are evident. Even pruning cuts that are completely closed over with wound wood can create weakness from decay behind the wound wood. The goal is to relieve the load on the branches to reduce the likelihood of failure. This is accomplished by reducing the weight at the end of the branch.

For older, mature trees, focus primarily on reducing loads and risk with a few intentional pruning cuts.

REDUCTION CUTS VS. REMOVAL CUTS ON LARGER BRANCHES

The alternative to excessive pruning cuts on larger branches is implementing a series of reduction cuts. Reducing a larger stem or branch in increments—rather than completely removing the stem in one session—may be more desirable to prevent stress in the tree and improve wound recovery rates. In this case, a progressive pruning plan may be used to reduce the negative impact of a higher pruning dose. If appearance is a primary concern, this approach allows for corrective pruning without disfiguring the crown dramatically. However, it may take several pruning sessions over a few years to complete the task and meet pruning goals. Plan to subordinate or reduce the stem or branch—especially if it is larger in diameter and length—over multiple growing seasons until the branch is removed or the leader is twice the diameter of the codominant stem. The act of reducing the length through multiple sessions still benefits the tree by reducing the weight (load) on the weak attachment or codominant stem and improving stability until the union is no

TABLE 7.1 Structural Pruning by Age Group

Developmental Age	Pruning Strategy
Planting and early establishment	Select the most upright and vigorous stem as the leader. Eliminate competitors with a removal cut or use reduction and removal cuts to remove 50% or more of foliage on each competing stem. Review the following year and remove or reduce stems that continue to compete with the leader.
Young and established	Select the most upright and vigorous stem as leader. If the competing stem is approximately less than 3 inches in diameter, eliminate it with a removal cut. If the competing stem is too large to remove or if the customer will not tolerate the void, subordinate it with reduction and removal cuts to remove 50% or more of the foliage. Repeat subordination treatment annually until the stem is one-half or less the diameter of the leader.
Medium-age	Select the most upright stem as the leader. Subordinate the competing stem with reduction and removal cuts less than 3 inches in diameter removing about 40%–60% of the foliage. Repeat subordination treatment every 2 or 3 years until stem is one-half or less the diameter of the leader.
Mature and veteran trees	Use removal and reduction cuts to reduce loads (stress) on identified weak attachments. Repeat pruning treatment every few years as needed to control weight distribution. Remove no more than 10% of live foliage, if possible. Consider tree support systems when pruning isn't an acceptable approach.

longer at risk of failure. Often, arborists do not have the opportunity for session pruning and are required to accomplish pruning objectives in one visit. This requires informed decision-making, which balances the mechanical objectives with the biological needs of the tree.

PRUNING DOSES FOR OLDER, MATURE TREES

It is important to understand as the crown becomes thinner, so does the tree's ability to thrive and survive. The key to pruning on mature trees is to never do too much of anything! It is a delicate balance of needs between the client and the tree. Live foliage should be removed only for justifiable reasons. These food producing leaves are critical for older trees and any significant reduction in their leaf mass can compromise health. When thinking in term of amount, less than 10% of live buds and leaves should be removed. As best practices and pruning standards indicate, only prune what is needed to reach objectives.

SPROUTS ON TREES

Here's the thing about sprouts: sprouts grow naturally on trees. Remember, a tree doesn't grow what it doesn't need. Typically, the production of epicormic sprouts can be a product of excessive pruning or from wounds or stressful conditions. Sprouting develops when growth hormones (auxin) are unevenly moving to the underside of the limbs, stimulating dormant buds on top of the branch to produce new sprouts. In the case of storm damage or topping, some of these sprouts could be kept as a permanent part of the crown. When sprouts are young, numerous, in the way, or an aesthetic issue, remove or reduce them as needed and be sure they are spaced and placed to grow away from the trunk and interior parts of the crown. Sprouting can be managed by removing the less aggressive sprouts and shortening the more aggressive ones. Restorative sprouting is critical for recovery from storm damage. When developing a pruning strategy, think in thirds: remove one-third, reduce one-third, and retain one-third. Mind the spacing and direction of the growth, as these characteristics will influence the now permanent branches later.

FIGURE 7.32 A tree after receiving storm recovery pruning; remember, trees do not grow anything they don't need for survival.

CONIFERS

Some tree species naturally tend to self-maintain a single central trunk from which smaller lateral branches arise. This growth pattern creates a pyramid shape called an excurrent form. This type of structure tends to be strong and stable. Examples include conifers, bald cypress, and dawn redwood. On excurrent trees like most conifers, vertical spacing is less critical because the branches usually are only moderately vigorous and more horizontal and seldom compete with the leader. Minimal or no pruning is necessary except when an occasional limb becomes overly vigorous or competes with the leader.

FIGURE 7.33 Pinch the new growth back to control size on evergreens.

FIGURE 7.34 A new leader can be developed on evergreen trees with an existing shoot and splint applied.

SELECTIVE PRUNING, NOT SHEARING

First, avoid shearing evergreens. Instead, prune using a thinning technique, which keeps the evergreen's natural shape. The woody, non-green portions of needle-leaved evergreen branches do not normally produce new branches. So, do not cut evergreen branches back to the old wood. It's possible to thicken the new growth or control the size of some coniferous trees such as pines, spruces, or firs by pinching them in the spring. Manually break off or sever with secateurs half of the new growth (called candles) when it is approximately 2–4 inches long. Most importantly, never top or remove the central leader. This will destroy the apical dominance, losing it's excurrent form and replacing it with an unusual flush of uncontrolled growth at the top. If the terminal growth of a pine or spruce has been lost, you will need to help the tree grow a new terminal shoot. If not corrected, the plant will likely produce multiple leaders.

To form a new terminal, bend one of the youngest lateral branches near the terminal into an upright position. Secure it to the dead terminal stub or insert a small stake for rigidity. Over time, this branch will take over as a terminal leader. Once it assumes the leader role, remove the stub or stake, and the plant should resume its characteristic growth habit. Conifers that produce side buds (lateral buds) on shoots and branches, such as spruce and fir, can be pruned by cutting the tips back to the desired length in late winter when buds are dormant. The recommended place for cuts is just above a side bud or side branch. Pines do not produce buds along new shoots or existing branches, or from cuts made in older wood. Consequently, pruning must be confined to the current year's growth if tree size is to be managed. Unfortunately, it's too late to start corrective pruning on pines once they've outgrown their site space.

Arborvitaes, yews, and junipers are best pruned before new growth begins in the late dormant season so any unsightly cuts can be covered quickly by new spring growth. Some landscape managers prune these species again during the growing season for appearance. Fall and winter pruning should be avoided because it often results in tip die-back of the pruned branch. Arborvitaes, yews, and junipers respond best to the "thinning out" pruning method. A branch is cut off at its point of origin on the parent stem or back to a lateral side branch. This technique promotes a more open plant with a natural, unpruned appearance, without stimulating excessive new terminal growth. Shearing, on the other hand, stimulates a proliferation of growth at the branch tips that shades the shrub interior and favors the formation of undesirable "dead zones" (areas devoid of foliage). New growth will not develop unless there are surviving green twigs and needles.

PRUNING DOSE OR AMOUNTS

Arborists and horticulturalists are often perplexed with determination of the amount of pruning needed to "get the job done." However, best management practices indicate very clearly the pruner should "prune only what is necessary to achieve objectives." Establishing the objective before pruning begins can help with this decision.

The pruning dose is the estimated percentage of live foliage and buds removed during pruning compared with the total amount on the tree prior to pruning. According to the ANSI A300 Part 1 pruning standards, the percentage and distribution of foliage to be removed should be adjusted according to objectives, the tree's species, age, health situation, and environmental conditions.

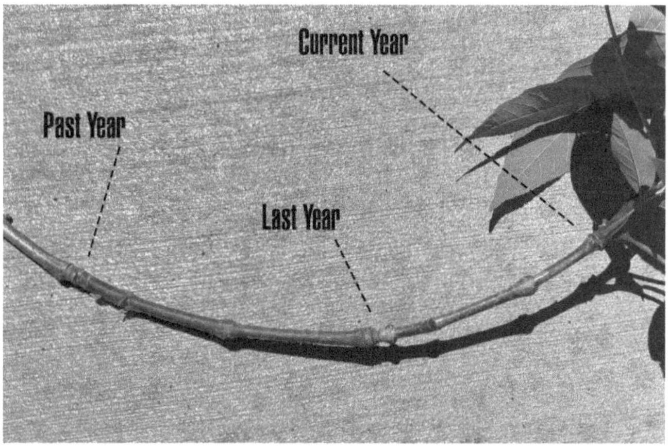

FIGURE 7.35 Health conditions can be determined by observing the growing distance between terminal bud scar locations.

A helpful aid to help determine tree health is the examination of the terminal growth. The terminal bud represents the current year's growth on the tree. Each twig or branch makes only one terminal bud annually and the previous year's terminal bud location will be marked with a bud scale scare that encircles the twig. This measurement can be used to get an idea of growing conditions over the past 3–4 years.

When searching for bud scars, they can easily be confused with leaf scars, which are the scars left behind after leaf drop or abscission. They differ from the terminal bud scar because

TABLE 7.2 A suggested pruning dose based on developmental age.

Development Stage of Tree	Pruning Dose (maximum % of total foliage removed at one pruning)
Young, newly established	50%
Medium-aged	25%
Mature	10%

A large debris pile may not be the most desirable outcome.

– Dr. Edward Gilman

FIGURE 7.36 Pruning debris isn't a good metric for pruning dose.

they don't encircle the branch and there are more of them. This is where the leaf petiole was attached and the arrangement can be either alternate or opposite. Greater distance between the terminal bud and bud scars indicates normal growing conditions. Smaller increments may indicate stress such as drought conditions, limiting apical meristematic growth and elongation.

There are situations when a larger pruning dose may be necessary and suitable for the tree and meeting objectives. Young, healthy trees tolerate greater amounts of live branch removal than do mature or unhealthy trees under stress conditions. Sometimes it may be necessary to remove more than 25% of the foliage to correct a significant structural defect such as multiple branch attachments or to reduce growth rates of competing branches.

There are several ways to express pruning amounts or doses mentioned in the ANSI A300 Standards. Arborists can specify the length to remove in feet, the percent of foliage to remove on a branch or branches, the height for elevating the crown, the vertical clearance from the ground to the lowest foliage or branch wood in feet, or other expressions to indicate the amount. The percent of foliage removed is traditional but inconsistently interpreted in actual practice. The percent of foliage removed is problematic because *arborists differ greatly in this subjective estimate*. It is nearly impossible to determine how much foliage was removed if the pruning debris is disposed of quickly or not stacked and visible. These are challenges which question whether the actual desired dose was achieved. The number of cuts, the type of cuts applied, and the diameter of cuts on the branches are proven to be the most accurate and efficient methods for quantifying pruning amounts.

Much larger doses can be applied to certain branches or sections of the crown to improve structure or manage growth. For example, about 60% of the foliage and buds on a medium-aged tree may be removed from a large codominant stem to subordinate it relative to parts of the tree that are not or only lightly pruned. Pruning a branch slows growth on that branch in proportion to the pruning dose. When a nearby branch is pruned, resources are reallocated and directs future growth by invigorating parts of the tree left unpruned. Reducing large branches instead of removing them not only preserves the trunk from a large pruning wound and ensuing decay, it allows the pruning dose to be distributed across other parts of the crown. Removal of too much foliage and buds at any single pruning reduces the amount of stored carbohydrates the tree needs for growth and defense. When pruning trees, the arborist and tree owner or manager must agree on the objective, with the dosage rationale explained. It may be necessary to plan for multiple pruning sessions scheduled over successive growing seasons or years to minimize stress by reducing the number of wounds and live tissue removal.

TIMING THE PRUNING SESSION

The best time to prune depends on the biological needs of the tree and client needs. If a tree is healthy and vigorously growing, timing is not as critical as when a tree is stressed and constrained by site resources, or when timing may produce unwanted late season growth, subject to winter injury. Unfortunately, most trees in the urban forest have the challenges of surviving in less than favorable conditions, which makes the timing of pruning even more important.

Winter and dormant pruning is considered an unfavorable pruning season for some tree species. During the winter months, a tree's reactions are significantly weaker than in other seasons, especially for those considered to be poor compartmentalizers. Much of this timing is related to the activity of parenchyma cells. These are highly active cells that store and transport specialized chemicals which aid in several life processes, especially wound healing. This occurs when these specialized cells are most active, during the growing season. This is considered the most beneficial time for trees to recover. However, it may be necessary to prune during these inactive biological seasons to reduce the transmission of tree diseases such as the various wilts.

BLEEDING TREES

Some trees ooze or bleed from pruning cuts made in the late winter or early spring, just before or during active growth. These may include species such as *Populus, Acer, Betula, and Carpinus* to name a few.

During the dormant season, the transpiration process virtually ceases. However, the roots do not enter dormancy and become active whenever the soil temperature rises above freezing temperatures. In evergreens, shoot growth may cease but water will continue to be lost from the leaves through transpiration. It is important to remember roots must be able to absorb moisture in order to replace the water lost from the leaves. This movement of water from the soil into the roots continues until the pressure of the water column within the tree is the same as the osmotic pressure exerted by the roots assisted by the capillary pressure in the xylem vessels.

The pressure of water within the roots ensures the xylem vessels in the tree are recharged with water needed for bud, leaf, and shoot expansion when favorable conditions occur.

This positive water pressure in the tree will create the potential for bleeding should there be any damage in early spring. The injury caused by pruning may release the internal pressure and allow sap to escape. This has been described as bleeding or oozing. The flow may continue for many weeks since liquid loss from the wound will release the internal pressure and again allow water to enter the roots from the soil. The condition typically stops when there is adequate leaf expansion, which means transpiration is occurring, releasing the water pressure within the tree.

According to research, recovery from pruning wounds best occurs after leaf expansion in late spring and early summer.

This is not necessarily harmful to the tree. It can be alarming or unsightly for some tree owners and even invite pests to the tree. To minimize most sap leakage on these trees, prune after full leaf expansion in early to midsummer. Some arborists also find that pruning in the fall minimizes bleeding on susceptible species.

PROPER PRUNING CUTS

Proper pruning technique is essential for recovery and aesthetics when pruning trees. Before any type of cut is made, it is critical for the arborist to be capable of identifying two very important components: the branch bark ridge and the branch collar.

The *branch bark ridge* (BBR) is a raised strip of bark at the top of the branch union where the growth and expansion of the trunk or parent stem and adjoining branch push the bark into a ridge structure. There is an inner partition of compacted xylem vessels extending along the angle of the ridge. Typically present on every branch union, the BBR is an important identifying feature for determining tool placement in pruning.

Branches on trees arise from lateral buds present in leaf axils. As branches become shaded naturally by crown expansion, photosynthesis is reduced in that location and growth slows to a lesser rate than the parent or main stem. A swollen area or collar develops at the junction of branch and stem. This swollen area is commonly referred to as a **branch collar** (BC)

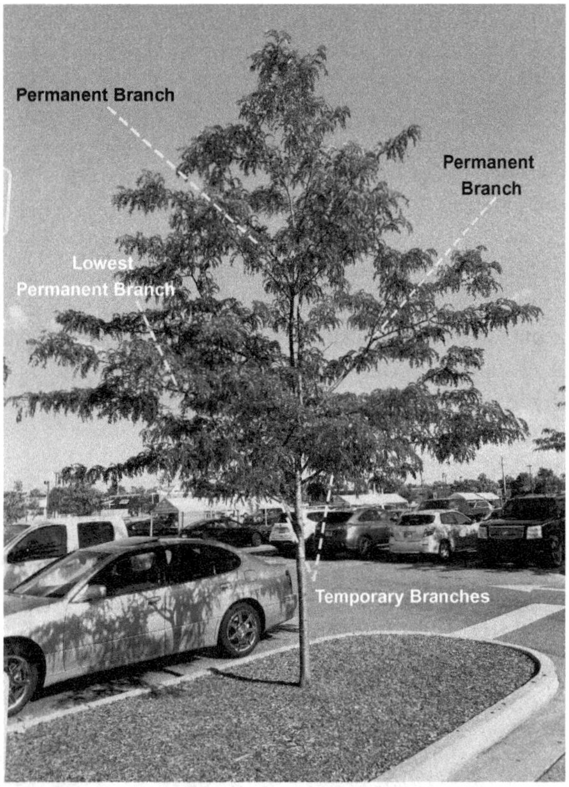

FIGURE 7.37 Branch components should be identified prior to making any type of cut.

FIGURE 7.38 Well-defined branch bark ridge and branch collar on a small branch.

and often presents in many branches on the underside of the branch. However, these branch collars are not always present, especially on codominant stems and branches with and unfavorable aspect ratio.

BRANCH PROTECTION ZONE

The combination of the branch collar, branch bark ridge, and the overlap between the branch and stem are the physiological components forming the branch protection zone (BPZ). The importance of the BPZ is limiting parent branch or trunk infection after injury or the removal of a branch. The BPZ is generally recognized as a three-dimensional cone-shaped area of morphologically and physiologically unique wood at the base of a branch. The xylem within the BPZ can become occluded with decay-resistant compounds. The BPZ xylem is structurally distinct from normal xylem with thicker cell walls, more parenchyma, and shorter and fewer vessels. The BPZ separates decayed and discolored wood in the branch stub from healthy sapwood in the trunk.

This region of specialized chemical compounds helps resist the spread of disease in the tree and facilitate wound sealing. If the branch collar is damaged or removed as in the case of a flush cut, the branch loses the ability to defend against invading diseases. As a result, disease organisms are more likely to invade the wounded area, creating decay. The amount of discolored and decayed wood resulting

Before the equipment comes out, remember these basic concepts:

- Each pruning cut has the potential to change the tree forever.
- The removal of living buds and branches and affects the tree's ability to gather sunlight for food production.
- Large limb removal can impact form and symmetry, affecting stability.
- The indiscriminate removal of large branches can leave the tree susceptible to decay and decline.

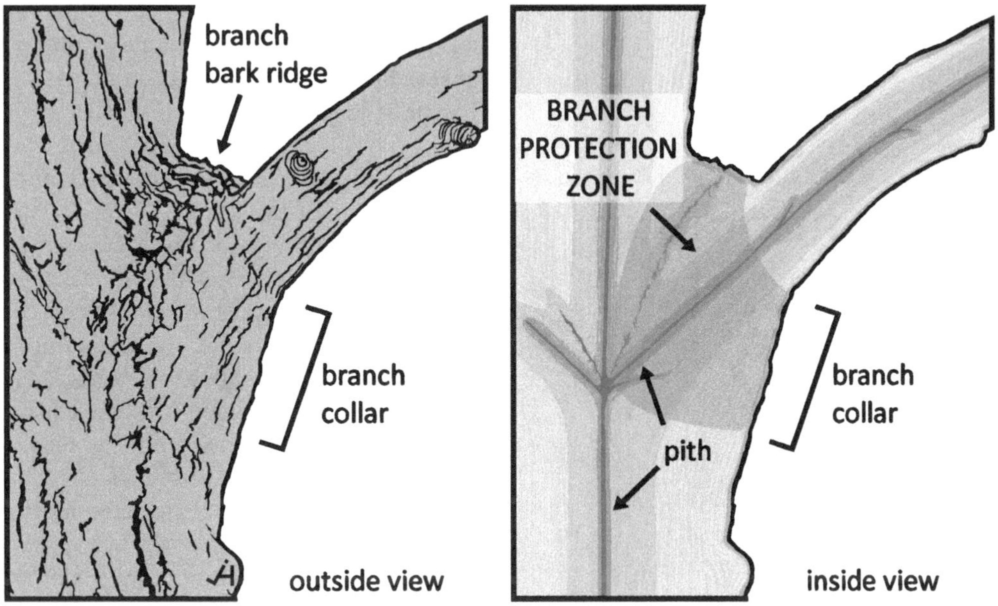

FIGURE 7.39 The branch protection zone contains special properties to facilitate wound recovery.

from pruning cuts appears to be minimal when removing small-diameter branches. Removing larger branches creates a greater volume of dysfunctional wood. Therefore, trees should be pruned to minimize the diameter of branches that will eventually require removal as they age. This is accomplished through using reduction cuts, during the early establishment of the tree. The diameter of the cut branch may affect the development of discolored and decayed wood after branch removal, particularly in large branches that contain heartwood. This infers that there is faster healing and pruning recovery when there is a greater amount of sapwood compared to heartwood in the branch.

Pruning wounds that expose hardwood are more likely to decay.

TYPES OF PRUNING CUTS

There are three types of pruning cuts: reduction, removal, and heading. *Reduction cuts* shorten a branch by removing the terminal portion back to a lateral branch of equal or smaller diameter. The cut should be made just beyond the lateral branch and the remaining branch should be one-third or larger than the size of the branch removed. Reduction cuts are often used for improving crown structure, directing growth, correcting branch defects, or decreasing tree size. This is a preferred method for reducing risk and directing growth by suppressing branches with large aspect ratios. The weight on branches is lessened as well, for situations with weaker branch attachments.

Removal cuts eliminate the target branch back to the main stem or a larger branch, just outside the branch collar and branch bark ridge. If the BC cannot be identified, position the pruning tool so that the resulting cut is as small in diameter as possible. Removal cut examples include removing a branch from the trunk or severing a smaller branch from the larger one. A removal cut should retain the branch collar if one is present. There is no standard angle to the trunk or parent branch at which to make a proper removal cut. The angle should be determined by the location and of the branch collar (if a collar is present), the location the branch bark ridge (which is usually present on a branch attachment). Most cuts are made at a downward angle, away from the trunk or branch. A general guideline used by arborists is to perform the cut that takes the shortest path through the wood, which is almost always a cut nearly perpendicular to the top surface of the branch. Typically, this leaves smallest amount of wood exposed for the tree to seal. One of the biggest challenges for pruners is judging cut locations if no collar is present. If not present, the final removal cut should be made at a 30- to 45-degree-angle from the stem, with angle of stem considered. Visualize an imaginary line, as seen in Figure 7.43, with the stem or branch to get the cut angle. A different option is to determine the cut angle from the location of the branch bark ridge. Included bark is another challenging situation. The recommendation is to remove the branch or stem with included bark by cutting as far down into the union as possible without

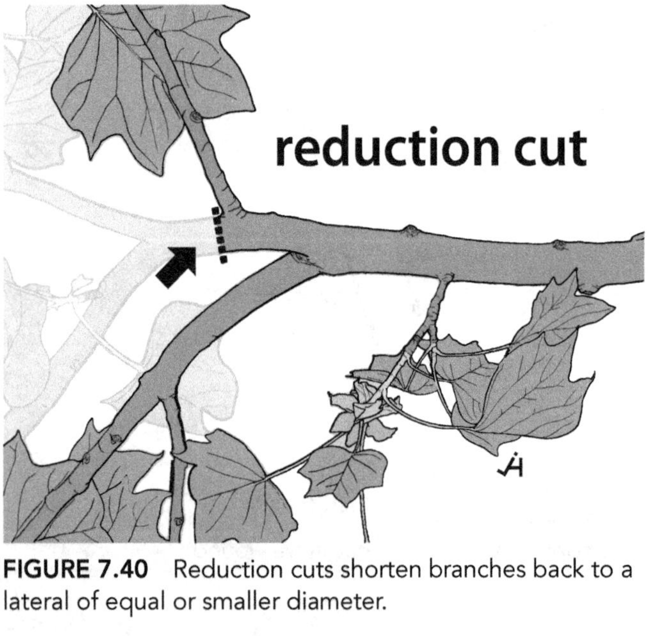

FIGURE 7.40 Reduction cuts shorten branches back to a lateral of equal or smaller diameter.

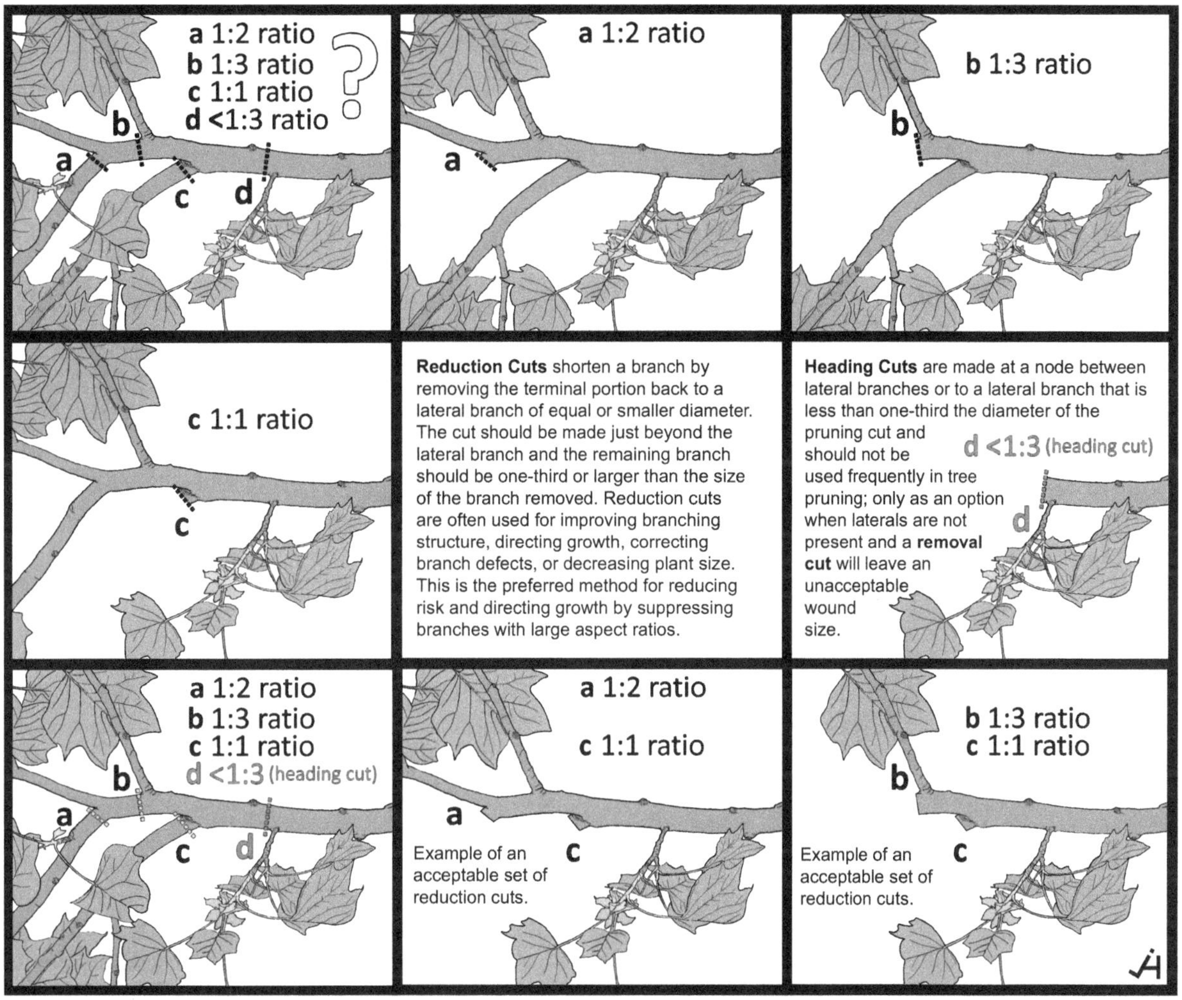

FIGURE 7.41 Pruning cuts should be made to laterals at least 1/3 or larger than the size removed.

TABLE 7.3 Pruning Cut Comparison

Reduction Cuts	Removal Cuts	Heading Cuts
Shorten branches with reduction cuts (subordination) is one of the most effective ways to suppress growth on larger branches.	Improves clearance by removing the lower branches, especially with young trees.	Sometimes necessary, especially when repairing trees after a storm.
Reduction cuts are preferred over heading cuts when reducing the crown because they result in less wood decay behind the cuts and less sprouting.	Helps direct future growth into another stem by reallocating resources to the preferred stem.	Allows for large branches to be progressively reduced without making large removal cuts on the trunk.
Reduction cuts allow more sunlight to reach branches growing from the leader above the cut, stimulating growth.	Eliminates undesirable vertical growing branches down to the parent branch.	Can create excessive sprouting and decay at the cut ends.

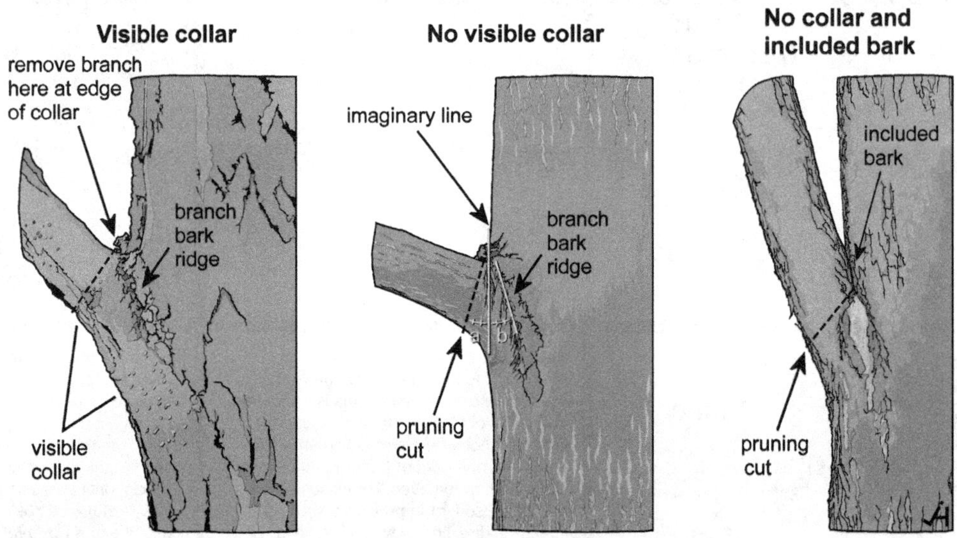

FIGURE 7.42 Removal cuts sever a branch back to the trunk or larger branch.

injuring the retained trunk/branch. Cut down as close as possible to where the double ridge ends or merges together at the bottom of the inclusion as shown in Figure 7.43. The important consideration is the smallest wound possible is always the best practice.

Heading cuts are made at a node between lateral branches or to a lateral branch that is less than one-third the diameter of the pruning cut. This also describes cutting a larger branch back to any size lateral branch. Often, these cuts are used when restoring trees following storms or when managing tree size where there are no suitable laterals to execute the cut. This is considered a better alternative than cutting back to the trunk and leaving a larger wound, especially on mature and veteran trees. Heading cuts should not be used frequently in tree pruning and only as an option when laterals are not present and a removal cut will leave an unacceptable wound size. A heading cut is made perpendicular across the axis of a branch, or at a slight downward angle away from the retained branch or bud. Internodal cutting between buds is not an acceptable option because sprouts may emerge from buds some distance behind the cut surface, leaving a dead stub. This is the type of cut made when topping trees.

FIGURE 7.43 Heading cuts reduce a branch back to any size lateral branch.

Although reduction cuts are preferred, heading cuts may be necessary on young trees because they usually have few lateral branches on their main stems. Properly executed heading and removal cuts gradually or quickly close from the outer wound margin (bark) inward. Heading cuts can be appropriate for damage restoration from storms and occasionally when elevating the crown is a critical pruning objective for clearance.

PERFORMING THE PRUNING CUT

The pruning cut depends on the size of the branch to be cut and if a simple, single cut can be made with hand pruners, loppers, or a hand saw. If the branch is too large to support with one hand, a specific cutting technique will need to be executed, most likely with a hand saw or chainsaw. Before making any cut, remember to identify the branch components to ensure proper alignment of the pruners, loppers, or saw during the cut. When removing any branch or limb, always make the cut just outside the BBR and the BC when present.

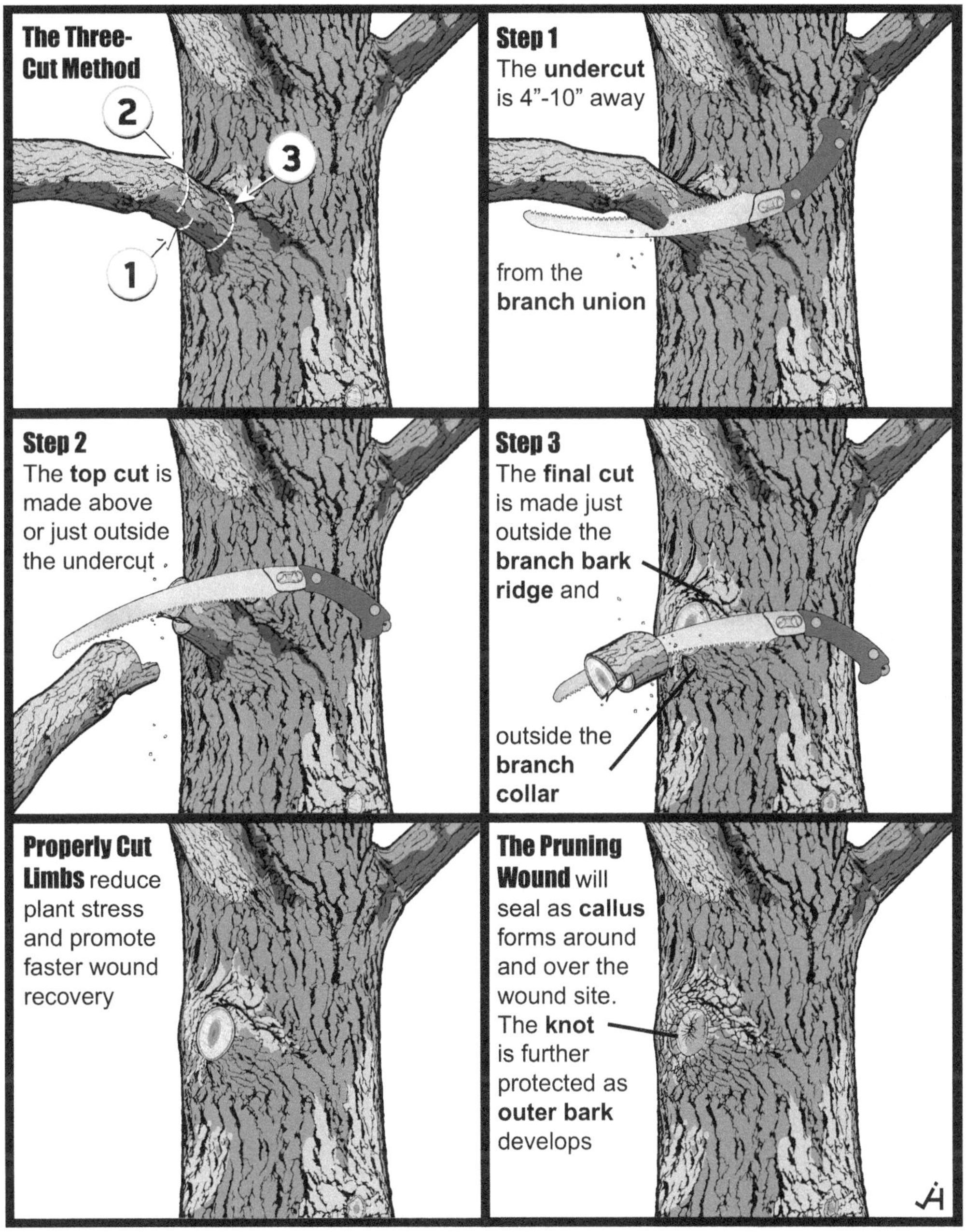

FIGURE 7.44 The three-cut method requires three distinctive pruning cuts to sever a branch.

FIGURE 7.45 Rip cut.

FIGURE 7.46 Stub cut.

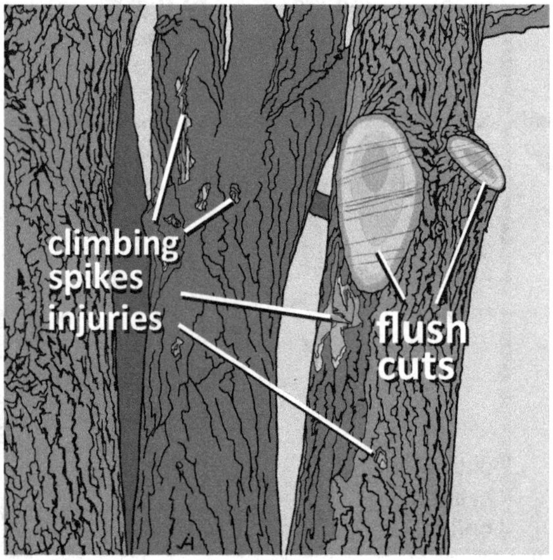

FIGURE 7.47 Flush cut.

FIGURE 7.48 Dead pruning stubs should be removed to the swollen collar without damaging the live tissue.

Branches too large to be supported by hand should be removed using the *ternary method,* also known as the three-cut method, to avoid tearing or splitting the bark and damaging the branch protection zone. (The triple-cut method was formerly called the "double-cut," which is a misnomer, because it actually takes three cuts to finish the process, rather than two as the name implies.)

In the triple-cut method, the first cut, called the *undercut,* begins on the bottom of the branch anywhere from 4 to 10 inches away from the branch union. The second cut, the top cut, is made above or just outside of the undercut; proceed with the saw from the top of the branch moving downward. This is the cut that allows the branch to be cleared completely. As the saw moves through the wood the branch will naturally fall as gravity takes over. This top cut will soon meet the plane of the previous undercut, stopping it and preventing the bark from ripping. After both cuts have been made, the branch should easily fall and be removed. However, the pruning process is not finished. Make the third and final cut just outside the branch bark ridge and outside the branch collar on the bottom side of the attachment. Now a proper cut has been made, let the sealing begin.

Determining the two points on the branch to align the cut can be challenging at times, especially when the BC may not be readily visible. Branches with a defined collar should be removed outside the swelling at the base of the branch (mostly in a slanted angle to the stem). Now, for branches without a collar, the branch should be removed outside the branch bark ridge, but with a cut more or less parallel to the stem to avoid the formation of a dead stub at the lower margin of the wound where the branch was cut. This is not a flush cut, because the branch bark ridge remains at the stem, and the cut is outside the branch or stem. The wound is oval shaped and is a lot smaller than the flush cut. The cambium tissue at the wound edges receives assimilates, allowing the tree to wall off the wound directly from the wound edges.

The primary aim of any pruning cut is to create the smallest possible wound, allowing for the most rapid closure possible. Recent research has found that the smaller the wound diameter, the less the potential for decay. This suggests decay would be less by making a cut which takes a shorter path through the wood. Cutting parallel to the branch bark ridge exposes too much wood surface area, and results in greater decay. The objective is to find an angle and position for the cut that minimizes the surface area of the cut and minimizes the amount of bark and cambium dieback behind or under the cut. The appropriate angle can vary among species.

Poor pruning cuts, such as leaving rips, stubs, or flush cuts, create many issues detrimental to recovery. Pruning without damaging the branch collar and branch bark ridge encourages the formation of wound wood, which protects the tree. Never make a pruning cut through the collar or through the branch bark ridge. This will damage the branch protection zone within and leave a significantly larger wound. Making a removal cut too close to the trunk or parent branch is a flush cut. Implications of making a flush cut become evident as the affected tree grows. If you could look behind a flush cut made some years earlier, you would find cracks, wood discoloration, and in many cases, decay. Also, never leave the stub cut outside the branch collar area. This leaves woody fiber with no support from leaf tissue; it will soon decay and provide a conduit for disease to spread into the remaining branch or stem.

Dead branches are a common occurrence in living trees. There is little evidence that retaining dead branches in trees is detrimental to tree health. Dead branches are often removed to reduce risk near targets, such as buildings, people, and vehicles. Collars on dead branches can protrude well beyond collars on living branches. The removal of the swollen wound collar growing around the dead branch stub is not recommended. This stub contains living tissue associated with the trunk. Leaving the collar intact helps prevent the spread of decay in the trunk and will eventually be occluded.

RESPONSE AND RECOVERY

What should be expected after pruning relative to how the tree responds to this intervention? Trees respond to pruning in various ways, again, depending largely upon their developmental age, health, and vigor. Phytochemicals are initiated in the pruned area to help facilitate healing and recovery. Proper pruning can be used to encourage growth in the portion of the tree needed to become dominant. The placement of the pruning cut can affect the direction and amount of growth. Pruning to subordinate or remove a competing stem allows the tree to reallocate food and water resources toward the remaining branches and stems, promoting improved development. Codominant stem removal increases the penetration of more light energy to the favorable leader left unpruned, stimulating growth and elongation in the remaining buds and smaller lateral branches. This eventually balances out the crown. If pruning objectives require removing large amounts of living tissue on codominant stems, it can leave large voids in the crown. This isn't necessarily harmful, but it may create an unbalanced appearance for a time. However, patience should remedy the situation as the crown develops and fills in with new growth.

TREE RESPONSE TO PRUNING

Most arborists and landscape professionals understand pruning as an important arboricultural practice. How a tree responds to this intervention is mostly unknown to many practitioners. Pruning elicits amazing reactions internally and externally. Because pruning is an important practice for arborists and landscape professionals, it is important to understand these actions create reactions from the tree.

When trees are pruned, removing live green tissue and living buds, changes occur within the tree. At the very least, we are modifying the physiology and morphology of the tree when pruned. The negative aspect of pruning includes the loss of food production and carbohydrate reserve capacity especially from excessive pruning. Pruning live tissue limits tree processes, because there are fewer shoots, lower carbohydrate supplies, less storage, and less nutrient uptake. This can lead to decline and dieback in the tree and predisposition to pest and environmental issues. Professionally executed pruning strategies balance pruning objectives with biological needs.

PRUNING DISRUPTS CHEMICAL ROUTES

Let's look at the hormonal response created by removing living tissue. First, nearly all aspects of tree growth and development are regulated by phytohormones (plant hormones). A plant hormone is an organic compound synthesized within the plant in very low concentrations,

FIGURE 7.49

causing a physiological response. It may be transported to another site or may be active at the site of synthesis. According to the definition, only naturally occurring compounds can be described as plant hormones. However, manufactured chemicals that mimic the structures and functions of hormones can also have significant effects on plant growth and development. Such substances are usually called plant growth regulators and are often analogues of endogenous hormones.

Plant scientists indicate there are five major groups of endogenous plant hormones affecting over 2,000 plant processes. Tree biology and management typically focuses on the following groups or substances considered to be the major hormonal groups having an impact on plant growth and development: auxins, gibberellins, cytokinins (sometimes called kinins), abscisic acid (ABA), and ethylene.

FIGURE 7.50 Excessive removal of branch terminal buds will result in uncontrolled growth.

Hormones are simply chemical messengers often produced at one location and translocated to another location where they go to work. Pruning trees creates a disruption in the flow or routes for the many chemicals found in trees, including hormones. The long, connected pathways of xylem and phloem not only transport food and water, but also the phytohormones signaling how the tree should respond when wounded, damaged, or pruned.

Severing branches during the pruning process interrupts or disconnects the chemical pathways that influence root and shoot growth as discussed in the tree functions chapter. The downward flow of auxins, which are produced in the green shoot tips, stimulate root and shoot growth. The cytokinins produced in the root-growing tips stimulate crown growth. Cytokinins produced in roots encourage the emergence of lateral buds, while auxin produced in the apical meristems and new foliage inhibits cytokinin activity through a process which results in apical dominance. Branch buds are initiated in the shoot apex and their flushing action is activated by cytokinins.

Pruning cuts that remove growing tips sever the auxin-cytokinin pathways in the tree by removing the auxin producing shoot tip. The removal then triggers the release of dormant growing points such as lateral and epicormic buds along the trunk and branches due to the accumulation of kinins. These once unseen dormant buds move outward with the vascular cambium as the tree grows.

When removing branches interrupting the pathway, buds may emerge and grow into branches, attempting to restore the broken pathway from the released sprouts to the roots. A proliferation of lateral buds can develop into branches below the location of the pruning cut where the apical meristem on the branch was removed. However, not all pruning causes sprouting.

Basically, the cytokinins are transported upward in a branch until the tip is removed through pruning. Due to the recently pruned and now missing apical shoot, which produces auxin, the latent buds on the truncated branch are wakened and produce new

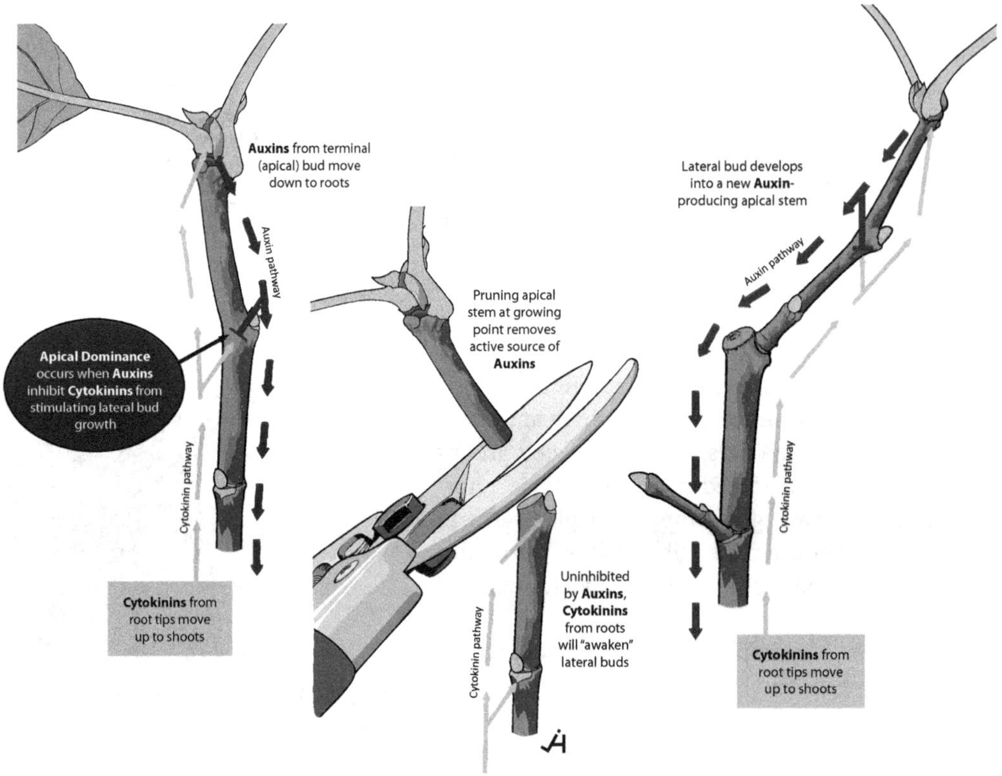

FIGURE 7.51 Effects of pruning and new growth.

shoots. The concentration of cytokinins increases in this area, flushing new growth below the location of the cut. Consequently, all this new growth requires resources to survive, such as food and water. This newly formed expansion of growth is often at the expense of the root system and the health of the tree as the allocation of resources has changed unfavorably. More resources are required in greater quantities for this newly expanding growth.

PRUNING DISRUPTS APICAL DOMINANCE

Apical dominance varies within tree species which influences plant shape, as well as the length and number of lateral shoots. Trees with strong apical dominance are more upright in growth (excurrent crowns). Those with weak apical dominance have more horizontal and spreading growth (decurrent crowns). Pruning can alter the shape and direction of growth of both by manipulating branch habit and influencing the hormones involved in development and expansion.

The apical buds in the tree structure produce auxins that are actively translocated down the shoot. As the auxins pass the lateral buds along the shoot, the buds maintain inactivity and are unable to sprout and grow. If the growing tip is still there, auxins flow and these lateral buds remain inactive. However, if the growing point is removed, auxin production is interrupted (chemical pathway). This interruption allows the cytokinin present in the stem to activate growing points and allows for flushing of new growth and lateral branches. The process of heading back or making reduction cuts on an excurrent tree form can alter the natural shape of the plant.

Pruning focuses on improving structure, clearance, and appearance. Many tree care professionals overlook or fail to understand that when pruned, trees do respond biologically. However, when pruned properly, there is an overall favorable response, leading to improvements in structure and stability.

MINIMIZE DAMAGE, MAINTAIN BALANCE

Dr. Alex Shigo, considered the father of modern arboriculture, once said that to be a good arborist, first, you must know your tree. This can include recognizing the health, vigor, and developmental age of the tree before determining the pruning dose or the amount of living tissue removed during any single pruning episode. Removing dead, damaged, or dying tree parts often referred to as unbeneficial plant parts, such as basal sprouts and perhaps epicormic sprouts, shouldn't significantly contribute to the pruning dose calculation.

The response from a wound such as a pruning cut stimulates the production of plant hormones involved in wound recovery. Specific phytohormones are necessary for cell division, which facilitates compartmentalization. The compartmentalization of decay in trees process (CODIT) is an important tree strategy for surviving the damage of pruning maintenance. Walls are developed internally and wound wood is created externally which "compartmentalizes" the wound by sealing off the damage and attempting to

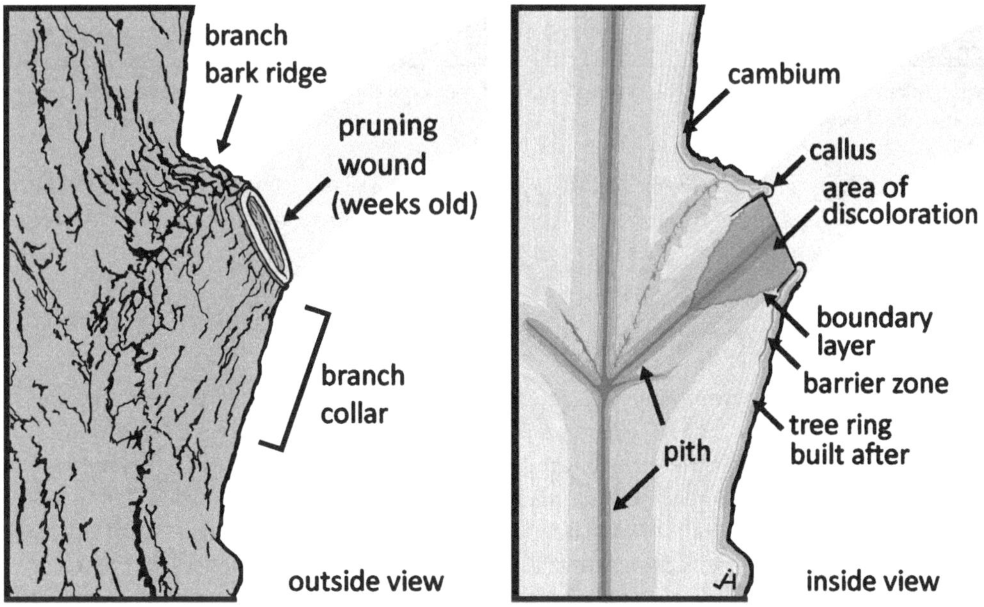

FIGURE 7.52 Wound reactions of bark, cambium, and wood.

TABLE 7.4 Tree Species and Their Comparative CODIT Abilities

Strong Compartmentalizing Trees	Weak Compartmentalizing Trees
Aesculus	Acacia
Acer rubrum	Acer saccharum
Betula	Castania sativa
Celtis	Carpinus
Cercis	Gleditsia
Eucalyptus	Juglans
Erythrina	Lysiloma
Fagus	Quercus macrocarpa
Ficus	Q. petrea
Magnolia grandiflora	Q. germinate
Malus	Q. robur
Persea	Q. rubra
Pinus virginiana	Robinia pseudocacia
Populus	Tilia
Prunus	Tabebuia
Quercus laurifolia	Ulmus americana
Q. nigra	Ulmus alata
Q. laevis	Ulmus parviflora
Q. shumardii	
Q. palustris	

Note. Adapted from Lonsdale, 1999, Dujesiefken, 1991, personal experience, and anecdotal evidence. Citation in references.

minimize pathogen movement and decay. Tree species have varying abilities to seal off these wounds depending upon health, conditions, and their resource allocation strategies as a species.

TIMING AFFECTS RECOVERY

Current research contributes largely to determining the optimal times to prune to reduce the impact on the tree and facilitate faster wound recovery. If we want the trees to recover quickly, we should prune when the trees naturally recover quickest. This timing is when they are actively growing and their physiological processes are functioning best, which is during their growing season. Additionally, severe, heavy pruning, especially at times of stress, such as during drought conditions, can result in poor compartmentalization and health consequences for the tree.

Also, it is important to recognize which trees facilitate wound recovery more effectively. Some trees are considered strong compartmentalizers; some are weaker in the process. This will help when deciding on the amount of branch removal and green tissue elimination required. Examples are provided in the table below.

CARBON IS CURRENCY

It's really all about that carbon. It's a plants currency. Typically, tree death and decline may be summed up as the result of two conditions: carbon starvation and hydraulic failure. These can happen alone and in combination. Trees decline and eventually die when carbohydrate storage becomes depleted and they lack the capacity to create energy to support metabolic functions. Hydraulic failure occurs when the water can no longer be delivered to the leaves due to xylem embolisms. Embolisms can occur anywhere there is xylem damage or interruptions to the water supply. These are gas bubbles within the tracheid and xylem vessels. At this point, photosynthesis stops and the ability to move assimilates is reduced. The respiration function is reduced and the decline spiral begins.

This relates to pruning as the most obvious concern is the reduction in leaf area, which can lead to a reduction in carbon fixation and assimilates. At the same time, resource allocation is needed for recovery and defense just when the photosynthetic area is reduced. This leads to an increased competition for carbon for other processes, such as growth and reproduction. Live tissue amounts should always be considered as part of the pruning strategy, considering the health and condition of the tree.

The vital concepts of pruning practices are to

- minimize the impact of the wounds;
- Do not prune trees experiencing stress conditions;
- Do not exceed the pruning dose, while considering developmental age, species, health, and environmental conditions; and
- always avoid damaging the BPZ, the area within the branch collar and branch bark ridge.

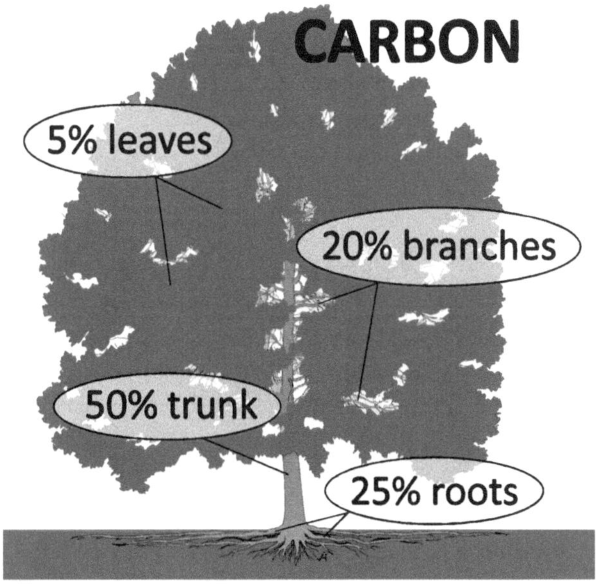

FIGURE 7.53 Removing living buds and foliage reduces carbohydrate production and reserves.

TREE WOUND APPLICATIONS

When trees experience an injury, such as pruning, the objective is to manage the wound to facilitate quicker healing. Formerly, this meant the application of a tree wound paint or dressing. Tree wound dressings have been around for decades and frequently used by arborists, fruit tree growers and tree owners as a post-operative step in pruning. However, research has confidently proven these chemical "bandages" are not recommended for the tree and slow natural wound closure. It has been believed the application of these petroleum-based paints would help deter decay and fungal activity. However, research has demonstrated tree wound dressings do not prevent decay and provide limited benefits regarding wound closure. Many arborists current with research and best practices have known this for a long time and have stopped using them.

The best way to understand the implications of prophylactic dressings requires a review in tree biology and the tree recovery process. Trees isolate damage from injury by forming wound wood in the affected area, such as a pruning cut. This new tissue around the wounded area is highly lignified with fungistatic properties which forms on the perimeter of the wound, gradually sealing it off. The pruning cut or injured area forms a specialized wall

FIGURE 7.54 Good wound wood development on a pruning cut.

around the wound to help reduce the spread of decay. This wall helps prevent the wood decaying organisms from establishing itself in the tree leading to advanced decay and damage.

Although the seemingly legitimate advertisements on the wound dressing packaging make claims such as these:

- "A clean, easy, simple way to aid in healing cuts and protecting tree wounds, pruned-edges and graft unions of roses, trees, and shrubs"
- "Insure your trees, shrubs, and vines against decay, insects, and fungi in any kind of weather"
- "An artificial bark for treatment of wounds . . . made of ALL NATURAL biodegradable materials"

The reality is petroleum-based products do not inhibit decay but limit the oxidative processes necessary to compartmentalize the wounded area. It has been thought a wound dressing can be used to prevent wood rot and the entrance of decay organisms. When in actuality, tree wound dressings seal in moisture and decay which can make problems worse. In fact, the ANSI standards state, "Wound treatments should not be used to cover wounds or pruning cuts, except when recommended for disease, insect, mistletoe, or sprout control, or for cosmetic reasons. Wound treatments that are damaging to tree tissues shall not be used."

There may be some benefit for wound treatments with certain diseases such as Oak Wilt along with a pesticide application or alone to the wound. Entomologists recommend the application of water-based latex paints if wounds cannot be avoided during times of increased risk for pathogen transfer. When pruning is necessary on susceptible trees, prune when in dormancy or when there is less risk for transmission by vectors, the chance of infection is greatly reduced, and wound treatment is unnecessary. If pruning trees prone to disease such as fire blight, sterilize pruning tools with a dilute alcohol solution to help reduce the spread to other trees. The best treatment for wounds especially from pruning, is a proper cut using the proper tools at the appropriate time. When pruning trees prone to disease such as fire blight, sterilize pruning tools to help reduce the spread to other plants.

PALMS

Palms are pruned mostly for appearance or safety reasons, not for health or strength. Most palms can be left unpruned and remain perfectly healthy and strong provided their nutritional requirements are met. One of the main reasons tree owners want palm trees pruned is to remove yellow and dead leaves, so preventing these leaves from turning yellow reduces the pruning requirement. Fertilizer containing 100% controlled (slow) release nitrogen, potassium, and magnesium should be proactively applied to palms, especially the non-native species. Another common objective of owners is to remove developing fruit before they fall and make a mess. This must be timed to correspond with the development of the flower stalks.

Many planted palms are needlessly overpruned. There is no biological reason to remove green leaves from a palm. Removing green leaves likely slows growth rate and can cause a narrowing of the trunk if performed over a long period of time. Insect pests are also attracted to some palms that are overpruned and unnecessarily pruned. Overpruning, also known as "hurricane pruning," has been associated with increased damage in hurricane force winds, not less. The critical component missing is the strong, lignified older leaves which protect the palm bud from damage in strong winds. The high lignin content in old leaf bases that surround the tender young leaves and bud provide the support required to prevent most damage in strong winds. Removing this protection contributes to failure in storms.

FIGURE 7.55 Comparison image of proper and overpruned palm.

Palm pruning should primarily remove dead fronds. When live fronds are removed, it should only be those that are below those above with a more horizontal orientation. Leaves (dead or alive) can be removed by cutting back to the trunk or near the trunk without causing damage to the trunk. Do not remove yellow fronds, they are still contributing to overall plant health. Retain the base of a booted palm, such as cabbage palm, because the saw can easily damage the trunk tissue during removal. On palms with a crown shaft, petiole bases can be carefully cut at their base to remove the petiole entirely. Leaves can be cut in half or to any length to clear an obstruction such as a building or utility; the retained portion of the leaves will remain green for the normal life span of the cut leaf.

CHAPTER 8

TREE APPRAISAL AND VALUATION

LEARNING OBJECTIVES

- Define tree appraisal and explain how it is applied to urban trees and landscapes.
- Define the different appraisal approaches and techniques used for a tree appraisal.
- Recognize the primary depreciation factors to consider when preparing an appraisal.
- Apply the systematic process to develop and communicate an assignment result as either a cost or a value.
- Understand the applications of the cost approach in appraisal and identify the appropriate technique for appraisal assignments.
- Demonstrate the application of the trunk formula technique for a tree appraisal assignment.

FIGURE 8.1 Trees are capital assets with economic value.

Trees provide many benefits and value to property owners in functional, aesthetic, social, environmental, and even economic ways. In appraisal terms, value may be defined as the present worth of future benefits. According to the Council of Landscape and Tree Appraisers, value means the monetary worth of a property, good, or service to buyers and sellers at a given point in time and not to be confused with cost or price. Many of these benefits can be quantified using money, and it is the responsibility of an appraiser to assign the monetary value. An appraisal is considered an opinion of value based on industry procedure. The valuation of trees and living landscape components is often complex, requiring training, expertise, and experience. A qualified consulting arborist must be current with appraisal methodology in the event of contentious or litigious situations.

Consulting arborists who prepare appraisals are usually focused on trees as components of a maintained landscape such as on residential properties, in parks, and in public spaces. Trees growing in forests may be appraised for their value as forest products, such as sawlogs and pulpwood. Trees grown in commercial nurseries for the purpose of selling and transporting in their entirety or as Christmas trees, as well as trees that produce crops such as fruit and nuts, are appraised using a different set of methods and techniques. This chapter focuses on the appraisal of landscape trees. Arborists who are not properly trained in other aspects of appraisal should not use landscape tree appraisal methods for trees that have highest and best use as forest products, or as income-producing trees.

DEFINITIONS OF MONETARY RELATIONSHIP

There are distinctive differences in appraisal terms for "worth" and for everyday use. Price, cost, and value of things are often used interchangeably in everyday language; however, there is a big difference when it comes to appraisal. *Price* is the amount of money asked, offered, or paid for an item or service. Price may or may not relate to value or cost and is a final fact or numerical rate. For example, the price of a tree is stated as a number on the item, like the price of goods in a department store.

Cost is the amount of money required to create, produce, or obtain a property or service. It can be applied to the cost of a functional replacement of a tree or landscape element. The term is used as an estimate of current, future, or historic reproduction cost of something. It is the amount of money actually paid or will have to be paid for the appraised subject. Cost may be equal to value in some situations, but cost is not synonymous with value. Unlike value, which is an opinion, cost is a fact.

Value is a difficult term to define and is a general concept best described as the monetary worth of something at any given time. Also, value can be understood when it can be measured, such as by relating value to the current prices for comparable items, plants, or to amounts that were actually paid for comparable items which were sold or purchased in the past. Again, the term, "value" is expressed as an opinion and not as fact.

Identify the Appraisal Problem

The purpose of an appraisal is defined by the clients' needs. The arborist is contacted by a client with a specific need for valuing trees or plants. These needs may include tort claims, insurance claims, tax deductions, real estate assessments, and proactive planning.

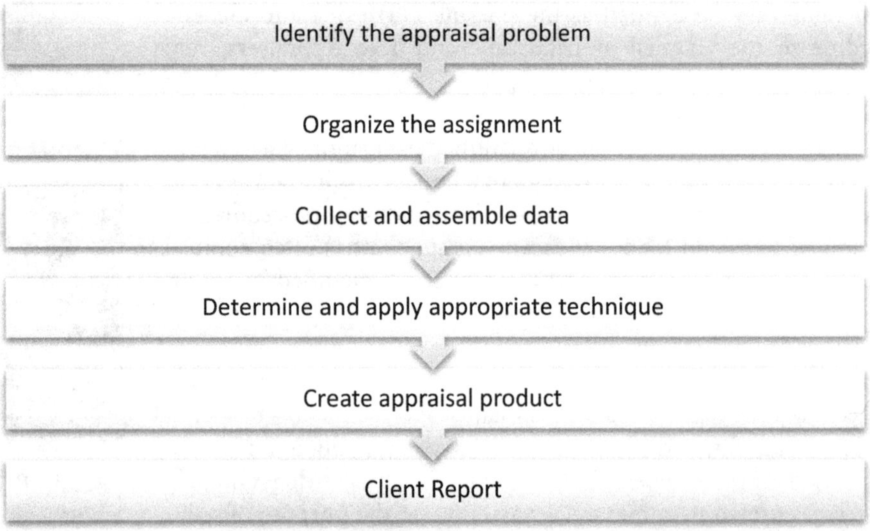

FIGURE 8.2 Tree appraisal development.

Typically, appraisals are requested by insurance companies, attorneys, and homeowners, among others. Also, city foresters may use tree appraisals to communicate the value of the trees in the urban forest. This helps with assembling budgets and acquiring the necessary resources to maintain the trees in the community.

Besides client needs for an appraisal, the next question the arborist must seek answers to is the intended use of the appraisal. This will help determine the appraisal method and technique to develop the final appraised value or cost. Additionally, before the assignment can be defined, the subject of the appraisal needs to be determined. Trees aren't the only landscape element appraised as shrubs, hardscape, or a collective landscape valuation may be required. For example, a distracted driver may leave the road and hit a tree, causing severe damage, or they may drive through the homeowner's landscape, damaging not only trees, but shrubs and other plants in the landscape.

Organize the Assignment

The terms of the assignment, sometimes called the scope of work, should be defined, and agreed upon between the client and the arborist conducting the appraisal. This is where appraisal development begins. The final plant appraisal can be an opinion of the cost or the value of the items, plants, elements or services provided in the process. Key components of the assignment include, at minimum, the appraisal problem (or what will be appraised), the type of report to be presented as the final product, the timeline or due date of the final appraisal, and a fee schedule of services as well as the anticipated cost of the final appraisal product. More detailed assignments may include additional information, such as comprehensive relevant data and detailed appraisal methodology. The assignment should be in writing, agreed upon, signed by the arborist and client or agent of the client.

In many cases, it is recommended to obtain a retainer from the client. Retainers are payment agreements between a client and a service provider, and then the client pays a specified amount of money to the consulting arborist. Typically, consulting arborists will use the pay-for-work model of retainer. A recurring fee is set for a predetermined time. The client pays a partial payment upfront and then the consulting arborist is expected to work throughout that time to achieve specific deliverables, such as an appraisal. Retainer agreements clarify the roles and expectations of both the service provider and the client. Moreover, this legally binding document protects both parties financially and legally. Typically, appraisal assignments are conducted on a time and material basis. Payment for appraisal services should never be related to, or a percentage of, the appraisal assignment result.

Collect and Assemble Data

An appraisal estimates the defined value of personal or real property, including plants. The tree appraisal process is used to develop a supported estimate of current value. Unfortunately, most appraisals are done after trees have been removed or damaged. Site review and inspection protocols are used to research the relevant issues and any site information. Often, additional investigation is necessary and might include determining pre-casualty value or sampling similar trees nearby. However, the best time to conduct an appraisal is prior to any incident or damage, but this is rarely the case. If available, previous site records, tree assessments, site reviews, images, and even a witness can help determine the tree's pre-damage condition. The appraiser should document all activities related to the process, from initial contact with the client, including establishing the background information on the tree, to inspecting the site and formulating values. With all the facts gathered, it is the duty of the appraiser to determine the appropriate method and technique to provide an unbiased appraisal.

Determine and Apply the Appropriate Technique

The methods used to value trees are published in *The Guide for Plant Appraisal* authored by the Council of Tree and Landscape Appraisers (CTLA). The guide has various editions used by plant appraisers with the most recent being the 10th edition. The guide describes the various appraisal processes and gives examples of each method and technique. Although it is a good tool for the valuation of plants, it is considered a guide and the procedures involved require appraisal awareness and experience. For more detailed information on conducting plant appraisals, consult the guide provided by the council.

The most commonly applied approach is the cost approach. This approach analyzes the costs of goods or services to estimate (1) the cost to repair the item, (2) the cost to replace the benefits provided by the item, or (3) the cost to reproduce the item. If the subject tree can be replaced with a tree of the same size, then a direct cost technique may be applied to the assignment. For subject trees larger than what can practically be replaced with a similar tree, a different approach is applied. A popular technique within this approach is the trunk formula technique. This appraisal process extrapolates the costs to purchase the largest commonly available tree to the size of the tree being appraised. It's about proportionated comparative values applying the trunk cross-sectional area measured at the tree's DBH. Often this is considered a "starting point" to determine a value for many arborists. Of course, the technique applied is highly situational and requires a skilled determination to arrive at a reasonable and defensible appraisal.

Independent, Objective, and Impartial

Frequently, appraisals are involved in litigation where one party has damaged or destroyed another party's trees. Lawyers for each party—the plaintiff and the defendant— are responsible for advancing their respective client's motives and objectives. The role of the consulting arborist responsible for completing the appraisal assignment is quite different. It is necessary for the appraiser to be resolutely independent, objective, and impartial. The appraiser must not be involved with the emotions and objectives of the disputing parties. The appraiser collects the relevant data, analyzes, and evaluates the information collected, and then prepares the appraisal product, usually in the form of a report. The professional appraisal product is handled and presented in the same manner, regardless of which side of the litigation is paying for their services. Maintaining consistent integrity and ethical conduct is essential for preserving a professional reputation. While a consulting arborist/appraiser is conducting themselves independently, objectively, and impartially, it is possible at the same time to be an advocate for the tree and their client. This is accomplished by providing reasonable and defensible results, reviewing documents during the discovery phase, and providing advice to the client and/or their lawyer for the litigation process.

Ultimately, the appraiser's responsibility is to provide an independent, objective, and impartial result without discrimination or accommodation of personal interests. The client provides information and context to the assignment but should not direct the appraiser's work product. If an appraiser is perceived as having a personal interest, bias, or a conflict of interest, it may be necessary to step away from the assignment or, at a minimum, to disclose the conflict at the onset of the assignment or as soon as it is dis covered. The CTLA strongly discourages advocacy and other practices that are not rooted in economic reality and empirical data.

Create the Appraisal Product

The next step of the process is assembling all the information into the agreed upon appraisal product. This is how the arborist conducting the appraisal will communicate the results of the assignment either as a cost or value. The written report is the preferred way to deliver the information to the client. The length, content, and details of the report can range from a single page to numerous pages, including tables, images, research, and other relevant data. The final result will either be a cost, such as the replacement cost of a damaged tree, or a value, in the event of a situation like a tree that has been removed.

Prepare and Submit the Report

After the report is created, the final step is delivering the results to the client. The preferred delivery method includes a written report with a verbal presentation providing an explanation of the contents. It is important to be clear about whether the appraisal is a cost estimate or a value estimate, communicating the methodology, appraisal process and data supporting the appraisal. Verbal reports are acceptable when delivering the results to the client; however, it is recommended to follow up with a formally written product.

FIGURE 8.3 Accurate measurement is required for appraisal process.

Appraisal Factors

Regardless of the appraisal method used, there are primary elements to consider that consider depreciation factors. The major elements involved in properly assessing the value of a landscape tree are size of the tree, condition rating, functional limitations, and external limitations. A thorough understanding of each is crucial for a well-developed opinion of the appraisal subject. Without an in-depth knowledge of these trees and landscape plants, and their contribution to the site, the appraisal will lack credibility and significance.

The *size* of the plant is measured using common tools and industry standards. A diameter tape or tape measure can be used to measure trunk size and is typically recorded as the diameter at 4.5 feet above grade, or shoulder height. There are a considerable number of variables in measuring DBH on a tree, including trunk configuration, single stem or multi-stem, branch location, such as low bifurcations, and even the tree's orientation on the ground. Slopes and uneven topography present different procedures for accurate measurement. Consult with the appraisal guide on proper sizing practices.

FIGURE 8.4 Tree health, structure and form are considerations for the condition rating.

The condition rating is an aggregate metric which depreciates the tree based on an assessment of the health, structure and form of the tree or plant. Condition refers to assessing the tree's overall health, looking at vigor and vitality. It is an assessment of the tree's structural integrity and health at the time of appraisal. Thought should be given to rooting, branching, health, any damage or wounds, or evidence of pest infestation. It is important to note the current condition as the most reasonable gauge for the rating determination. Methodology here should be consistent and verified with current practice. Each tree has a species profile, often with unique characteristics for consideration in the condition rating. Also, the developmental age of the tree figures into the condition as well. Does the tree have a good form for the species? Each species has a typical, genetic form or "normal" traits, which should be representative of the species. See the tables later in the chapter for more information on rating the condition.

The *location* factor involves the landscape value of the site and the placement of the tree on the property. Consider the location of the property and the overall quality of the landscape, hardscape, and related elements. Recognize the tree's contribution to the site and the functional and aesthetic values to determine how effectively the placement of the tree provides these benefits. The impact of the location on natural plant growth should be considered as well. These values will be recognized in the function and external limitations section. See Table 8.2 for more information on location ratings.

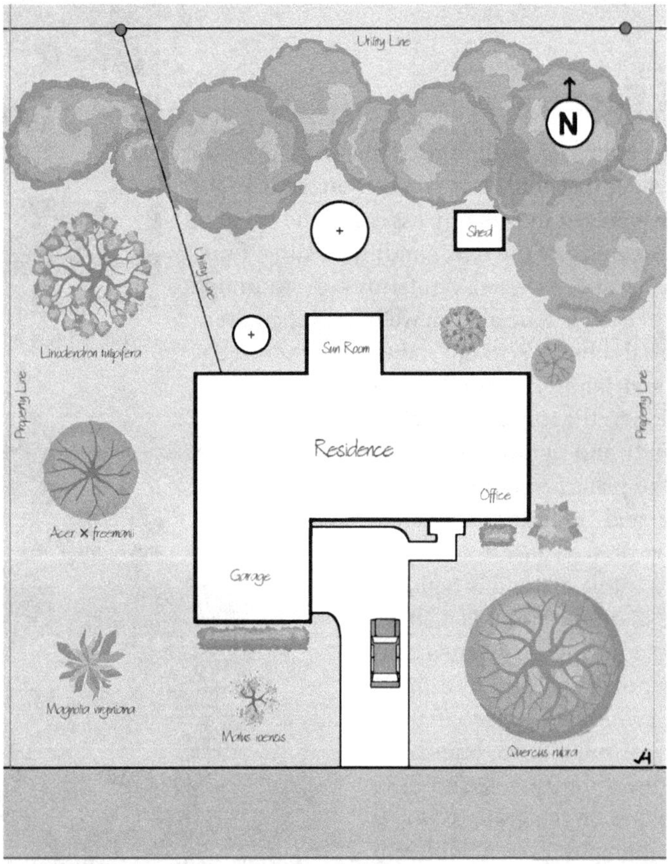

FIGURE 8.5 Tree location is an important variable in the appraisal process.

TABLE 8.1 Condition Rating for Landscape Trees

Condition Rating	Tree Structure Consider root condition/formation, trunk condition, and branch assembly and arrangement	Tree Health Consider crown indicators including vigor, density, leaf size, quality, and stem shoot extensions	Formula Values
Excellent	Root plate undisturbed and clear of any obstructions. Root flare has normal development. No visible trunk defects or cavities. Branch spacing/structure and attachments are free of any defects.	Perfect specimen with excellent form and vigor, well-balanced crown. Trunk is sound and solid. No apparent pest problems. Normal to exceeding shoot length on new growth. Leaf size and color normal. Exceptional life expectancy for the species.	1.0–.90
Good	Root plate appears normal; only minor damage may be found. Possible signs of root dysfunction around trunk flare. Minor trunk defects from previous injury, with good closure; less than 25% of bark section missing. Good branch habit, minor dieback with some signs of previous pruning. Codominant stem formation may be present. Minor corrections required.	Imperfect canopy density in few parts of the tree, 10% or less, lacking natural symmetry. Less than half normal growth rate and minor deficiency in leaf development. Few pest issues or damage, controllable. Normal branch and stem development with healthy growth. Typical life expectancy for the species.	.90–.75
Fair	Root plate reveals previous damage or disturbance and dysfunctional roots may be visible around main stem. Evidence of trunk damage or cavities with decay or defects present. Less than 30% of bark sections missing on trunk. Codominant stems are present. Branching habit and attachments indicate poor pruning or damage, which requires moderate corrections.	Crown decline and dieback up to 30% of the canopy. Overall poor symmetry. Leaf color somewhat chlorotic with smaller leaves. Shoot extensions indicate some stunting and stressed growing conditions. Obvious signs of pest problems contributing to lesser condition. Some decay areas found in main stem and branches. Below average life expectancy.	.75–.50
Poor	Root plate disturbance and defects indicate major damage with girdling roots around the trunk flare. Trunk reveals more than 50% of bark section missing. Branch structure has poor attachments, with several structurally important branches dead or broken. Canopy reveals signs of severe damage or topping, with major corrective actions required.	Lacking full crown, more than 50% decline and dieback, especially affecting larger branches. Stunting obvious with little evidence of growth on smaller stems. Leaf size and color reveals overall stress in the plant. Insect or disease infestation may be severe. Extensive decay or hollow. Life expectancy is low.	.50–.30

Note. Each plant can have any combination of the above health or structural issues, and others. The expression of symptoms and signs is subjective. The appraiser should consider the individual species characteristics and use existing circumstances as a reasonable scale for condition determination. This table is a general representation to assist in formula values.

TABLE 8.2 Location Rating for Landscape Trees

Location Rating	Site Position The general appearance and quality of the site in relation to the market value in the area; overall site appearance	Site Contribution The functional and aesthetic attributes the plant has on the site overall in terms of appeal and purpose	Site Placement The effectiveness of realizing benefits and the plant's impact on contributing to the site	Formula Values
Excellent	Arboretum Well-maintained residential area Historical district Designated parks and recreation areas	Air filtration Water purification Noise abatement Erosion control	Specimen trees in a functional landscape design Single, historic or specimen tree Outstanding aesthetic value in the landscape	1.0–.90
Good	Suburban residential areas Golf course School/corporate campus Green spaces/memorials Cemetery Scenic parkways	Windbreaks Shade/cooling effects Specific plant aesthetic factors Structural accents	Considerable element in the landscape for design quality or function Plants in a windbreak, screen or other integral planting Planting allows maximum functional benefits	.90–.75
Fair	City streets/boulevards Rural residential areas Urban streets Industrial/commercial areas	Framing views Space definition Privacy Traffic management	Well-spaced planting site Tree installation in planting pits or lawn strips Mass, unplanned plantings on a site	.75–.50
Poor	Streets/roadways in rural areas Woodlots, managed Freeways/interstates Countryside, naturally occurring woodlands Woodlots, unmanaged	Create vistas Screening	Trees with utility interaction Improper spacing with infrastructure conflicts Species with fruit or leaf litter issues Invasive species	.50–.30

Note. The table above illustrates representative factors that influence functional limitations and provides formula values for various sites and applications.

DEPRECIATION

An accurate appraisal will require the application of depreciation factors. Appraisers use depreciation to justify differences between a "perfect tree" compared to the appraised tree. Also, it may be that the tree is not placed or located optimally for the site and situation. This will account for less-than-ideal tree characteristics, placement in the landscape or the site it occupies. The three elements or variables figured into depreciation include the actual condition of the tree, functional limitations, and external limitations. Remember, conditions place a focus on the health, structure, and form of the tree, whereas functional limitations are relative to the tree's growth and function and external limitations are likely issues outside of the tree owner's control. A cost estimate may be the desired result of the assignment, regardless of depreciation factors. In either case, the appraiser focuses on cost. In other cases, the appraiser may use a cost estimate as the basis for estimating value, in which case depreciation should reflect the difference between cost and market value. The type and objective of the assignment will determine which methods are appropriate.

FUNCTIONAL LIMITATIONS

Functional limitations are a type of depreciation that primarily apply to the tree itself or the location. These are factors that restrict or constrain growth or functionality of the tree due to poor placement, limited space, or quantity. These factors limit the future growth, development, and overall health of the tree. Consider how site conditions and placement, such as proximity to utility lines, could limit full development due to necessary pruning for clearance. Infrastructure conflicts with neighboring buildings or other landscape elements are included. Finally, investigate for any genetic limitations related to the genus and species itself, including pest pressures.

Species rating is a depreciation factor assigned to a given tree species based on the recommendations of the Regional Plant Appraisal Committee (RPAC). This local group of professionals determines a species rating for common trees in a designated region for plant appraisals. This is a depreciation schedule accounting for the functional limitations of the species. This subjective rating is based on individual qualities and traits, which may vary geographically within the state because of local climate and environmental influences. The arborist must consider usability and the urban applications to rate specimens. RPAC information and data is a baseline for each species; yet it is the responsibility of the appraiser to determine ratings and wholesale values based on availability within a region. The 10th edition of the *Guide for Plant Appraisal* recommends species rating be considered as a component of functional limitations. Earlier editions of the guide recommends species rating be a separate percentage depreciation unto itself.

FIGURE 8.6 Analyze the location for conditions that limit future growth and sustainability.

EXTERNAL LIMITATIONS

External limitations factors include issues outside the control of the tree owner or manager affecting sustainability, structure, health, or form. Examples include environmental issues, such as water availability or limitations and threat of significant pests, or legal restrictions, such as rights of way, or utility management easements, where there are impending conflicts between power lines and the tree.

The appraiser will assign a numerical rating to each of the depreciation categories (condition, functional limitations, and external limitations), which then will be applied to the basic cost. The basic cost is multiplied by each of the three categories to estimate the depreciated value, which would be the final functional reproduction or replacement value using the trunk formula technique. This appraisal technique is only one way of appraising large trees. It is important to realize these estimates of value may not be proportionate to the value of the property or what would actually be paid for the tree. Every credible appraisal must pass a litmus test; it must be reasonable and defensible. The ultimate decision is not made the appraiser, but often by the courts, counsel, clients, or opposing litigants.

METHODS FOR ESTIMATING COST

After the primary plant and site assessments have been completed, it is necessary to determine the proper approach. *The Guide for Plant Appraisal* details three basic methods for plant valuation: cost, income, and sales comparison approaches. The approach chosen depends on the kind of property, the purpose of the appraisal, and the quality of information available for analysis. It may be necessary to use other methods, or a combination of approaches, in the appraisal process to verify the results.

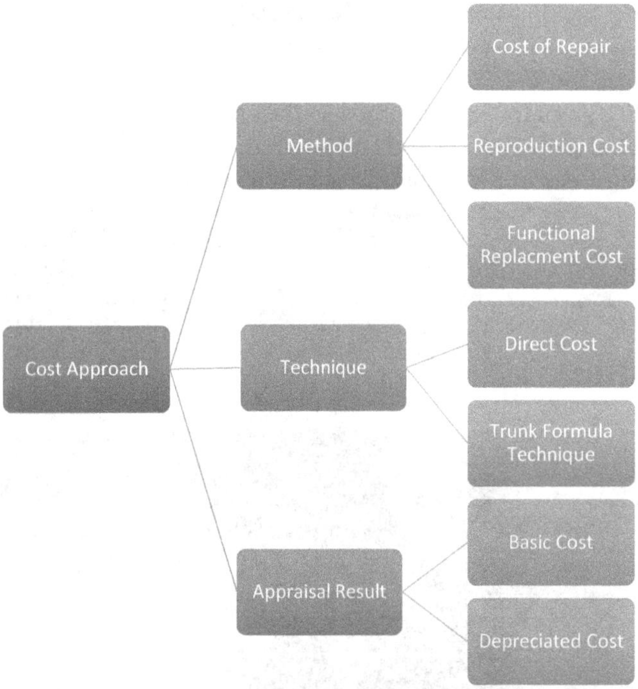

FIGURE 8.7 Appraisal methodology within the cost approach.

The *cost approach* considers the amount in dollars to either repair damages to the tree, cost of a replica of the tree, or the cost to restore the benefits the tree provided before the damage or destruction. There are various methods to use within the cost approach, including:

Repair cost considers the cost of correcting damage to a tree and preventing further loss. If a large branch in a tree was broken by a backhoe, it would need the attention of an arborist to prune and perhaps tend to the wounded area to facilitate wound recovery and prevent further damage by ripping bark down the trunk. Another common example would include wound repair from some mechanical damage on the trunk.

Reproduction cost estimates the cost of replacing a tree in the same location with the same species and, if possible, of a similar size. Usually, the tree has been removed or damaged beyond repair. The cost estimate is the amount needed to replicate or reproduce the same tree in function and utility. However, in most cases the tree is too large to be replaced by a single tree and still be of equal value and utility. In those cases, appraisers may designate several smaller trees collectively to be equivalent to the value of the original tree. This becomes a matter of calculating the value of the replacement trees and their associated costs. Depreciation factors are applied during the appraisal process and are found to be an accurate determination of market value for the tree, resulting in depreciated reproduction cost.

Function replacement cost is a substitution for the appraised tree or perhaps a group of trees that provides equal utility and functional benefits. For example, if a tree is destroyed by fire, a replacement is figured into the cost estimate that would function in the landscape as the original tree. It can be the same tree, or it can be a group of trees which might be considered as serving the same purpose. For example, it may be reasonable that a hedgerow once used for privacy may be replaced functionally by a privacy fence. Depreciation factors may be applied in situations for the imperfect conditions or location.

The *trunk formula technique* (TFT) defines value in comparison to other trees of the same species. However, it does not necessarily provide an adequate market value of the tree. This technique falls within the cost approach, applying the reproduction or functional replacement methodology. It is an extrapolation procedure that considers that a mature tree may not be reasonably replaced and compares the cost to the largest commercially available transplanted tree from a nursery or grower.

The TFT is often used when the tree is too large to be replaced with typical nursery stock and can be a good representation of overall value to the property. The basic value of a tree is the sum of two factors: the cost of transplanting the largest normally available tree of the same or comparable species, and the increase in value because of the larger size of the tree being appraised compared to the size of the replacement tree. The formula to determine the basic value of the tree requires a *unit cost*. This is the cost per square inch of trunk cross-sectional area measured at DBH. The unit price is determined by either using the RPAC designated unit price or from calculations derived from available trees at local nurseries, whichever is most applicable. The RPAC is comprised of industry experts who are typically associated with their local chapter of the International Society of Arboriculture. This committee gathers data based on statewide information to determine unit costs for commonly available trees, the trees' obtainability, and their functional limitations in the area. The results of the appraisal using this technique is considered a depreciated cost. Also, additional costs may be added to the appraisal for ancillary work such as clean-up, maintenance, stump grinding, etc. An example of RPAC unit cost from the state of Indiana surveys have determined that for the computations needed in cost techniques, the largest commonly available, transplanted deciduous tree would be 3.5 caliper inches with a unit tree cost of $37.71.

The trunk formula technique is only one method to appraise large trees. It is important to realize that estimates of a tree's value may not be proportionate to the value of a property or what would actually be paid for a tree. Studies estimate that trees may account for 8–20% of a residential property's value. However, much lower values could result due to various factors, such as location. For example: An ideal, mature sugar maple with a 24-inch diameter at breast height may be valued at more than $15,000, but the home it resides next to may be worth $35,000. This is an unrealistic tree value that is not reasonable in a professional appraisal situation.

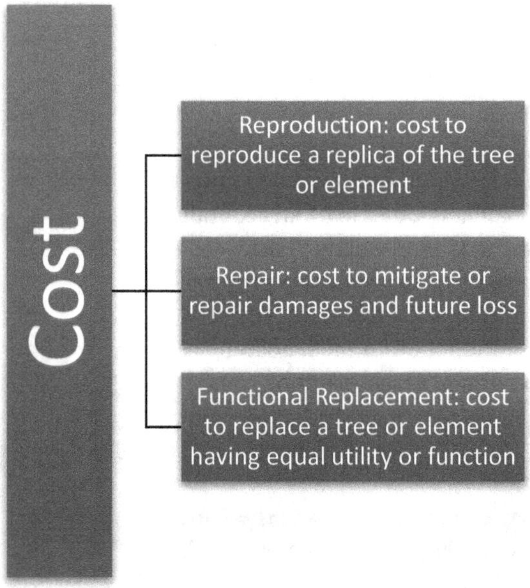

FIGURE 8.8 Cost approach techniques.

Calculating a tree's value using the TFT begins with multiplying the cross-sectional area of the tree by the unit tree cost. To determine a tree's cross-sectional area, first calculate the tree's diameter at breast height (DBH). DBH is measured using a special diameter-measuring tape wrapped around the tree at 4.5 feet above the ground. If a diameter tape is unavailable, a regular tape may be used to determine tree diameter. First, measure the circumference by taking the tape around the tree. Now that you know the circumference of the circle, divide it by π to get the diameter. π is equal to approximately 3.14, but you should use a calculator to get the most accurate results.

Using the results to find the diameter, calculate the cross-section area. This is found by using the formula for the cross-sectional area of a cylinder (π x R2), where π is a constant (3.14159265). The square of the radius (half of the diameter) multiplied by π will give you the value for cross-sectional area of the tree.

For example: If a tree has a diameter of 20 inches, the computation would be 10 x 10 x 3.14—equaling 314 inches. Multiplying 314 by a unit cost of $37.71 equals $11,841, which is the overall basic cost. However, this is not necessarily the "value" of the tree. Additional factors will affect a tree's value, which is why a professional appraiser is recommended for an accurate value.

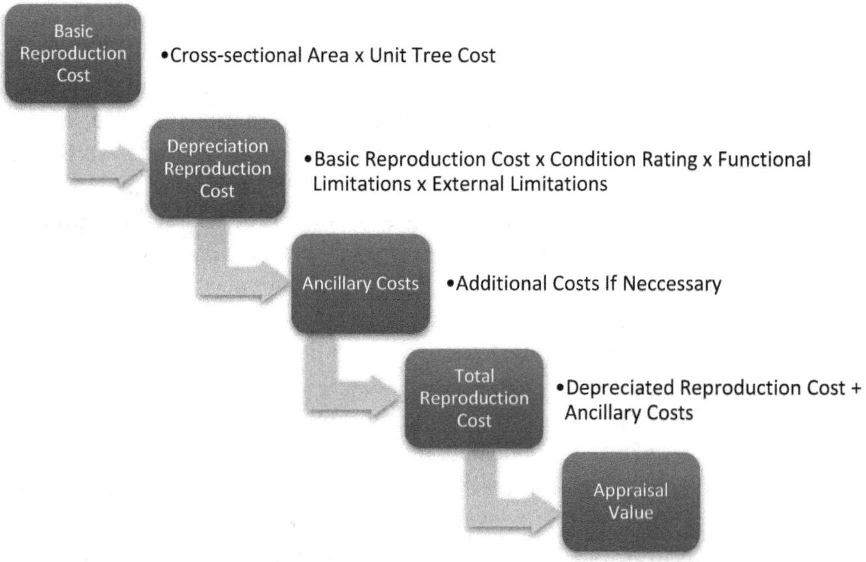

FIGURE 8.9 Trunk formula technique process.

TREE APPRAISAL SCENARIO

Let's put the Trunk Formula Technique to work with an example for a typical suburban landscape.

FIGURE 8.10 Maple tree in front lawn.

Example

A residential site in an Indiana neighborhood has a sugar maple (*Acer saccharum*) on the front lawn, shading the front of the home. Measuring 14 inches at breast height, the tree is in good condition and in a proper location. The tree's owners wish to have it appraised to determine the value of the tree on their property.

The appraisal calculation method would be as follows:

1. Basic Reproduction Cost = CSA x UTC
   CSA = Cross-sectional area of the subject tree
   UTC = Unit tree cost, determined by the Regional Plant Appraisal Committee (RPAC) or local wholesale cost
2. Depreciated Reproduction Cost = CR x F x E x BRC
   CR = Condition rating
   F = Functional limitations rating
   E = External limitations rating
   BRC = Basic reproduction cost
3. Total Additional Costs = cleanup, installation, maintenance
   These other costs would be included if there is a loss that requires the removal of a tree and the installation of a new tree, along with post-planting care costs for a determined time.
4. Total Reproduction Cost = DRC + TAC
   DRC = Depreciated Reproduction Cost
   TAC = Total Additional Costs
5. Appraisal Value = Total Reproduction Cost rounded to the nearest thousand.

Now, back to our example using the steps above:

1. Basic Reproduction Cost (BRC): 154" CSA x $37.71 UTC = $5,807
2. Depreciated Reproduction Cost (DRC): 1.0 CR x .80 F x 1.0 E x $5,807= $4,645
3. Total Additional Costs (TAC): $0 (not applicable in this example)
4. Total Reproduction Cost (TRC): $4,645 DRC + $0 TAC = $4,645
5. Appraisal Value: $4,600 (TRC rounded to the nearest thousand)

Following the calculated steps, the reproduction value of the tree would be $4,600. This value is the depreciated reproduction cost of the tree if it were damaged or removed. In other words, it is the cost of replacing the tree with an exact replica. As mentioned earlier, depending upon appraisal situations, there are other approaches, methods, and techniques beyond the trunk formula technique used to estimate costs and tree values.

CHAPTER 9

TREE RISK MANAGEMENT AND ASSESSMENT

LEARNING OBJECTIVES

- Recognize how managing tree risk is an important component of the overall urban forestry management plan.
- Describe how duty of care and standard of care apply to tree risk managers and assessors.
- Identify the components of the tree risk assessment process and the factors which affect likelihood of failure.
- Recognize common tree-related conditions of concern as well as site factors influencing tree failures.
- Explain how common wood decay organisms contribute to likelihood of failure.
- Identify targets and target zones and how they affect the risk assessment process.
- Describe risk mitigation and the various strategies used to reduce risk to an acceptable level.

TREE RISK MANAGEMENT CONCEPTS

Trees provide many benefits for our homes, businesses, and properties. However, there are associated risks with trees wherever there exists human activity. All tree owners and managers should take reasonable steps to protect themselves and others with knowledge and understanding of tree risk assessment and management. In many jurisdictions, it is important

Risk management is about managing uncertainty to reduce the potential for tree-related consequences later.

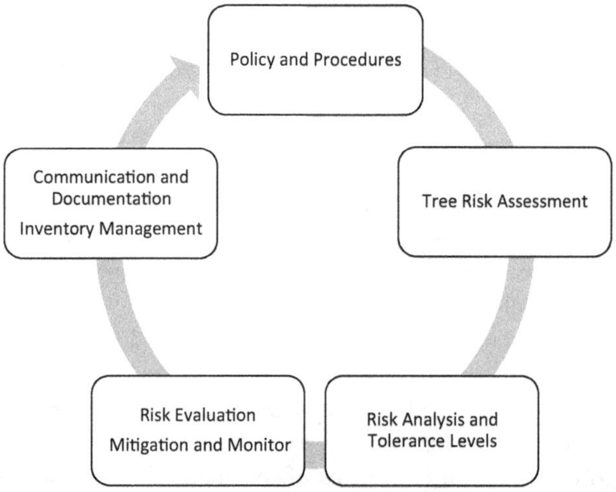

FIGURE 9.1 Risk management process.

to recognize tree owners and managers have a duty to inspect and maintain their trees and understand the protocols. The intentions of this portion of the book are to create an awareness of the management process and provide some insights on the procedures involved in tree risk assessment, mitigation, and communication of those risks. This document should not be interpreted as guidance or legal advice on risk assessment or risk management policy. Many concepts found in this chapter are provided by ANSI A300 Part 9 standards and ISA best management practice Tree Risk Assessmentand should be applied to prescribe the specified process and provide more details for determining tree risk.

Tree risk management is an important element of any urban forestry program. In many communities, it is a component of the overall comprehensive risk management platform. This collaboration creates a more forceful emergency response in the event of seasonal storms or other emergency situations in the municipal application. Additionally, tree risk management policies can reduce the liability for communities as well. One of the main objectives of a tree risk management strategy is to create policies reasonable for the situation while demonstrating an acceptable standard of care in the process.

FIGURE 9.2 Tree owners and managers have a duty to inspect and maintain their trees.

FIGURE 9.3 Standard of care are usually based on guidelines that are research-based and referred to as "best practices."

Recommendations

- Develop and implement an inspection program.
- Create and enforce a policy for risk trees.
- Develop a standard of care suitable for the situation.
- Develop and maintain records or case history.
- Make it all "attorney- approved".
- Keep current with BMP's and protocols.

FIGURE 9.4 Recommendations for tree risk management programs.

In general, most laws obligate landowners to periodically inspect their property and take reasonable care in maintenance to reduce risk to an acceptable level. If landowners fail to identify and remedy dead or dangerous trees or tree parts and injury or damage results, they can be held liable for that harm.

The standard of reasonable care may be defined as the watchfulness, attention, caution, and prudence a reasonable person in similar circumstances would exercise. If a person's actions do not meet the standard of care, then their acts fail to meet the duty of care which all people (supposedly) develop toward others. Failure to meet the standard when there is harm is negligence, and any resulting damages may be claimed in a lawsuit or tort claim by the injured party. The challenge lies in the fact that the "standard" is often a subjective issue upon which even reasonable people can have differing opinions. In the end, the courts decide what is reasonable during litigation.

Negligence, as identified earlier, is a failure to behave or act with the level of care someone of ordinary prudence would have exercised under the same or similar circumstances. In the event duty holders are found negligent in meeting their duty of care, such as in the case of a tree inspection where a significant risk existed, it does not automatically follow that they would be liable for any harm that arises. Liability will only flow from that negligence if it can be established that a competent check within a reasonable timeframe of the incident would have identified an unacceptable risk of harm and resulted in mitigation that would have prevented that harm from occurring. If a defect resulting in failure would not have been found during a competent assessment, then, irrespective of any negligence from not carrying out the inspection, a duty holder is unlikely to be held liable for the consequences of the failure. This is a common scenario and often results in a court examination focusing on the competence of assessors and whether causes of harm could and should have been discovered before the event.

Additionally, a standard of care should be proportional to the situation and the duty holder as well. Duty holders are expected to consider the risks posed by their trees and manage those risks in a reasonable and *proportionate* way. Proportionate means control measures should be in proportion to the risk. For example, if an assessment identifies negligible risks, it is unreasonable to introduce measures requiring a stringent and costly management strategy.

Typically, standards of care, or expectancies of care, increase with the scale of the property and the resources available to the duty holder. In short, it is expected that duty holders act in a practical and sensible way, according to the overall size of their land holding and the

A property owner, tree owner, or manager is negligent and liable if a claimant can prove that:

- the owner had a legal duty to the claimant;
- the owner breached his duty; and
- the owner's breach was the legal cause of the claimant's injury or property damage.

In a court case involving tree failure, fundamental questions come up:

- Was the manager negligent?
- Was there an inspection system?
- Was it prudent and performed to standards?
- Was it documented?
- Could the cause for failure have been detected in a reasonable and prudent inspection?

availability of resources. There is case law upholding the principle that the standard of the duty of care varies according to the resources available to the duty holder (i.e., large landowners and managers, such as municipalities or large public places, would be expected to apply a higher standard than smaller property holders, such as homeowners).

It is important to recognize arborists, tree risk assessors, or anyone working on and around trees owe a duty of care and standard of care as well. The duty of care for a tree risk assessor is to use the generally accepted standard of care (as defined in applicable standards, best management practices, qualifications, and training courses) when performing work. Developing a tree risk management and assessment process and adhering to the prescribed procedures creates a clear framework for managers, supervisors, and staff to follow and implement due diligence, which forms the basis for articulating a proactive and defensible program.

RISK RATINGS

The process of assessing trees for risk is complex and requires training and specific skill sets to be confident and competent. A tree risk assessment can frequently be misinterpreted as the risk management program, but it is only one component of the program. The overall program includes the establishment of policies, procedures, risk analysis, mitigation, and monitoring protocols and tree inventory management that includes communication and documentation. Tree assessment provides information to determine the potential risk of specified trees and the management plan determines mitigation strategies for trees at the current risk rating.

Each type of condition of concern and defect presented can be assigned a distinct range of indicators identifying its severity and the potential for failure. There are four risk ratings used which follow the International Society of Arboriculture best management practices for tree risk assessment and the ANSI standards: extreme, high, moderate, and low. These risk ratings are used to communicate the level of risk to the client and support recommendations for mitigation. They are a combination of rating the likelihood of failure and impact as well as the consequences of failure should the tree or its parts affect the target.

The Process of Risk Determination

Extreme-risk situations are those in which failure is imminent, there is a high likelihood of impacting the target, and the consequences of the failure are severe. Mitigation should be implemented as soon as possible. In areas where there is constant occupancy, or high value targets, restricting areas within the target zone is recommended.

High-risk situations are those in which likelihood of failure is very likely, and consequences are considered significant. The decision for mitigation and timing of treatment depends upon the risk tolerance of the tree owner or risk manager.

Moderate-risk situations are those in which consequences are minor and the likelihood of failure is likely or somewhat likely. The tree risk assessor should recommend mitigation and timing dependent upon the risk tolerance of the client.

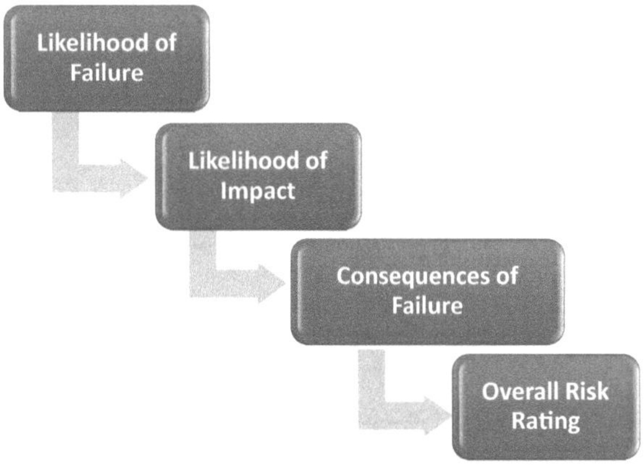

FIGURE 9.5 The process of risk determination.

Low-risk situations are those in which consequences are negligible, and the likelihood of failure is unlikely; or consequences are minor, and the likelihood of failure is somewhat likely. Some trees with this risk level may benefit from mitigation or maintenance measures to reduce risk to an acceptable level, but immediate action is not usually required.

The primary responsibilities of the arborist as a tree risk assessor are to assess for the likelihood of tree failure and determine a risk rating. Then, they should advise on the consequences of failure and recommend appropriate mitigation options to reduce the associated risk. Along with communicating tree concerns, it's essential to understand the client's acceptable level of risk to meet their needs. The overall risk rating represents the level of risk posed by the tree part or parts after thorough investigation, analysis, and evaluation of the situation. The acceptable risk level determined by the client identifies what level of risk they are willing to assume.

The process of implementing a tree risk assessment can be associated with a significant amount of uncertainty. The assessor, therefore, provides opinions given the best information available and the ability to investigate and analyze trees. First, for a tree to be considered a risk issue it must be defective either in some part or as a whole, with risk for failure and a threatened target. Trees receive an elevated risk rating when the assessment requires measures to prevent a failure from adversely affecting a target. To understand the implications of this concept, there are some common terms to understand.

A *target* is people, property, or activities that could be injured, damaged, or disrupted by a tree failure, according to the standards. When an assessment is conducted, the assessor must carefully assess the area around homes, playgrounds, roadways, sidewalks, and parking areas. Is the target static, moveable, or mobile? What is the occupancy rate of the target or

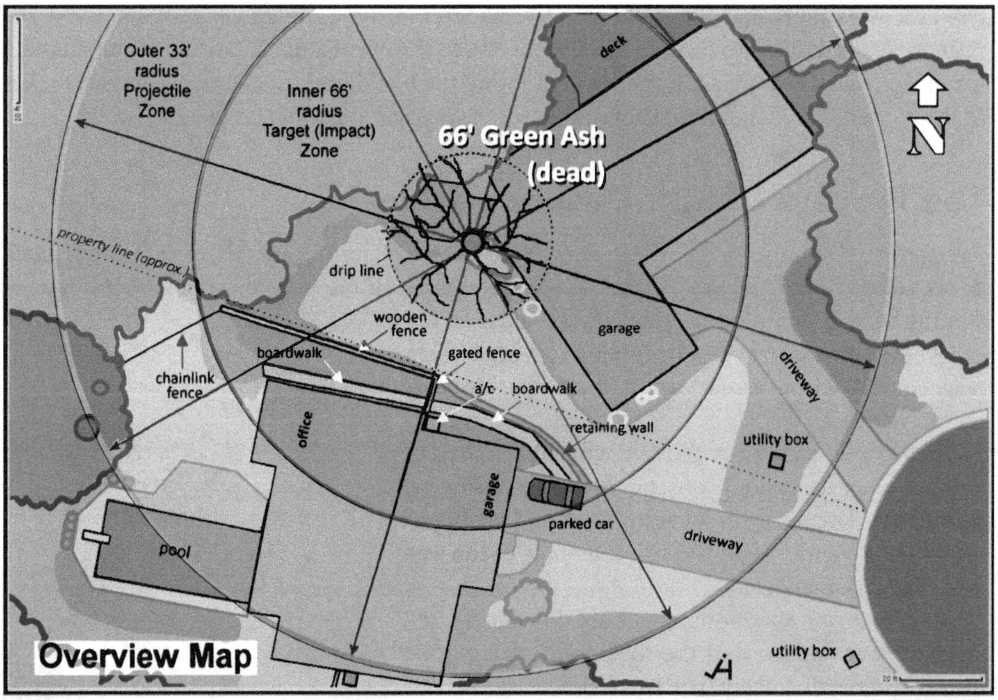

FIGURE 9.6 Targets include people, property, or activities that can be injured, damaged, or disrupted by a tree failure.

rather, how often is the target within the target zone of the tree? Consider whether people can be kept away or separated from the target area, either temporarily or permanently. Also, assess the target's value in relationship to the site and client. Review the target zone, which is the area where the tree or a branch is likely to strike when it falls. Finally, determine the consequences should the tree or tree part fail. The target zone should include the areas inside a circle around the tree, which is at least as wide as the total tree height.

Risk is the combination of the likelihood of an event and the severity of the possible consequences of that event. In reference to trees, risk is the likelihood of a tree failure occurring either with whole tree failure or perhaps just a part, such as a branch, adversely affecting a target, and the severity of the consequences. Typically, a tree failure impacts people, property, or disrupts activities in some way. Every tree has the potential to fail; however, only a small number of failures cause injury or damage. It is impossible to maintain trees free of risk and some level of risk must be accepted by the owner.

Often the term hazard is used for tree risk, especially with the utility vegetation management industry. When a tree is labeled as a *hazard,* this indicates that it poses a likely source of harm, as the tree part or parts will affect the target zone. For example, an entire tree or a single branch could be designated as a hazard. Hazards are identified through tree assessments, and tree owners are required to take steps to minimize the risk of damage from failure. It is recommended not to use this term when following the current standards.

TREE FAILURE AND BIOMECHANICS

Trees fail at their weakest point and those with serious defects will have a higher likelihood of failure than trees without defects. Determining the failure potential of a tree requires knowledge of biomechanics and is the most challenging component of the process.

Biomechanics studies trees as mechanical objects using engineering and physical principles to understand the structural properties of trees and how they interact with their environment. Biomechanics, (from Ancient Greek: βίος "life" and μηχανική "mechanics"), is the application of mechanical principles to living organisms, such as trees and is a field of study with a very long history. It is only recently in the last 40 years biomechanics has been applied directly to tree failure.

Trees are naturally engineered to withstand normal weather conditions for their area. Normal weather is an accepted term used to describe weather

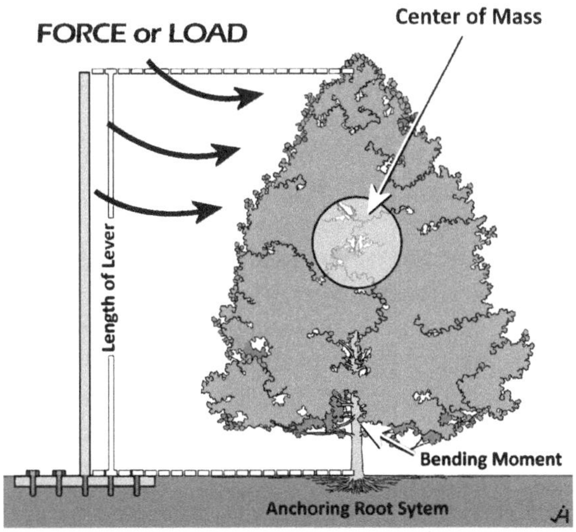

FIGURE 9.7 Forces or loads on trees are important concepts in risk assessment.

Bending moment (**M**) = Lever arm length (**L**) x Force (**F**)	Force is magnified by lever arm length.
Lever arm length is determined by location of center of mass.	Loads on trees lead to internal stressors.

FIGURE 9.8 Biomechanics concepts.

FIGURE 9.9 Common stresses within a tree.

FIGURE 9.10 Torsion crack with wound wood present.

based on a location's average temperature, wind, and precipitation during the past 30 years. Extreme and abnormally extreme storms are less clearly defined. Normal storms occur often during a defined time frame and these include thunderstorms, snow, and light accumulations of freezing rain in areas that are subject to those conditions. Abnormally extreme may include tornadoes, hurricanes and typhoons.

The assessing arborist is challenged with the determination of assessing the effect of loads as mentioned above on trees to determine their likelihood of failure (Figure 9.5). Load is a general term, which can refer to force, bending moment, or weight. Failure occurs when loads exceed the tree's load-carrying capacity, typically from gravity, wind, static, and dynamic forces. A static load is simply the weight of the tree's own wood fiber, branches, and stems. A static load can increase with the additional weight of ice, water, snow, and parasitic plants, whereas a dynamic load results from wind forces applying pressure to the crown, branches, and stem. Climbers and rigging from tree work can exert dynamic loads as well. In most areas, the common loads affecting trees are wind loads. The related force on the tree as a system is called "drag." Essentially this can be thought of as wind resistance and the amount of resistance is determined by the crown volume, density, and wind speed.

Internal stressors experienced by the tree are the forces or loads applied to the tree structure. This can include trunks and branches. There are four basic stressors examined during the assessment process.

1. Tension, which is the stretching or pulling of wood fibers.
2. Compression, which is the squeezing or crushing of wood fibers.
3. Torsion, which is simply a twisting force on the wood structure.
4. Shear, which is the force created by tension and compression forces in opposing directions.

Loads and stress factors can occur alone or in combination when it comes to impacting the tree's capacity to withstand these forces. For example, as wind blows a tree back and forth, each bending motion either compresses the wood fiber or causes stretching or tension within the stem. Also, it is building fiber as well! A leaning tree experiences both forces as well. On the side of the lean, wood fibers are squeezed together and compressed. On the opposite side of the lean, the fibers are stretched longer, similar to when someone bends over to pick something up. In this analogy, the abdomen experiences compression, while the back is under tension. In between the zones of compression and tension is a layer where these two forces come in contact and the result is fibers under tension pulling away from fibers under compression. This is a shear force. Torsion or twisting is a common response from wind impact on an asymmetrical crown. A tree is much like a sail and will respond more where the largest area of the leaves (mass) is located on the crown. If a tree has been cut in half due to a structural conflict with nearby utility lines, the tree will not bend uniformly but rather will twist in the wind, causing torsional stress and separation of the wood fibers. Think of the orientation and configuration of sails on a boat. Tree crowns can be similar to sails in response to wind.

Wind is energy or simply air in motion. Usually in meteorology, when talking about the wind, it typically pertains to the horizontal speed and direction of this moving air. Although we cannot actually see the air moving, we can measure its motion by the force it applies on trees. This wind energy needs to be dissipated to prevent failure. This is accomplished by the moving leaves and branches and as it is transferred through the branches, down the trunk and roots into the ground. The dissipation of wind is called mass damping. Basically, the movement of leaves, twigs, and branches diminishes the force on the tree and its parts by changing its shape. The crown and leaves become more streamlined in the face of wind. Through this morphological modification, this energy becomes barely discernable. Mass damping can be observed on any tree during a windy day. This is why maintaining crown biomass actually helps the tree reduce the impact of wind as a loading force. The opposite of this is sparse crowns or those that have been improperly thinned, topped, or lion's tailed, which reduces the tree's ability to dampen the load, resulting in branch and possibly trunk failure.

DEFECTS, CONDITIONS OF CONCERN, AND LIKELIHOOD OF FAILURE

There are many ways a tree can fail, either in part or as whole tree. An ice storm can overload branches on a tree, a blustery wind can blow a tree over if its roots are damaged, or a cracked tree can fail just under its own weight. Figure 9.11 below describes the most prolific failures.

A tree develops an increased risk of failure when the woody structure is weakened by one or more defects. Most defects can be linked to previous wounds and decay, pest infestations, severe storms, or to growing conditions limiting the root system. Given that defects and old injury sites are structurally weaker than uninjured wood, the tree is predisposed to fail at the location of the defects.

Trunk Failure

- failure just above the root flare on the main stem below the primary main branches.

Windthrow

- typically, whole-tree failure at the root plate where roots are lifted, snapped, or dislodged at the soil surface.

Codominant Stems

- frequently fail due to the lack of a dominant leader system. Included bark will exacerbate the likelihood of failure.

Branch Failure

- results from cracking and splitting often from excessive loads and poor branch aspect ratio.

Sprout/Shoots

- fail due to their poor attachment to the parent branch. They are not connected to the branch with overlapping anatomy and typically attached only on the outermost xylem and cambium tissue.

FIGURE 9.11 Common tree failure types.

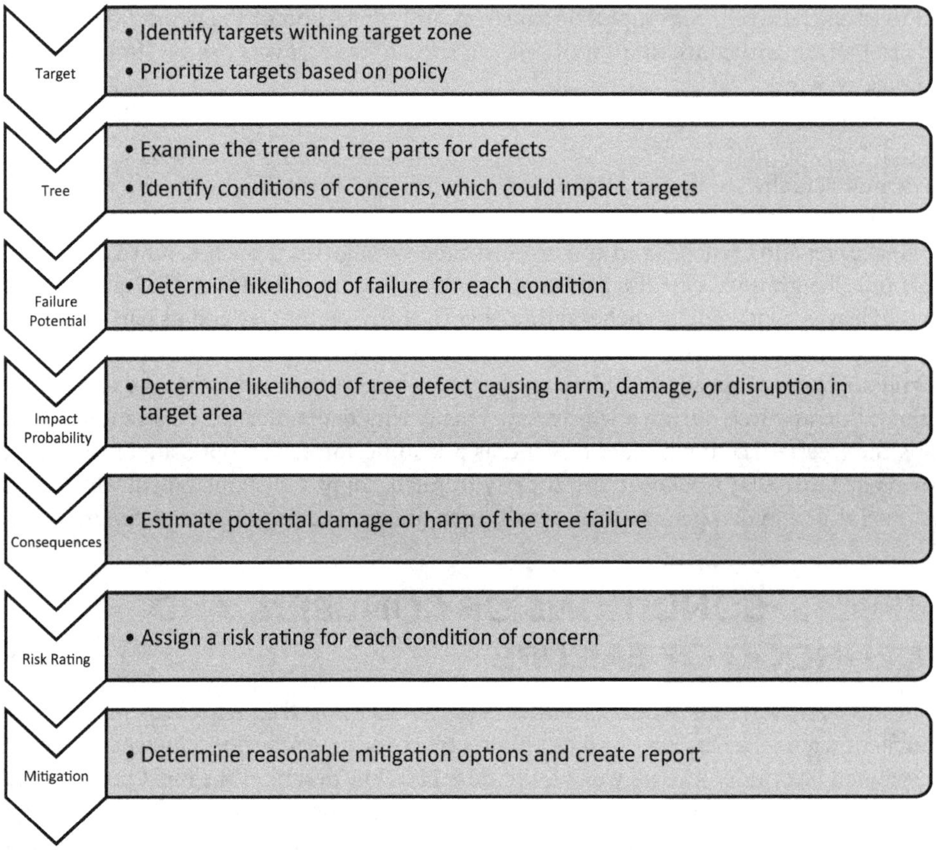

FIGURE 9.12 Tree risk assessment considerations.

Healthy, well-maintained trees growing on the appropriate sites are capable of minimizing the impact of wounds and the spread of decay. Trees located on stressed spaces likely have reduced energy reserves and limited water resources, and therefore, a compromised ability to recover from wounds and limit ensuing decay. Urban trees are more likely to be site-stressed due to several factors. Most urban trees grow in soils that have been altered by construction, which can lead to compaction, poor drainage, high clay, sand, or gravel content, high alkalinity, or the presence of construction debris. Quite often, these trees are growing in root-confined spaces, such as narrow boulevards or limited volume sidewalk planting pits. Additionally, many urban trees are subjected to chemicals such as deicing salts, herbicides, and fertilizers commonly used in landscape maintenance. Poor tree maintenance aggravates the stresses placed on trees by the above factors. These cumulative stressors affect health and structural integrity, increasing the risk of failure.

Tree failures can be predictable, if visible, because defects show us where the tree is likely to fail, at its weakest point. We can identify some of the more common categories of concern and defects, including decayed and missing wood, cracks, root problems, weak branch unions, cankers, poor tree structure, and dead or dying branches and stems. Consideration should be given to tree species and conditions as well in the assessment process. Stress can produce notable weaknesses, and some tree species are more prone to stress such as wind loads. These include trees with poor structure, weak branching habits, or previously topped trees. Failures occur as a result of excessive loads such as wind damage, poor branch attachments and arrangement, or damage from a disease pathogen. Any condition recognized as a factor in tree health and stability should be examined carefully to identify the most appropriate mitigation strategies as shown in Figure 9.12.

TREE RISK ASSESSMENT: THE PROCESS

There are distinctive differences between tree risk management and tree risk assessment, and it is important to recognize these differences when creating and implementing policies and procedures. Simply stated, tree risk management is the macro-level system of stratifying, classifying, and prioritizing risk. Risk assessment is a micro-level system within risk management. It aims to break down threats into identifiable categories and define consequences and impacts of each risk. When it involves trees, risk assessment is the systematic process to identify, analyze, and evaluate the risk presented by the tree. Additionally, the management process should include procedures to mitigate the risk to an acceptable level, communicate the risk, and monitor the tree, or system of trees, as well.

In assessing and managing trees for risk, it is important to create a balance between the inherent risk a tree poses and the benefits provided by the tree. As trees grow and mature, ecosystem services provided increase as well, becoming important assets to any site, when managed appropriately. However, as trees get larger, parts get larger and its loads increase as well. Trees are dynamic, living organisms that respond to external stimuli, such as environmental influences, pests, and people. So, any defect or concerning condition that compromises the strength or stability of the tree will increase the likelihood of failure under the right conditions, such as wind or ice. Tree failures during normal weather conditions are often predictable and preventable. However, not all conditions can be seen within the tree and belowground with its root system. However, any tree with defects or weaknesses visible or not will fail if the forces applied exceed the strength of the tree or its parts. While every tree has the potential to fail, only a very small number actually affect something or someone. In fact, the odds of a tree falling and causing injury are relatively low. According to the UK Health & Safety Executive, the chances of being killed by a tree in a public space is about 1 in 20,000,000. However, it's important to note that these odds can vary depending on factors such as location, weather conditions, and the health of the tree. Also, the frequency of targets within the target zone. All trees have some risk associated with them, but it is important for the arborist to articulate that the benefits far exceed the risks from tree failures.

> Risk fluctuates and can be ambiguous when considering the dynamics of the tree and its environment.

A thorough tree risk assessment protocol includes not only the inspection, but also events leading up to and after the assessment. For example, it's essential to consider the maintenance levels and inspection protocols that were in place prior to the assessment, as well as the mitigation options and monitoring protocols required after the assessment.

TREE INSPECTIONS

Inspections should include a thorough examination of each section of the tree. A systematic approach using standardized evaluation methods aids the assessment process. Review the crown, branches, and root zone to check for conditions of concern. Read the body language of the tree to determine how the tree has responded to previous failures, current conditions, and environment, aboveground and belowground. Inspections should take a systematic approach, assessing the potential for tree failure, understanding the impact of such failures, and outlining a plan of action to mitigate the risk presented at the time of inspection. Arborists assessing trees should understand the site factors and characteristics of individual tree species. There are several possibilities that influence tree failure potential and risk. There are external factors such as site conditions and weather playing a role in the evaluation. Soil texture and moisture saturation, wind exposure, pest damage, unfavorable growing conditions, and poor pruning such as topping all should be identified during the inspection.

LEVELS OF INSPECTION

Three levels of tree risk assessment may be used for tree inspections. They include:

- Level 1: Limited Visual
- Level 2: Basic
- Level 3: Advanced

Each level has its specific application based on the scope of work or assignment. Also, if risk cannot be adequately assessed at a certain selected level or conditions of concern cannot be sufficiently inspected, the arborist may recommend a higher level of assessment. Typically, this means a more detailed inspection of the tree and its parts, and advanced research.

Level 1: Limited Visual Assessment

The Level 1 assessment is a visual assessment from a specified perspective of an individual tree or a population of trees. This can include street trees, park trees, or rights of way. A limited visual assessment focuses on identifying trees with an increased likelihood of failure. This inspection is a one-dimensional look at the tree, such as a walk-by, drive-by, or even an aerial view with a UAV (unmanned aerial vehicle). Limited visual assessments are the fastest but least thorough means of assessment and are intended primarily for large populations of trees after a storm event or utility rights of way.

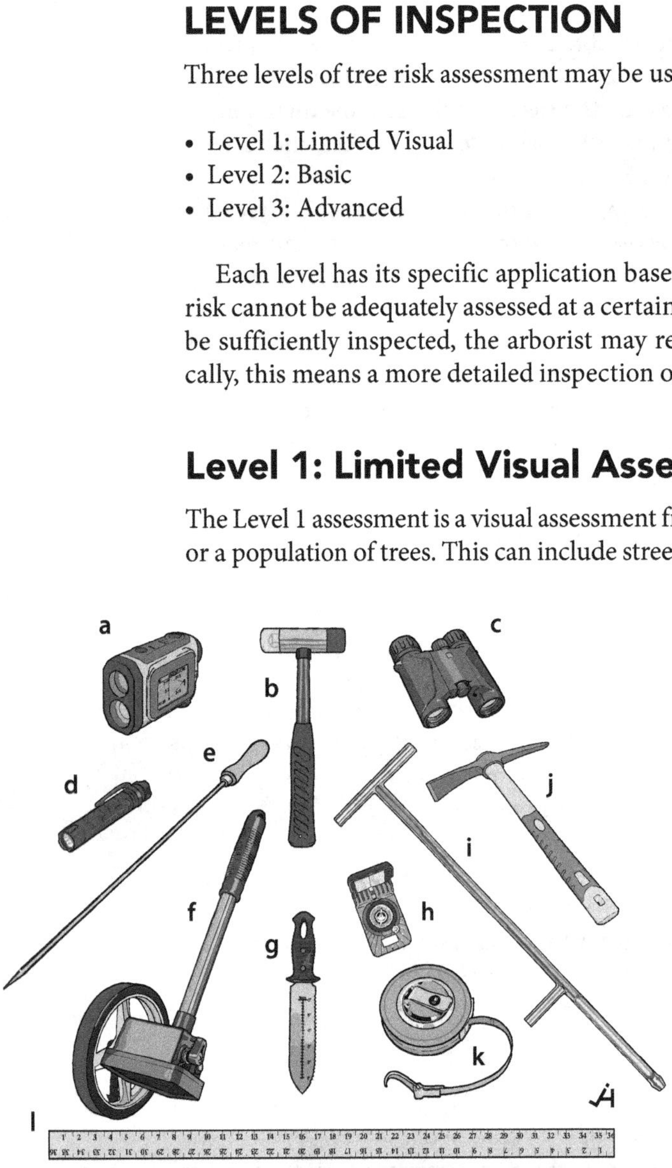

A Selection of Basic Assessment Tools: a) Rangefinder, b) Soft Face Hammer, c) Pocket Binoculars, d) Pocket Flashlight, e) Tree Cavity Probe, f) Telescoping Measuring Wheel, g) Soil Knife, h) Mirror Compass, i) Soil Core Sample Probe, j) Mattock, k) Tree Diameter Tape, l) Yardstick

FIGURE 9.13 Basic tools help identify conditions of concern.

Level 2: Basic Assessment

Typically, most of the tree inspections performed for any risk issues are going to be some form of Level 2. A Level 2 or basic assessment is a detailed visual inspection of a tree and surrounding site. It is a ground-based, noninvasive, 360-degree look at the tree crown, trunk, and root zone. It requires the assessing arborist to walk completely around the tree carefully looking at the root flare, buttress roots, trunk, and branches. A basic assessment should include the use of simple tools to learn more about the tree and any conditions of concern.

The basic tools needed include a diameter tape (D-tape) or tape measure to measure the tree. The usual inspection tools include binoculars, a sounding hammer, a trowel, a garden shovel, and a probe.

Level 3: Advanced Assessment

When a tree requires a more detailed assessment or when there is additional evidence necessary for decision-making, advanced assessments are necessary. Advanced assessments are performed to provide detailed information about specific tree parts, defects, targets, or site conditions. They are usually necessary after a Level 2 is performed and not enough information was found at this level to make a sound decision on risk level and mitigation. Specialized equipment, data collection and analysis, and expertise are required for advanced assessments. Technical equipment, such as a sonic tomograph or resistance recording drill, is used to determine the extent and location of the decay. Also, it may be necessary to climb or ascend the tree with a lift for a better view of the crown. A Level 3 assessment is often utilized when the decision is between retaining or removing the tree for the client, of which there are a lot of options.

FIGURE 9.14 Specialized equipment can improve decision-making with mitigation options.

WHEN SHOULD TREES BE INSPECTED?

Inspections should be performed prior to any tree care activity and in consideration of any nearby targets. Any tree care activity, including plant health care, support system installation, pruning, or removal, should be preceded by some level of inspection for any defect that could endanger the tree owner, manager, or other target, such as a vehicle. Something as simple as a broken branch could be dislodged and fall on someone while they are working beneath the tree. Ideally, proactive inspections occur as recommended by the consulting arborist. However, most inspections leading to assessment happen after a storm or some other disturbance that concerns the client, tree owner, or manager.

Inspection intervals are based on the level of risk presented at the time of an initial inspection. Intervals can be as often as every 6 months to a year. Typically, the range for inspection intervals is from 2 to 5 years; however, the developmental age of the tree, conditions, and other factors will affect frequency. Inspection timing can vary according to the client's needs. It is especially important to inspect trees after major weather events and before events that bring many people to the vicinity of a tree or trees, such as a tournament on a golf course or a concert in a park.

The frequency of inspections is ultimately the decision of the property holder and dictated by predetermined acceptable levels of risk. For example, trees in active-use areas, such as public spaces, parks, and greenways, could be inspected annually for conditions leading to failure. Often, it is recommended to conduct inspections after storms to search for damage which can lead to elevated target risks. Also, be sure to frequently monitor and inspect trees with a history of failures or those with problematic structure.

Every tree inspection should be recorded and kept safely filed for future reference. There are management considerations and legal implications for documenting the review of a tree or trees on a site. Records of past inspections can show how a tree has changed in its health and structure over the years. These past inspections can aid in determining management strategies and risk mitigation for the tree owner. Also, written evaluations can help

FIGURE 9.15 There can be several conditions of concerns and defects affecting likelihood of failure.

FIGURE 9.16 Loads from climbing and rigging can cause failures. Inspect the tree before any tree work is implemented.

to minimize liability if a failure occurs and a claim is filed against the tree owner or the company providing services. A good database and archive of tree assessments is a prudent strategy in the overall risk management plan.

Although we can readily recognize many defects in trees, there can be root problems and internal conditions not easily detectable. These trees may require an increased level of inspection with more in-depth assessments and the use of specialized diagnostic tools. Secondly, defect severity can and does change with time and season. Finally, trees are adept at disguising potential problems and surviving these issues for some time. Consistent and timely inspections will help to successfully manage risk associated with trees.

Individual tree features must be considered when assessing for conditions of concern, defects, and failure potential. What is considered a condition of concern or a defect on a tree? A tree defect is an imperfection or abnormal condition that impairs quality or reduces physical strength. This could include an injury, included bark in a codominant stem, decay, or other condition that reduces the tree's physical strength. A condition of concern is a situation or an atypical or abnormal condition that could lead to a defect or worsening condition requiring closer examination. This might be poor branch aspect ratio, stem bifurcation, or clustered branching. Defects are flaws, weaknesses, or the absence of important features that impact failure. Some common defects are injuries, poor attachments, and decay in trees. Additionally, it is important to observe the severity and just as importantly, the location of the identified condition of concern or decay.

In addition to being attentive to the structure and health of the tree relative to risk of failure, the tree worker must also understand the forces applied to the tree during tree work, and how these forces can be magnified or diminished depending on how rigging operations are implemented. Ultimately, three questions should be thoroughly investigated before work on the tree begins and throughout the work assignment:

1. Where are the strengths and weaknesses within the tree?
   a. A systematic inspection of the tree canopy, branches, trunk, root collar, and root zone is performed before work begins.
   b. Focused attention during work operations continues to identify conditions that change once forces are applied to the tree.
2. How must the work plan take advantage of the strengths of the tree while avoiding the weaknesses?
   a. Can the work be accomplished from the ground?
      i. If not, is there functional access for an aerial lift?
      ii. If not, are there appropriate tie-in points?
      iii. If not, is there an adjacent tree with appropriate tie in points?
   b. Can tree parts be allowed to free fall?
      i. If not, are the branches, trunk, and roots strong enough to support rigging?
      ii. If not, are there adjacent trees that can support rigging?
      iii. If not, is there functional access for a crane?
3. Do we have the appropriate resources to accomplish the work plan?
   a. Do we have the right people for the job?
   b. Do we have the right tools and equipment for the job?
   c. Do we have adequate time to safely complete the job?

Tree workers should understand that if after conducting a tree risk assessment and devising a safe work plan it becomes apparent the appropriate resources are not available to safely accomplish the work plan, they have a right to refuse to do the work until appropriate resources are available.

There is no aspect of tree risk management more important than tree workers understanding the factors leading to strength loss in trees and the forces that are applied to trees — particularly those forces associated with tree work—that can trigger failure of trees or tree parts. Be sure all tree workers are informed of risk situations for each tree.

WHERE DO I LOOK FOR DEFECTS?

During the inspection, look at the tree systematically from top to bottom. Read the body language of the tree and inspect each section of the tree, including the crown, branches, trunk, and root zone, for indicators of past or potential failures. Identify the structural, mechanical, and

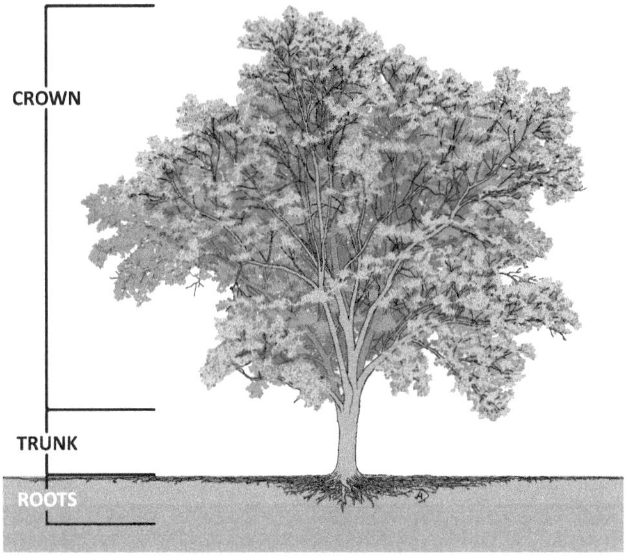

FIGURE 9.17 Systematically inspect the tree roots, trunk, and crown for conditions of concern.

biological factors that create a condition of concern. These factors can occur alone or in combination, such as:

- Dead, diseased, dying, or broken branches because of poor growing conditions, storms, or pest activity.
- Crown size, shape, and weight distribution. This is especially true in situations where a tree is exposed to windy conditions, is leaning, or has an unfavorable live crown ratio. Live crown ratio is the ratio of the vertical extent (height) of the live crown to the height of the entire tree.
- Thinning or poor crown health from environmental stressors or pests.
- An unstable branching pattern with overextended or weakly attached branches, such as some codominant stems or poor aspect ratio.
- Ripples in the trunk or decayed areas with large cavities present, such as nesting holes.
- Insect populations that can destroy the cambium layer through boring and feeding, such as emerald ash borer (EAB) or pine bark beetles.
- Cracks or seams through the bark and into the wood of the tree.
- Root and root plate issues, such as exposed or decayed roots, soil heaving, conks, or mushrooms or cracks in the soil around the root plate.
- Cut roots or soil compaction from construction can result in defects immediately or can be delayed for several years.
- Cankers or oozing around the root flare, which may indicate internal decay.
- Crown structural issues, such as poor branch aspect ratios. Included bark in narrow branch angles can create high-risk situations with wind, ice, or other dynamic loading.
- Plant health and vigor. This determines how likely it is that a tree might overcome wounding or pest infestations. Trees that have adequate moisture, nutrition, and soil conditions are less vulnerable to stress, reducing the impact of some defects.
- Decay, cankers, missing wood fiber, and other positive indicators of weakness in the roots, stems, and branches.

FIGURE 9.18 Pests, such as borers, can quickly weaken trees, increasing the likelihood of failure.

FIGURE 9.19 An unbalanced crown shape can lead to twisting, resulting in cracks and potential failure.

FIGURE 9.20 Mechanical damage injures the cambium layer, resulting in decay conditions.

Due to the variability of defects, it may be helpful to group them to aid in the inspection process or categorize types of defects by causal factors. Causal factors are a common term which is simply an unintended contributor to an issue, in this case, tree weakness. Typically, defects and conditions of concern are found to be the result of structural, mechanical, or biological factors.

These are visible indicators that contribute to the likelihood of failure:

- **Biological factors:** conditions caused by a living agent, or biotic cause. This would include decline and dieback of branches because of a pest infestation, decay, and cavities in the trunk, disease causing sap rot on a limb, insect pests such as borers, or fluxing from bacterial wet wood. Also, soil-borne fungal agents should be considered as an impact on failure potential as well.
- **Structural factors:** poor architecture, such as codominant stems or branches with unfavorable branch aspect ratio or branch arrangement (e.g., vertical branches, clustered branching, or lion's tailed branches). Asymmetrical crown shapes from utility pruning can create conditions leading to failure. Soil conditions are a part of this category as well. Soil failure is an often-overlooked contributor to whole tree failures.
- **Mechanical factors:** physical injury, such as damage from lawn equipment or vehicular impact, construction-related damage to the root system, limb breakage due to a storm event, pruning cuts that leave stubs or wounds that don't close, and even vandalism, can cause conditions leading to unfavorable conditions. It seems logical to introduce this category as well as the structural category as an abiotic or nonliving factor contributing to the issue.

WHAT ARE WE LOOKING FOR WHEN CHECKING THE TREE FOR CONDITIONS INCREASING RISK TO AN UNACCEPTABLE LEVEL?

There are many and varied types of tree defects and conditions, which can make an assessment complex. During a basic assessment, arborists walk completely around a tree and look for defects in all visible areas of a tree, including the roots and nearby environmental conditions. These assessments can include the use of simple tools such as a rubber mallet for "sounding" the tree and probes that can be used to evaluate cavities and wood fiber. The sounding process allows for listening for hollowness or soundness of wood fiber in the trunk. Using a probe can help the assessor determine the depth of a cavity, location of decay, and the integrity of the wood fiber affected. Another important tool is binoculars to help pinpoint potential issues in the crown. Leaf cover can camouflage branches that may have decay, posing a serious threat to the tree owner and any climbers who may enter the tree. During the basic, Level 2 inspection, the arborist can determine whether some aspect of tree structure or health indicates that a more comprehensive tree evaluation, called an advanced assessment (Level 3), is needed to evaluate tree condition and risk of failure more thoroughly.

WOOD DECAYING AGENTS

The wood-decay fungi are not so easy to classify as being pathogens or saprotrophs. Pathogens decay living cells in a tree and affect the health of the tree. Saprotrophs decay nonliving cells, such as leaf litter, on the forest floor or dead branches and are beneficial in recycling

FIGURE 9.21 Fruiting bodies from fungal pathogens are definite indicators of decay.

nutrients back into the soil. Saprotrophs feed by a process known as absorptive nutrition, in which the nutritional substrate (e.g., wood fiber) is directly digested by a variety of enzymes that are excreted by the wood decaying organism. The enzymes convert the nonliving organic matter into simpler molecules, which are then absorbed by the cells to feed the organism. These can include fungal and bacterial pathogens. It is considered highly advisable to identify the most common species of pathogenic decay fungi including *Ganoderma*, *Armillaria*, and *Ustulina* (the latter referred to as *Kretzschmaria* by some specialists). According to wood decay experts, species identification is challenging at the least and recognizing even the genus will assist with determining the mode of action for damaging the tree and assist with the tree risk determination.

The challenge for the assessor lies in the fact that as awareness of the role of wood decay in structural failure increases, there is a risk that practice guidelines may become overly cautious, causing practitioners to overreact to the presence of decay. We have learned from experts in engineering that even large volumes of decaying wood or even open cavities do not necessarily confer to a greater degree of risk. The extent of wood decay is important, but not the only important factor to assess tree condition. The challenge for the assessor is to accurately determine not just how much wood has been lost to decay, but to assess (1) the quality and quantity of the wood that remains and (2) the strength of the response growth of the tree to compensate for injury, infection, and decay. Arborists need to be aware that the presence of decayed wood or a cavity does not mean that a tree is likely to die in the foreseeable future or structurally fail. The presence of an active decay fungus does not necessarily mean the tree is compromised but does present the need to elevate the assessment of the tree's condition through accepted standards.

FIGURE 9.22 Wound wood development around a large cavity.

DECAYED AND MISSING WOOD

Wood decay is an internal process with often limited external indications. Positive indicators of decay are fungal fruiting bodies, such as mushrooms, conks, or brackets, and the presence of rotten wood or cavities. Cracks, oozing, in-rolled cracks, seams, and bulges in the wood are potential indicators of decay, compelling investigation. Advanced decay leading to cavities results in reduced structural strength and stability.

Typically, microorganisms causing decay are pathogens of opportunity, meaning they need some sort of opening or invitation to invade the tree system. These openings can be wounds caused by lawn equipment, storm damage, vehicles, excavation, pruning cuts, animals, or insects. These injuries expose the cambium, providing an opportunity for wood-decaying organisms to invade the tree.

Decay-causing agents present some of the most challenging conditions in determining their impact on the likelihood of failure. Also, mitigation strategies are limited, if not impossible for trees. This is due to the fact there aren't rescue treatments available. Advanced

TABLE 9.1 Positive and Potential Indicators of Decay

Positive Indicators of Decay	Potential Indicators of Decay
Conks and mushrooms	Bulges or swelling
Nesting holes	Cracks and seams
Cavities or openings	Sunken areas
Carpenter ants	Oozing through trunk
Missing wood	Termite trails

stages of decay that result in cavities, the presence of fruiting structures, and missing tree fiber are all critical defects to evaluate thoroughly. The informed arborist should be capable of identifying some of the basic wood-rotting agents and their modes of attack.

Decay and rot are general terms for the degradation of wood fiber. Causes include microscopic organisms, such as fungi, which colonize wounded or weakened areas of the tree. However, some diseases are more invasive, such as *Armillaria,* and can attack and colonize healthy tissue. These fungal organisms insert their hyphae into cells to take advantage of their nutrients and moisture. This results in cell death and decaying fiber. Often there is substantial wood strength loss before the decay is visibly evident.

There are indicators of disease issues called symptoms and signs associated with the fungal activities in trees. Symptoms of decay are the response from an infected tree to a causal agent. There are deviations in normal growth, health, and appearance. Then, there are signs that are evidence of the decay's causal agent. This is actually seeing the pathogen, such as the fruiting body including conks and mushrooms. These visual identification cues are helpful for identifying not only the wood-decaying organism but also its mode of action within the tree.

Understanding the way these decay organisms work can help predict the impact of the defect on tree failures. Before we can understand this, it is important to review basic wood anatomy. Trees are composed of primarily lignin, cellulose, and hemicellulose. The key load-bearing components in xylem cells are cellulose and lignin. The lignin is the concrete; it holds the cellulose fibers in place and keeps them from buckling. Lignin alone, like concrete alone, would crumble under the weight of a tree. The cellulose takes the role of steel reinforcement bars. Cellulose on its own, like rebar, would buckle. Together, however, cellulose and lignin provide unequalled strength: the whole being greater than the sum of its parts. Typically, rots lead to degradation of the woody materials, which leads to decay, attacking the cellulose, hemicellulose, and lignin.

FIGURE 9.23 Heart rot is internal decay reducing the strength of the tree.

Most decay-causing diseases occur from the inside out and are called heart rot. This is a condition of concern in any tree, but the extent and location become especially crucial when it comes to assessing a tree's likelihood of failure. This type of defect is common in many trees but isn't always an elevated risk level. Hollow trees resulting from heart rot are not all dangerous, requiring removal. Where is the living part of the tree? It is in the cambial tissue where the xylem and phloem are developed in the outer rings of the tree. Also, the secondary xylem vessels—containing cellulose and lignin—are essential for tree strength. The extent and location of decay are major determinants of strength or loss of strength when decay is present. Some guidelines recommend a minimum of 2–3 inches of sound living shell for every 6 inches of diameter. Regardless, the type of fungal organism the extent and location of the decay are critical factors to consider.

WOOD DECAY IN TREES

Researchers and practitioners typically categorize decay organisms into basic groups; white, brown, and soft rot:

- *White rot* breaks down lignin and, to various extents, cellulose in cell walls. A wet and spongy wood, often white or yellowish in appearance, remains. The flexibility of the wood that remains may allow the tree to flex, thereby stimulating the production of response growth. Response growth is adaptive growth or wood fiber that adds strength to weakened areas of the tree. *Ganoderma applanatum,* or artist's conk, is a commonly encountered white rot.
- *Brown rot* primarily breaks down cellulose, but generally leaves the lignin intact. This rot leads to major loss of bending strength and is often considered to be more serious than white rot due to the greater loss of strength in the decayed wood. Because brown rot is more rigid and does not flex, response growth tends to be less than in trees with white rot. *Laetiporus sulphureus* is a common brown rot decay fungus.
- *Soft rot,* such as *Kretzschmaria deusta,* typically decays cellulose first but may also affect lignin. It is generally localized and less aggressive in its rate of spread than white or brown rot, causing the softening of living tree tissue. The decay is not necessarily softer than brown or white rot, but modifies the wood fiber.

A canker is an area on a branch, stem, or root where the bark and/or cambium are dead. As the tree adds a new annual ring of wood each year, the affected area is unable to initiate wound

FIGURE 9.24 Ganoderma conk on oak tree.

FIGURE 9.25 Laetiporus species fruiting body.

FIGURE 9.26 Kretzschmaria on maple tree.

FIGURE 9.27 Sap rot attacks the cambium from the outside.

wood to occlude it. Large cankers or a number of small cankers in close proximity can predispose a tree to fail if there is not enough wood to support the tree at the location of the cankers. Stems and branches often fail on or near these cankers. Cankers can be caused by fungi, insects, lightning, or mechanical damage. Bark may or may not adhere to the canker face and be quite visible. Fungal cankers are long-term, tree-fungus associations that prevent normal wood formation at the canker location. Sometimes fungal cankers quickly girdle the tree, killing the stem and branches above the canker. Regardless of origin, research indicates cankers can lead to tree failure if they affect 40% or more of the tree's circumference. When decay is present, evaluate the shell thickness and the size of the opening caused by the canker. The rates and extent of this process are dictated by tree physiology, environment, developmental age of the tree, and available resources. All of these characteristics should be considered when making determinations on the likelihood of failure.

Sap rot refers to any fungus that erodes woody tissue from the outside in and can be considerably more dangerous than heart rots. It is critical for tree care workers to be capable of identifying this type of decay as it can affect the strength and structural integrity of stems and branches. This is of particular importance to climbers when selecting tie-in points or rigging locations.

Missing wood on the tree usually indicates decay is present and the wood has been weakened. Wood decay in trees on developed sites and urban landscapes is a common cause of tree failure, creating defects by reducing the mass and strength of the trees. Wood decay (wood rot) is decomposition of the structural components of wood in the tree. Decayed wood fiber is the result of the long-term interaction between a tree and decay-causing organism, which

FIGURE 9.28 Cracks in trunks and branches can result from several loading factors.

FIGURE 9.29 Shear plane crack.

can occur in branches, stems, and roots. Some tree species are more resistant to decay; others decay quickly and more extensively. This resistance and recovery rely heavily on the tree's species, vascular arrangement, and available resources.

Cracks are a separation of the wood fibers, horizontally or vertically, creating a split through the bark and into the wood. The majority of cracks are caused by wounds, the splitting of weak branch unions, and even lightning damage. Cracks can occur in branches, stems, or roots. The wood behind the crack may be sound, decayed, or missing (forming a cavity). Several types of cracks can be found in trees and, like other defects, the severity of cracks vary in their impact on stability. Vertical (longitudinal) cracks run with the wood grain along the length of the tree and may appear as shear cracks, inrolled cracks, or ribbed cracks. Horizontal cracks run across the wood grain as if to separate two halves of a tree.

FIGURE 9.30 Ram's horn or in-rolled crack.

Shear plane cracks, which are a type of vertical crack running parallel with the branch or stem, go completely through the stem and longitudinally separates wood fiber. As the tree branch or stem moves in the wind, one half of the stem slides over the other, along a neutral plane. Eventually the enlarging crack causes the two halves of the stem to shear apart. A shear crack can have a high risk of failure unless there is significant response growth in the wounded area.

Another type of vertical crack is an in-rolled crack, also called a ram's horn. The margins of this type of crack curl inward on each of its sides and forms inrolled bark and wood. The fissure of an in-rolled crack may appear open or closed. This is a form of response wood as well. Decay is typically associated with an inrolled crack because the margins rewound the tree each year allowing decay to spread. Trees with an inrolled crack, advanced decay, and another crack, all in the same stem segment, have a higher risk of failure. The potential for failure can be estimated by measuring the shell

FIGURE 9.31 Horizontal crack.

FIGURE 9.32 Ribbed crack.

thickness in a few locations around the tree's circumference, determining the width of the crack opening and looking for the presence of any other type of cracks.

Horizontal (transverse) cracks run across the grain of the wood. Horizontal cracks are rarely found because they develop just before the trees fail. Horizontal cracks can be a sign of imminent failure or the tree being in the act of failing, often requiring immediate action.

A tree with a ribbed crack has a raised rib of wood on its stem with a crack along the length of the rib. A ribbed crack has a higher risk of failure when associated with another crack or extensive advanced decay. Evaluate the thickness of the shell and the size of the crack opening. Ribbed cracks may also form at the base of weak unions or on large branches.

Seams are generally not indicative of increased failure potential, but they are considered conditions of concern. A seam is a vertical line in the bark and is considered a phase in the wound sealing process. As time passes, a solid shell of wood begins to form over the old wound, which strengthens the stem and, eventually, the seam may disappear. The wood inside the tree may be sound or decayed and it is often very challenging to detect with sounding procedures. If in doubt, evaluate the tree's shell thickness with advanced testing equipment, such as a resistance drill, to determine conditions.

Cracks are considered a risk when they compromise the structure of the tree by splitting the stem in two or when another defect, such as internal decay and a crack, do not provide enough sound wood in the outer shell to support the tree. The presence of multiple cracks and decay indicates a defective tree, requiring substantial mitigation.

SHELL THICKNESS AND STRENGTH

The undecayed layer of wood surrounding the compartmentalized decay column is called the shell. This is the area of living wood unaffected by the wood-decaying organism. If the shell thickness is thin relative to the size of the tree, the shell is likely to break under tension or buckle from compressive stress, causing the tree to fail. Studies have shown that if a tree

TABLE 9.2 Solid Shell Thickness

Tree DBH (Inches)	Shell Thickness (Inches)
6	1
12	2
18	3
24	4
48	8

Note. A solid trunk with no cavity or opening should have at least 1 inch of sound wood shell for every 6 inches of stem diameter.

TABLE 9.3 Shell Thickness Thresholds With Cavity

Tree DBH (Inches)	Shell Thickness (Inches)
6	2
12	4
18	6
24	8
48	16

Note. A trunk with a cavity or opening should have at least 2 inches of sound wood shell for every 6 inches of stem diameter. It should be less than 30% of the stem's circumference.

has less than one inch of sound wood in its shell for every 6 inches of stem diameter, then the tree has an increased likelihood of failure. Measure stem (or branch) diameter where decay is present. If possible, determine where the shell is the thinnest and take the measurement there because the tree is most likely to fail where the shell is the thinnest.

All trees do not have a sound shell of wood completely around the circumference. Some are likely to have openings or breaches in the column (trunk). An opening can be a hole, an old wound, a fire scar, a cankered area, or a wide crack. A tree can have internal decay and an opening and still be structurally sound, provided that the shell is thick enough relative to the size of the opening. If an opening in the stem is up to 30% of the stem circumference, then the shell needs to have at least 2 inches of sound wood for every 6 inches of stem diameter, according to research. Trees with larger openings or thinner shells are more likely to fail. These are recommendations based on research and apply to centrally located decay columns. The location of decay in the stem affects its strength and load capacity greatly.

Shell thickness can be affected by the presence of bulges or swellings on the stem. This is called response growth. Bulges in the stem are formed as a reaction to the presence of decay. Additionally, they could be burls. Bulges and swellings help strengthen the tree and can decrease the likelihood of failure due to the presence of decay with highly dense, packed cells in the affected area. It is important to note that it is not only the presence of decay that affects failure potential, but also the extent and location within the cross-sectional areas of the trunk.

A burl is defined as a deformed rounded growth on the shell of the tree. Burls can be small or large, and often have an irregular shape. Burls are typically found on the trunk or branches of trees, but they can also occur on the roots.

These unusual growths are composed of many small fibers growing in many different directions. This gives them a unique appearance, and it also makes their material wood strength very strong. Their origin can vary and typically when a tree is under stress, it will produce chemicals tilling the cells in its bark to grow in a different way. This can result in the formation of burls. These stresses can include:

- mechanical injury
- fungi
- bacteria
- viruses
- environment

Mostly they are not harmful and can provide strength in some areas of the tree, especially if a response to wounds.

RESPONSE GROWTH

Just as the term implies, response growth is wood fiber that is produced in response to or as a reaction to mechanical stress or injury. It is a strategic placement of new wood cells affected by the mechanical stress experienced by the cambium layer. This new growth includes reaction wood (compression and tension) and wound wood. The amount of new wood a tree produces depends on basically five things: species, developmental age, environment (water and nutrients), health condition, and available resources (energy, carbon reserves). The placement of new wood is determined by mechanical stress, but it is also directed by physiological and morphological conditions.

Reaction wood is wood produced in response to a stimulus or load. Trees produce more wood than necessary to remain stable under normal weather conditions. Much of the response wood development occurs in the areas of the tree that are affected by movement in response to loads being applied repeatedly. In a more simplistic comparison, moving under normal winds and breezes is a tree working out. Movement stimulates the tree to place new wood in areas of the tree requiring additional support.

Compression wood, which is more commonly found in conifers, consists of cells that have expanded in size. The cell expansion applies compressive forces on the lower part of the branch or leeward side of the stem. Tension wood is composed of cells that have contracted, applying tension on the upper side of a branch or windward side of a stem. In response to gravitational forces,

FIGURE 9.33 Trees experience loads that stimulate various responses.

FIGURE 9.34 Wound wood can add additional strength in weakened areas.

conifers produce compression wood and hardwoods mostly produce tension wood; however, some hardwoods may also produce both types of reaction wood.

The development of trunk taper and buttress roots larger at the base of the tree are examples of response growth. As mentioned earlier, burls may be attributed to response growth. Tree branches will develop enlarged areas at the attachment point to compensate for increasing branch weight. The main stem or trunk will develop larger buttress roots and flares in response to prevailing winds. A cavity or a crack in a branch or stem often reveals obvious response growth with modified growth near the wound.

Wound wood is produced in response to cambial damage and is typically denser than, and chemically different from, the normal wood. This tissue contains densely packed and highly lignified cells with fungistatic properties. These properties help resist decay and reinforce the wounded areas. Research has demonstrated wound wood can be up to 30% stronger than normal wood under loads.

The rate of wound wood development is dependent on the same basic properties described earlier with response wood. When assessing trees, note the presence or absence of response growth. This will have an impact on the likelihood of failure. Areas of increased growth indicate the tree is responding to strength loss from the defect. This compensatory growth is critical for reducing failure potential and wound recovery. Compartmentalization of the wound is a form of response growth with special characteristics discussed in earlier chapters.

ROOT CONDITIONS

Root damage is often obvious with activities such as trenching, paving, grading, or soil compaction. The impact of the damage to the root system will depend largely upon how much of the critical rooting area was damaged. Critical rooting area is defined by the critical root zone (CRZ). The CRZ is a circular area around the stem of the tree, sometimes larger than the area defined by the tree's dripline. A tree is adequately anchored when the roots inside the area defined by the CRZ are healthy and undisturbed. Some tree species are more susceptible to issues associated with root damage than others and should be considered during the assessment.

FIGURE 9.35 Root decay and weakness can be expected when fruiting bodies are visible.

When a tree has extensive root damage, the whole tree has the potential for failure. Roots are the anchorage system of the entire tree. If the roots can no longer provide adequate anchoring, the tree is vulnerable to windthrow and whole tree failure. Roots can be damaged or removed due to excavation, trenching, soil compaction, grading, paving, fungal decay, or environmental stress, such as drought or flooding.

Severe root problems can become apparent if a tree develops a new or abnormal lean. In these cases, a portion of the root system has failed, and the tree exhibits a tilt but is stabilized, at least temporarily, by the remaining root system. Newly leaning trees are often accompanied by soil heaving, soil cracking, root lifting, or root breakage within the root plat and on the opposite side of the lean. A tree with a newly formed lean and compromised root system indicates an increased risk of failure.

STEM GIRDLING ROOTS

There are several reasons why roots begin growing in an encircling pattern around stems. Stem girdling roots (SGRs) are roots that are growing around the trunk or across the major lateral roots. Often it begins as either a pre-existing condition in pot-bound trees grown in the nursery or they develop when trees are planted in extremely compacted soil, planted too deeply, or when their roots hit solid obstructions in soil. Trees suffering from SGRs express common symptoms, including stunted growth, scorched foliage, abnormal leaning, lack of a characteristic trunk flare, early leaf coloration and leaf fall, and vulnerability to secondary problems such as insects and diseases. Symptoms are slow to develop and often elusive where seemingly healthy trees can suddenly fail during wind events.

FIGURE 9.36 Stem girdling roots affect tree health and stability.

Investigate the root system for decay on visible surface roots and at the root flare where possible. A tree root disease is one that originates in roots and remains in the roots and basal area of the tree. Those decaying organisms of concern are generally root and butt rots, specifically decaying wood in the roots and lower stem (at the root flare). Don't rely on visual aids such as fungal fruiting bodies. Mushrooms growing under the dripline of the tree are indicators of a potentially compromised root system, but they are not always present. Mushrooms are only present at certain times of the year, dependent on the decay organism. In trees on developed sites and urban landscapes, root diseases are often of critical importance, creating defects that increase the likelihood of failure.

FIGURE 9.37 Fungal fruiting bodies in the root zone indicate defects, which include decaying roots.

Root diseases damage trees primarily by decaying the wood fiber, specifically the lignin and cellulose in tree roots. This results in reduced water and nutrient uptake and slowed tree growth, predisposing trees to failures and death. As with any disease, root diseases can develop over a long period of time without any visible symptoms in the tree. Many root diseases are caused by fungi and some water molds (e.g., *Phytophthora spp.*). The most common root decay caused by fungi is Armillaria root disease. Armillaria root disease is caused by the *Armillaria mellea* complex (shoestring fungi) that readily infects stressed trees and is often associated with declines. Sudden oak death caused by *Phytophthora ramorum*, and a little leaf disease caused by *P. cinamommi*, are a few examples of tree root diseases caused by water molds.

FIGURE 9.38 Previous wounds are evidence of mechanical defects which may increase likelihood of failure, depending on the development of wound wood.

STRUCTURAL DEFECTS

Structural defects include characteristics of the tree itself and its growth habit. Whether it's a leaning tree or poor branch configuration, these should be considered during the assessment. In most cases, poor structure is a product of poor management, growth habit, or damage to the tree.

Excessively leaning trees are a common example of poor architecture. Many trees lean to some extent. In some cases, tree lean is a new or recent condition and is due to partial windthrow. In other cases, the tree has leaned for a long time and is well anchored and balanced for its load. Some situations warrant mitigation, however. Trees with an established, stabilized lean are often termed "corrected leans." These leans reveal a less than vertical angle at the stem origin, but the crown has grown more upright. An example may include a tree at the edge of a woodlot, seeking light, exhibiting phototropic response. This is not necessarily a condition of concern and typically has a low likelihood of failure.

Other common structural defects with increased failure potential are codominant stems and poor branch attachments. Codominant stems lack a dominant leader and present a condition where two or more stems about the same size arise from the same location. Such a configuration is prone to splitting under loading forces. Another important characteristic is the attachment of the branch to the parent stem or trunk. Branch attachments with similar aspect ratios are structurally weaker than those with a 2:1 branch aspect ratio.

Strong branch unions are those considered to have a favorable branch aspect ratio. Aspect ratio is the diameter of the branch relative to the diameter of the trunk, both measured immediately above the union. Branches with a small aspect ratio are very well attached to the trunk; those with a large aspect ratio separate more easily from the trunk. The goal should be to maintain a low aspect ratio between the trunk or stem and branch size of less than 2 to 1. This means that the size of remaining branches should be half that of the supporting stem or even smaller. Codominant stems and weak branch unions, especially with V-shaped bifurcations, are notably more prone to failure. Included bark in stems and branches are at an even greater risk of creating a dangerous situation. The branch union is even more weak when bark is included or trapped in the attachment. This defect is prone to failure and becomes a concern not only for the tree owner, but for climbers as well. These attachments are unreliable and unsafe tie-in points and rigging points for climbers and are likely to split with dynamic loading.

Dead and broken branches are structurally unsound because of preexisting defects or rapid decomposition of the wood. Failed branches lodged in the crown may fall at any time. Live trees most often fail first at their weakest point. Dead trees, however, can fail anywhere, unpredictably at the ground line, just above the stump, just below the lowest branch, or anywhere in the crown. As time passes, the probability of failure increases with the further degradation of the wood fiber. Dead branches may remain attached to live trees for an extended amount of time or may drop suddenly. Dead branches and broken, hanging limbs usually break off near or at the living part of the stem as the tree sheds the decaying component. Branches on dead trees usually decay and fall first, leaving a slowly decaying main stem that may remain attached for many years. A broken branch caught up in the tree's crown has an elevated risk level because it has already failed and only waits to be dislodged by the wind or by the failure of the supporting branch. Dead trees within striking distance of a target should be removed as soon as possible.

Mechanical injuries are arguably the simplest of the three categories to detect. Typically, these are recognizable injuries, wounds, or damage from an external source. Such defects include construction damage in the form of physical damage to the roots, trunk, or branches. Trenching within the dripline of a tree or cutting roots to make room for new infrastructure such as walks, or curbs can lead to serious defect issues for a tree. Inspect the ground below the dripline for roots exposed from wind throw or construction impacts. However, some root weaknesses may be hidden from a previous disturbance and not so easily identified without searching for clues in the top of the crown. Sparse or undersized leaves, chlorosis, necrosis, or dieback can signal problems in the root zone.

FIGURE 9.39 Topping trees leaves them vulnerable to health concerns and increases the risk of failure.

When a tree is repeatedly wounded from inrolled cracks, included bark, canker-rot fungi, or equipment (such as mowers, plows, and weed whips), decay occurs in every annual ring of wood. These trees should be carefully inspected because their decay is never fully compartmentalized, and a sound shell of wood does not form. Sounding can also be unreliable in these situations. The tree is likely to fail at or near the location of the crack or wound because a large and ever-expanding column of decay is present there. Again, evaluate shell thickness and opening width to help determine the tree's potential for failure.

Poorly executed tree care can cause defects as well, which may result in elevated risk. Bad pruning cuts, such as flush cuts or stubs left behind, can result in disease and decay affecting tree health and strength. Adventitious branches, such as water sprouts produced after topping, are weaker due to the lack of overlapping tissue attaching it to the parent stem. Topping trees results in an identifiable defect, leaving behind weak branch attachments and decay due to excessive pruning and a disregard for appropriate cuts. These have risk levels that vary based on the severity of the issues.

Identifying and recognizing conditions of concern and defects can be a complicated procedure, especially from a ground-based perspective. However, the skilled arborist should e able to examine the tree and identify critical factors affecting targets and risk. The proper tools, resources, and knowledge can prepare the arborist for a well-informed inspection. The goal is to minimize the likelihood of failure and create a safer environment around the tree and for those working in trees. Identifying defects and conditions of concern is an integral component of any pre-work planning process and understanding their impact on arboricultural practices is critical.

TREE INSPECTIONS AND USE OF SPECIALIZED DIAGNOSTIC TOOLS

Basic inspections of individual trees are sufficient for detecting most concerns and arriving at a confident risk rating. Some conditions of concern, however, do not clearly express visible defects and their detection requires in-depth and often invasive examination and the use of specialized diagnostic tools.

Advanced tree assessments are necessary when a tree poses an increased risk level and presents defects that cannot be fully evaluated during the level 2 basic inspection process. For example, where stem girdling roots (SGR) are a suspected defect, soil excavation near the base of the tree may be necessary to confirm the presence of SGRs and the extent to which they constrict the stem. For trees exhibiting external signs of decay where the extent and location of internal decay is uncertain, it is advisable to measure the thickness of the outer shell of sound wood within the tree, and to determine if safe shell limits are met. This may require a level 3 inspection and the application of an increment borer or resistance drill equipment.

ROOT COLLAR AND STEM GIRDLING ROOTS ASSESSMENTS

Root collar examinations are performed to detect dysfunctional roots, root damage, or decay in buttress and primary support roots. The root collar is defined as the base of the tree where the primary roots first begin to branch away from the stem and normally appears swollen or slightly flared. The examination begins by probing into the soil near the root collar with a metal probe such as a small diameter screwdriver or surveyors pin, to determine the depth of primary roots and the presence of any encircling roots or other root dysfunction. Next, the soil is excavated outward from the root collar and primary support roots are examined. Gradually loosen and remove soil until the stem/root conflict or root collar is exposed. Hand trowels, soil knives, stiff bushes, and forced air can be used to loosen and remove soil around roots. Shallow examinations of smaller trees do not require the use of complicated equipment. However, portable air compressors are most effective with larger trees and examination areas. Do not use spades or shovels unless you are certain that no roots exist in the soil to be excavated. After the soil is displaced from the root area, look for the presence of any dysfunctional roots, evidence of stem compression, or other root damage. In the case of root decay, a metal rod may also be used to probe the roots farther out from the trunk and to test for the presence of advanced decay. If the roots are severely decayed, they will be soft and easily penetrated by the metal probe. The risk of tree failure can be elevated if girdling roots constrict more than 40% of the stem's circumference or more than 40% of the roots within the CRZ are damaged or decayed.

FIGURE 9.40 Soil knives and garden trowels are helpful assessment tools.

DECAY DETECTION EQUIPMENT

Internal inspections are helpful to determine the location and extent of decay present in a tree and whether the decay significantly impacts the structural integrity of the tree. Several devices are available to help detect decay and other defects in standing trees. Traditional, low-tech devices include the steel probe, plastic mallet, and increment borer. High-tech devices include resistance drills (Resistograph, IML-RESI PowerDrill), sonic and ultrasonic detectors (Sonic Tomograph, Arborsonic Decay Detector) among others.

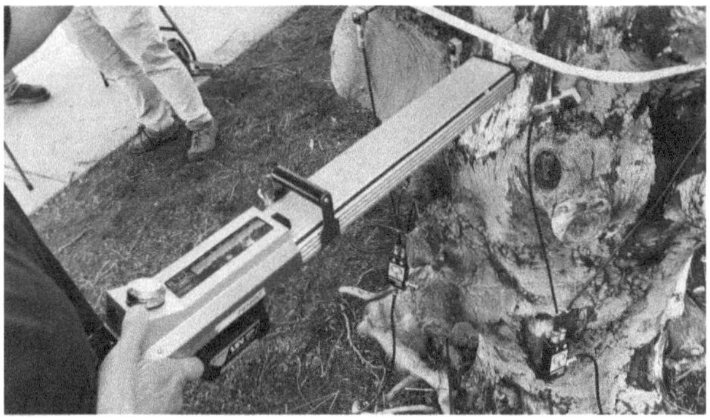

FIGURE 9.41 Resistance recording drills are advanced boring tools used to obtain detailed information on wood decay.

EXAMPLES OF DECAY DETECTION DEVICES COMMONLY USED BY ARBORISTS

Metal Rod or Probe

A metal rod or probe, like a screwdriver or chaining pin, may be used to probe woody tissues and roots to help determine the extent of cavities, the depth of cracks, or the presence of decay. If the stem or root tissue is severely decayed, they will be soft and easily penetrated by the metal rod.

FIGURE 9.42 Probing cavities can help determine the extent and location of decay.

Small Non-Metallic Mallet

A plastic or nylon mallet can be manually struck against the bark or exposed wood surface, and, with experience, an arborist can interpret whether the resulting sound indicates hollowness or severe decay. This method of sounding is highly subjective and is dependent on the arborist's experience and interpretation skills. Care should be taken not to mistake the sound emitted by striking loose bark as the presence of decay within the wood of the tree. There can be false positives and false negatives in response to sounding. This is especially true when sounding around cracks and seams on the trunk.

FIGURE 9.43 Sonic tomography provides graphic detail on the extent and location of decay.

FIGURE 9.44 Sounding the trunk can help reveal decayed wood.

FIGURE 9.45 An increment borer can provide information on age and wood fiber.

Increment Borer

The increment borer is a specialized boring instrument that consists of a hollow tube with an external screw thread at one end. It is twisted into a tree which allows for the removal and examination of a wood core approximately 5mm in diameter. The core can be examined for the presence of discoloration and decay along the cross-section, and the presence of decay can be manually mapped along the length of the core. Multiple samples around the stem circumference will provide a measurement of the thickness of the outer shell of sound wood and the location of decay when measured and mapped. Increment borers are inexpensive, easy to carry, and require limited maintenance. However, this method of assessment is invasive to the tree, and creates the largest diameter wound of the decay detecting devices commonly in use. In trees with internal decay, an increment borer can break an existing barrier zone within the tree and may allow decay to progress into healthy wood. Finally, if the coring tip becomes dull, the tool may get stuck and removal from the tree is difficult. Always lubricate the borer shaft with beeswax to help prevent this sticky situation.

A FINAL WORD ABOUT DECAY DETECTION DEVICES

Decay detection devices should be used with discretion because most are invasive in their mode of application and cause some degree of injury to the tree. Wounds, created when probes are drilled into the tree or when the bark is removed to attach sensors, may serve as entry points for decay organisms. Invasive methods may also allow existing decay to spread internally by interfering with the tree's ability to compartmentalize the decay. Although it is uncertain to what extent these small diameter wounds contribute to the development of decay within trees, the injury caused by most decay detection devices should not be overlooked. It is advisable to restrict the use of decay detection devices to situations when additional information about the location and extent of internal decay is critical to assembling the probability of tree failure.

Risk Mitigation Strategies

When unacceptable risk levels are identified, a mitigation strategy is communicated to the client, property holder, or manager. The ethical arborist conducting the tree risk assessment demonstrates competence by providing options, not opinions. Regardless of the situation, it is the responsibility of the assessor to provide expertise and mitigation options (Figure 9.46), and the decision to act on those recommendations rests entirely upon the responsible client.

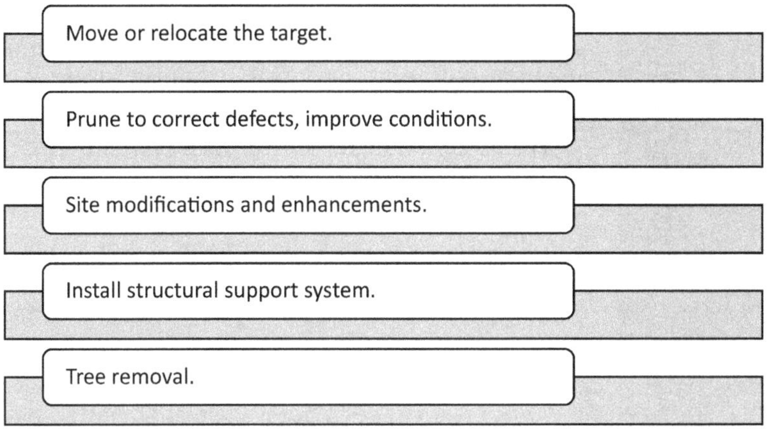

FIGURE 9.46 Tree risk mitigation options.

Those responsible for the tree should make every effort to reduce risk with proper pruning and health care practices. Consider all options before deciding on removal. If possible, move or remove the target. This can be a temporary solution until appropriate action can be taken or a permanent solution if it's feasible for natural areas. Install structural support systems when trees are good candidates for this strategy. Modify site conditions to improve the immediate location around the tree. This includes improving drainage, uncovering buried root collars, and fixing other detrimental tree-based issues. While evaluating trees for risk, the inspector should note any tree maintenance needs to further improve conditions. The assessment can determine the timing and priority of work needed as well. The work should be prioritized first for safety and then for tree health. Consider potential risk, activities around the tree, level of acceptable risk, and the needs of the client. Risk mitigation for trees considered to pose an unacceptable level of risk presents logical solutions to protect the target and preserve the tree where possible.

REPORTING

It is essential to communicate the findings to the client regarding the tree risk identified in the assessment. This can be accomplished either verbally, in writing, or through a combination of both to effectively communicate risk and mitigation options with the client. Written reports are usually preferred for documenting details, risk ratings, and mitigation strategies. Verbal reports are often preliminary discussions with the client or manager on the site or

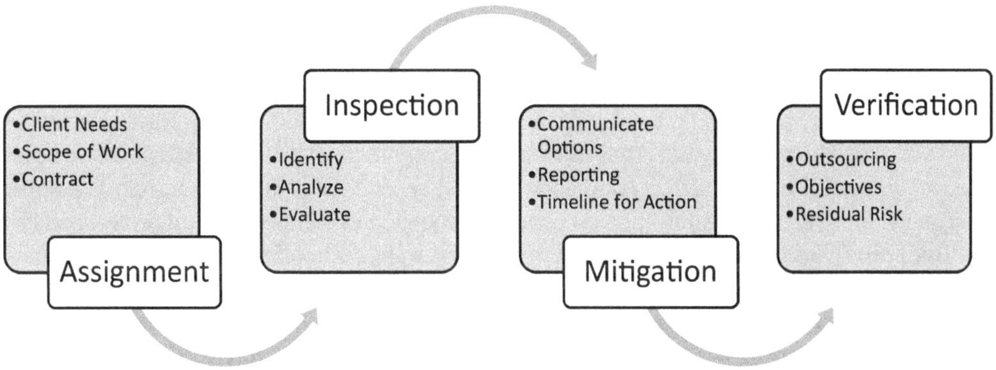

FIGURE 9.47 Tree risk assessment procedure.

FIGURE 9.48 The arborist assessing a tree for risk must identify, analyze, and evaluate the tree, reporting findings and mitigation options to the tree owner.

through telecommunications, describing initial findings and concerns. It is highly advised to follow up with documentation on the investigation, analysis, and evaluation of the tree assessment along with mitigation options.

The details of the report will vary based on the scope of work and client needs. Some situations require only minimal results for the assessment, while other investigations can be quite detailed and complex. Regardless, the communication should be clear and concise including important facts such as the date, the parties involved, the location, and the tree. The detailed written report includes several items. The level of details for each component is as variable as each tree assessed.

Key components of a tree assessment report:

- scope of work
- level of assessment
- methodology used during the inspection
- tree, target, and site assessment
- facts used to determine likelihood of failure
- explanation of the overall risk rating
- options for risk mitigation
- discussion of residual risk remaining after mitigation
- recommendation for reinspection options
- limitations preventing complete execution of the assessment

It is helpful to provide supporting documentation when necessary or required in the scope of work. Video segments and photography can be used to supplement reports. Often graphic and visual aids help to communicate complicated tree issues to the client. Videos and photos should not be altered to change their content. Aspects such as contrast, exposure, and color balance can be adjusted if it helps to accurately depict the scene. Placing a common object alongside a defect can help illustrate the size of a defect or condition to the tree owner/manager. Often, the scale of objects is challenging to interpret in images.

The final report should be considered a professional work product and treated as such with diligence and care. Appropriate, well-organized information conveys a sense of importance and professionalism to the client and is a critical step in the inspection process. The results and recommendations provided should be science-based methodology based on current research and best practices.

Skills, experience, and education are required in tree risk assessment to confidently identify the most challenging part of the tree risk assessment process: the likelihood of failure. However, it is important to recognize there is uncertainty in our inspection and assessment. Being realistic, honest, and ethical in communicating the level of certainty or inspection limitations within the written report is vital. These include full disclosure on limiting conditions, for example, inability to access the tree or excessive groundcover concealing the root zone. A diligent and thorough assessment that aligns with current research and best practice demonstrates competency and professionalism.

Trees aren't necessarily dangerous and we don't want to instill a fearful attitude, but as trees grow older and larger, more risk is involved. Arborists should strive for continual learning about trees to gain an advanced understanding of biomechanics and physiology. It is the best way to break free from the limiting mindset of the "remove or retain, kill or keep" paradigm. There is a world of options between those two, and that world is known as arboriculture.

CHAPTER 10

TREE DIAGNOSTICS AND DECLINE

LEARNING OBJECTIVES

- Distinguish between biotic and abiotic agents affecting tree health and sustainability.
- Describe the diagnostic process for identifying causal agents affecting tree health.
- Identify the four categories of abiotic tree stressors.
- Recognize and contrast disease symptoms and signs of pathogens in trees.
- Explain the concept of decline in trees and the factors affecting decline progression.
- Explain how anthropogenic impacts affect sustainable tree canopy.

TREE DIAGNOSTICS AND DECLINE

Many insects and diseases imperil the sustainability of trees in the urban forest. Also included in the urban forest challenge are the anthropogenic impacts and pressures experienced by trees. In this chapter, common issues leading to tree decline will be discussed based on the primary categories of biotic and abiotic pressures. Additionally, recommended strategies for minimizing the impacts of these pressures will be offered to establish an integrated pest management strategy. This chapter will focus primarily on the abiotic nuisances which lead to decline, dieback, and death in trees. Pathogenic disease and insect guidance should be sourced from current research and regional land grant universities.

TREE DISEASE DEFINED

Tree disease is a condition in which the growth and appearance of the tree has been altered, with obvious symptoms and visible signs. A symptom is defined as a change in plant growth or appearance indicating a plant health problem. Examples of symptoms include yellowing, wilting, dieback, fluxing, or burls.

Symptoms alone shouldn't be used for diagnosis because a symptom, such as wilting, can be produced by multiple causes (drought, borer, canker, root rot, etc.) Additionally, symptoms on one part of the plant can appear due to damage on a different part of the plant. For example, leaf wilt can be caused by clearwing borer feeding on the stem, or a canker

FIGURE 10.1 Symptoms and signs.

TABLE 10.1 Biotic and Abiotic Causal Agents

Biotic Disease (Pathogenic)	Abiotic Disease (Non-Pathogenic)
Fungi	Moisture stress
Bacterial	Cold injury
Viral	Chemical burn
Nematodes	Mechanical damage

disease. In contrast, signs are actual physical evidence of the harmful factor—the actual cause of the problem that allows you to conclusively diagnose the problem. Examples of signs include egg masses, larvae, mushrooms, rust, and powdery mildew, just to name a few.

The causes of disease can be grouped into two basic collections. They are either abiotic (noninfectious) and biotic (infectious) as shown in the table.

NON-PATHOGENIC (ABIOTIC) PROBLEMS IN TREES

Trees and shrubs, whether in a managed landscape or in street rights of way, are vulnerable to stressors not likely found in the more native environment. Additionally, the lack of proper cultural practices and mismanagement of the aboveground and belowground shortens the lifespan of urban trees. Trees have basic needs for sustainable growth. All plants, including trees, require adequate light, moisture, air, and nutrients, plus favorable temperatures and suitable soils for optimum growth. However, trees cultivated in the urban forest require supplemental practices to compensate for their unmet natural needs. For example, supplemental nutrition and especially water must be provided to urban trees due to limited soils. Because trees are planted in less-than-favorable locations without considering mature growth, often excessive pruning is required as well, with negative consequences.

FIGURE 10.2 Anthropogenic impacts on trees.

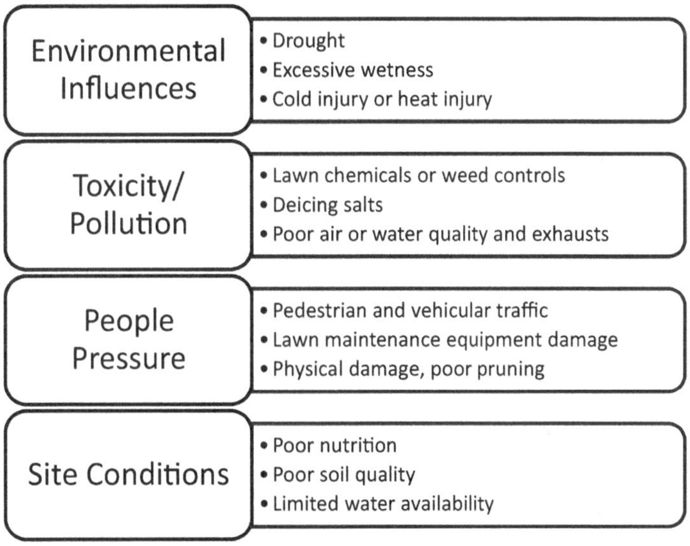

FIGURE 10.3 Examples of abiotic, non-pathogenic disease conditions.

The severity of symptoms for abiotic diseases can vary greatly, primarily based on environmental conditions existing before, during, and after symptoms appear, the intensity of the abiotic pressure, and the pre-existing conditions of the tree. There are several disease-like conditions, described as non-pathogenic conditions, that create adverse effects that interfere with normal plant processes. These may include toxic chemicals, pollutants, nutrient deficiencies, and people-induced problems or the *anthropogenic* impacts.

Pathogenic (Abiotic) Disease Diagnosis

The presence of characteristic symptoms associated with specific environmental or cultural factors sometimes makes abiotic disease diagnosis reasonably simple. Often, however, the symptoms of abiotic diseases are too vague for simple diagnosis. In such cases, accurate diagnosis requires careful examination and analysis of weather conditions before as well as during the appearance of symptoms. Other considerations include recent site changes, which can influence environmental and soil conditions in or around the area where the trees are growing. The cultural and management practices applied before the onset of symptoms should be taken into account.

FIGURE 10.4 Monitoring and scouting are critical actions for pest management.

The diagnostic process includes reading the body language of the tree, from roots to crown, discerning the cause of the plant problem and investigating the symptoms presented by the tree. The person inspecting tree conditions needs good observational and investigative skills. It is necessary to collect everything possible about the tree, site, and conditions to make an informed diagnosis. It is likely multiple causal factors contribute to overall tree decline. Typically, it is a combination of factors and this is especially true in the harsh, urban environment accompanied with elevated temperatures and poor growing conditions.

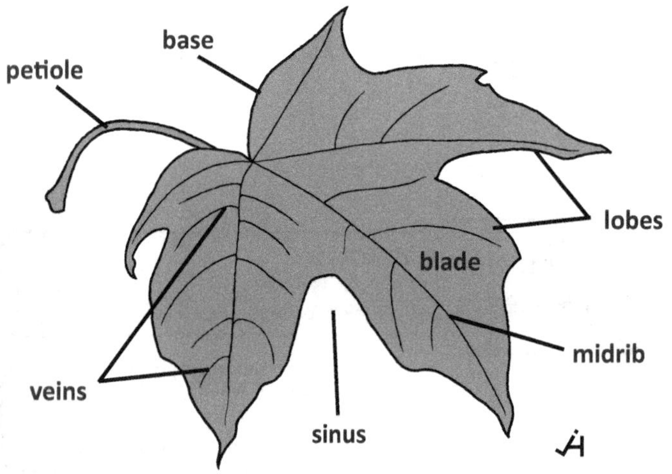

FIGURE 10.5 Parts of a leaf.

Be Certain of *Tree Identification.* Identification of the affected tree is the first step to making a proper diagnosis of the issue. In addition to knowing the common and scientific names of an affected plant, it is important to know the specific variety or cultivar whenever possible. There can be tremendous variability with how a tree responds to a specific disease within individual plant species and even different cultivars. Occasionally, an identification of simply the genus is all that may be required (i.e., two-lined chestnut borer on oak). In contrast, in other situations like with apple scab, it would be necessary to identify the specific cultivar to determine whether treatment is needed, based on its level of resistance to the disease.

The ability to recognize what is normal for the identified tree is important in the diagnostic process. Every species has distinctive characteristics that can be recognized through experience and training. Understanding the typical leaf shape, size, color, and other characteristics normally associated with the time of diagnosis can help narrow down the causal agent. Think about such fall needle drop in some conifers as a normal event. It is also important to remember appearance can vary with different cultivars. Certain tree cultivars, such as the sunburst honeylocust with naturally yellow to pale green leaves, may appear to present symptoms, leading the diagnosis towards nutrition deficiencies, pH issues, or other soil-related problems, when it is not the case. Some trees, such as weeping cultivars, are always grafted, which can lead to burls or faults in the graft and result in an abnormal appearance. Once the "normal" appearance of the specific plant is determined, several comparisons can be made between the tree being examined and similar species. Compare characteristics such as overall size, shape, and coloration; leaf shape, size, coloration, and distribution; root distribution and coloration; and bark, stem or trunk texture and coloration.

Observe How *Symptoms* Are Distributed. Another important step is to observe how symptoms are distributed across the affected tree area. This is a matter of distinguishing between random and uniform patterns. Are the symptoms distributed uniformly across an area or are they more randomly appearing in the crown? If there is a defined pattern to the distribution of symptoms, it is likely a noninfectious cause. As an example, does it occur on trees only along the edges of an agricultural field, next to roadways or driveways, or perhaps adjacent to nearby new construction? In contrast, there may be only one tree or even one branch expressing symptoms of poor growth compared to the other side of the crown. This distribution can be especially important when looking at the possibility of non-pathogenic problems, such as herbicide drift or soil contamination. A uniform pattern on an individual tree and uniform damage patterns over a population of trees are generally not associated with biotic agents but are usually due to abiotic agents.

Pathogenic diseases generally occur over time and are not associated with significant changes in tree health over a short period. However, some diseases can be very aggressive, particularly on vulnerable hosts. This includes oak wilt on red oak tree species. Oak wilt is caused by a fungus spreading locally from infected trees to nearby healthy trees. The red oak group, which includes northern red, shingle, Shumard, and pin oak, is primarily affected by the disease. Symptoms include sudden wilting, early leaf drop, discolored leaves, and vascular streaking in the sapwood. The white oak group is somewhat resistant, but still susceptible to the oak wilt fungus and may survive many years with an infection.

FIGURE 10.6 Deer damage.

The prevalence of a biotic pest in a particular region is an important factor as well. Investigate other trees in the area as well to determine if other similar species are affected. Generally, biotic agents causing disease will occur on a single tree or small sample of the population. However, it can spread and progress easily if favorable conditions prevail but rarely will all trees be infected or express the same level of disease severity at one time. If all trees in a location are affected, it is more likely due to abiotic factors such as soil issues (deficiencies or toxicities), adverse climatic factors (cold temperatures or drought), or toxic exposure (improper pesticide use).

If a plant problem is suspected to be associated with a disease or insect, or symptoms of a particular disease is observed, it is recommended to reach out to a land grant university plant diagnostic laboratory or other private lab to isolate the sample confirming if a disease is present and contributing to plant decline. Diagnostic questions include, "What has been the progression of symptoms on trees in the affected area?" If the symptoms all appeared at the same time with no further development, this indicates a possible acute episode, such as an extreme temperature change or off-target chemical application or drift. Conversely, if the symptoms started in a defined location, such as at the tip of a branch and progresses further into the branch to other tree parts with increasing symptom severity, it is likely a biotic agent, such as a fungal disease like needle cast. Needle cast is a common condition affecting various spruce species. The symptom begins with branch dieback on the lower limbs. If unchecked, it can progress over two to four years, causing needles to drop and a thinning appearance. By the third and fourth year, most of the lower branches are dead.

Review Environmental Factors. Information pertaining to the growing environment to which the affected plant has been exposed is another piece of the diagnostic puzzle. It is especially helpful to document changes in the environment. Environmental factors to consider include extreme temperatures (freezing and heat), rainfall, hail, lightning, prolonged drought, temperature inversions (important in possible air pollutant damage and pesticide drift), and prevailing winds. Any of these weather factors can be contributors. Investigate any external factors contributing to plant health or lack of plant health.

Analyze Management Practices. It is vital to question any maintenance activities that have been conducted around the affected trees. The problem may not be due to anything the owner has done; the problem could be related to neighboring activity. Most important are direct maintenance activities such as mulch type and levels, weed control, fertilization, watering, and pruning. The overapplication of fertilizers can lead to super abundant quantities of nutrients such as phosphorus. As a result, this can incite fungal disease issues, such as anthracnose outbreaks. Detailed information on any chemical application is essential for determining whether it is a suspected cause of decline. This information should include the product applied, application rate, date of application, and application method.

Site considerations for diagnosis
History of the property and construction
Light, soil, water, winds, other
Weather events
Changes in the property or site
Cultural practices
Environmental conditions
People pressure

Evaluate Site History. Site factors such as soil type, drainage problems, grade changes, and soil pH should be evaluated. Often, these site-specific factors are overlooked or not recognized as an inciting factor for many tree troubles.

Interviewing the site manager or tree owner can provide some insights on pre-decline activities around the tree such as any site construction which may have occurred previously. Even minor construction activity such as digging new water lines or sidewalk and driveway cuts can have an impact. Often it can be months or years before any expression of symptoms and decline. In some cases, errant applications or misapplication of herbicides, fertilizer, or other chemicals can be the hidden culprit and those involved may be unwilling to admit their error. Another example involves clearing unwanted mulberry tree seedlings or invasive honeysuckle, which often sprout from seeds within the canopy of other trees, such as an evergreen screen. Treating the cut-back seedling with the application of a stump killer can leach into the root zone of the spruce and cause it to die. Be sure to consider all factors and their relationship to the tree, environment, and situation.

DECLINE IN TREES

Decline is a descriptive term used to describe conditions in the tree that aren't considered normal and disrupt physiological functions. Often, the term decline is used to describe tree diseases with a complex etiology with several contributing factors: biotic, abiotic, or a combination of both. Misapplication of the term occurs when decline is used to describe what cannot be accurately defined and used as a specific disease category. Generally speaking, decline refers to the progressive deterioration and weakening of woody plants.

Symptoms of Decline Diseases
Slowed growth and development
Sparse, undersized leaves, which may be chlorotic
Early fall color during the growing season
Premature leaf or needle drop during the growing season
Dieback of twigs and branches, especially in the upper crown
Prolific sprouting following dieback in the crown

When referring to a tree in decline, it implies the tree is not normal in appearance, with a progressive condition leading to dieback. Typically, this condition is a combination of factors, not having a single causal agent. In most situations, it is the result of several agents acting together, reducing the vitality of the tree. Additionally, decline symptoms represent a progressive weakening condition due to stressors that accumulate over a period of years.

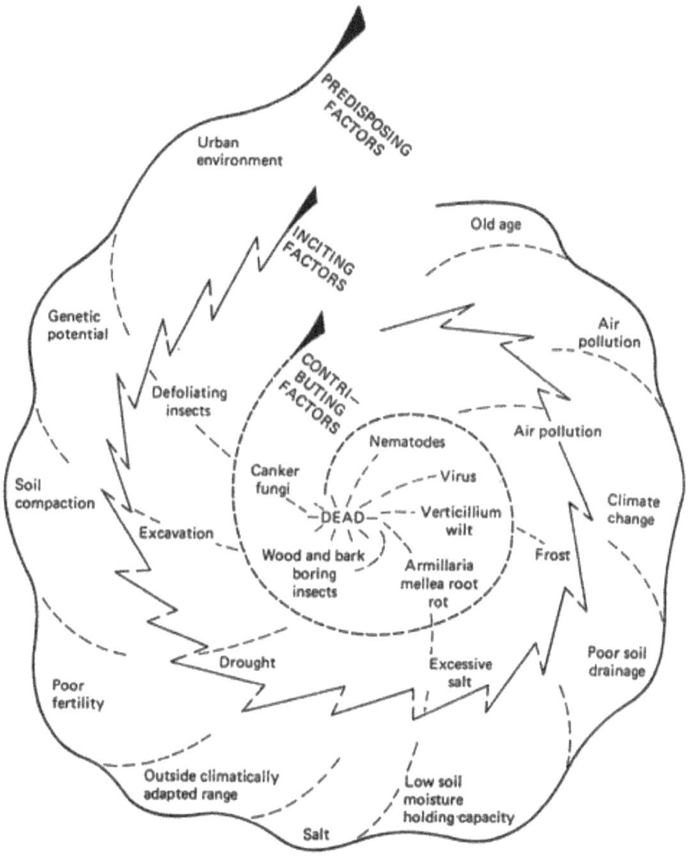

FIGURE 10.7 It is often difficult to pinpoint the factors leading to the decline of trees because of the complex interaction between organisms and the environment. Changes in the host plant, environment, pest, or pathogen, or a combination of changes, can lead to the emergence of novel disease epidemics or pest outbreaks as well.

Source: Manion, P.D. (1991) Tree Disease Concepts.

Generally, decline can be caused by several situations within urban trees. Without intervention, all of these situations contribute to the eventual death of the tree.

- Decline caused primarily by continued and chronic stress by one factor. This might include vascular wilts or other disease agents (i.e., oak wilt, *Verticillium*, *Xylella fastidiosa*, *Phytophthora*, or other infectious mechanisms).
- Decline caused by physical injury or mechanical damage leading to secondary stressors. This may include severe drought, storm damage, defoliation, or rodent damage, which reduces tree health to the point opportunistic pathogens and other insects invade. This is typical of trees in the built environment and is not likely to create serious decline in more natural environments.
- Predisposing factors in the environment, especially the soil. Inadequate soil volumes, poor belowground conditions, such as compaction or improper pH, reducing the ability of trees to obtain the water and nutrients they need to grow optimally and withstand biotic confrontations.

- Inciting factors, such as acute and sudden events like seasonal drought or defoliation, rather than long-term, chronic conditions. They result in the first symptoms of decline and a reduced ability to withstand additional stressors.
- Contributing factors where several causal agents plus insects and pathogens can contribute to decline, once incited. These include further defoliation by insects or cold extremes, poor soil conditions, and a wide variety of secondary biotic pests.

Basically, when a tree or any plant experiences stress especially in the presence of another pathogen there will likely be some level of decline. It may not necessarily mean tree death, as not all trees die immediately from decline. Often, it is a slow, chronic condition requiring intervention to slow or stop the downward spiral.

Tree vitality can vary widely from between sites and may fluctuate widely on the same site. As well as the fact that individual trees vary in exposure and responses to stressing factors. Repetitive minor damage or stress may have cumulative effects. A change in quality or quantity of the basic needs of a tree can cause stress. Inadequate moisture, poor air quality, or light intensity will likely cause stress. Chronic stress conditions will lead to decline and the accumulation of stress results in a weakened response to its environment, good or bad.

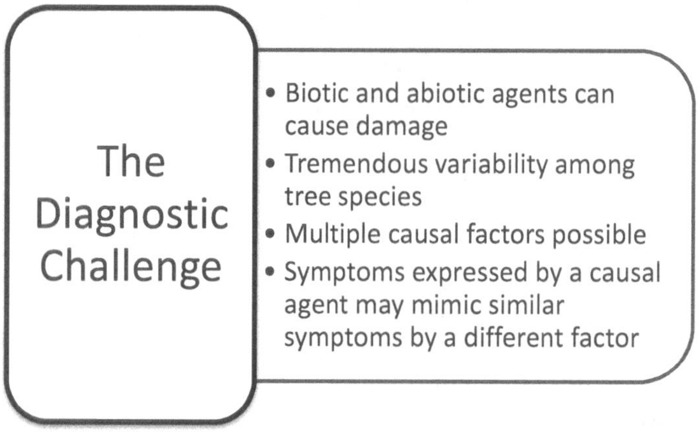

FIGURE 10.8 The diagnostic challenge.

SUSTAINABILITY AND SUCCESS

Diagnosing tree conditions includes identifying, analyzing, and evaluating conditions to determine causation. It can be challenging, yet it is an essential skill for yielding longer-lived trees in the urban forest. Diagnosis is essentially testing a theory postulation as to why the tree is in decline. It could be even more helpful to consider the application of differential diagnosis. A differential diagnosis is a method used to identify the cause of symptoms or a condition. It involves distinguishing a particular disease or condition from others presenting with similar characteristics. Basically, comparing the knowns to the unknowns relative to symptoms, signs, and conditions.

The process includes identifying the potential causal agents and seeking evidence through testing to confirm or refute the theory. The theory is generated through assessment of the tree, the environment in which the tree lives, and any relevant information from the tree owner or manager. When all information is collected, compare and contrast with other theories. Good detective work is required to gain essential details which allows for a more informed decision on the issue. All clues should be investigated to determine relevancy. Some clues will be helpful, and many will not apply in the process. Trees are dynamic living organisms and must be considered individually with each tree responding differently to stress. It is the compilation of the information that will ultimately lead to the most accurate diagnosis, which provides direction to the most appropriate mitigation strategy.

Decline and disease are most effectively managed through avoidance and management. In order to minimize stressful conditions selection, site preparation and proper installation are key to help the tree survive in the designed location, allowing it to reach maturity. Mitigation of stress factors, including cultural practices like irrigation during drought, and supplementation of nutrient-poor soil, will reduce the potential for diseases to develop and insects to attack. Plant health care professionals and tree owners and managers should focus on lessening the impact of predisposing stress factors. This basic strategy will help extend the life of a tree and reduce decline.

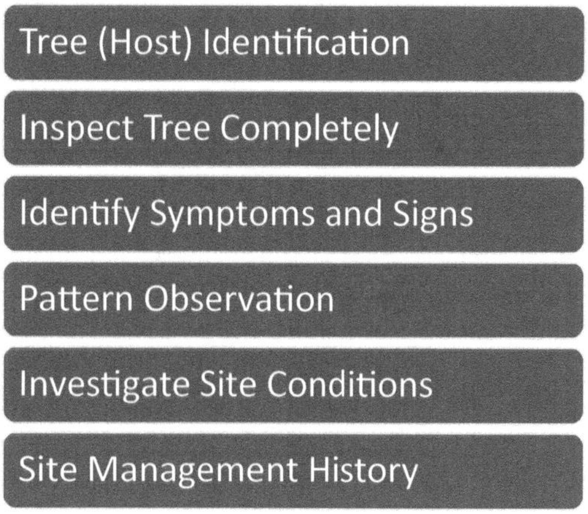

FIGURE 10.9 Tree diagnosis considerations.

INTEGRATED PEST MANAGEMENT

Integrated pest management (IPM) is the practice of using a variety of cultural, biological, and chemical controls to manage insects, weeds, and diseases. One goal of IPM is to reduce the harmful impact pesticides may have on beneficial insects, wildlife, or soil and water quality. IPM methods start with using the "right tree, right place" protocol following with biological pest controls, traps to monitor insect populations, scouting the landscape, and judiciously applying pesticides at the most vulnerable stage of a pest's life cycle.

Additionally, it involves the ability to recognize the threshold at which management strategies dictate the use of pesticides. This is often called the economic threshold. This level is the pest density at which management action should be taken to prevent an increasing pest population from reaching an unacceptable economic injury level. This is the level that leads to tree damage, lost production, and death.

METHODS OF MANAGING PESTS

Pest management methods fall into four basic categories: cultural, mechanical, biological, and chemical. Chemical controls will not be discussed in this text.

Cultural Control

Healthy plants free of stress can deter a wide variety of pest problems. Researchers find that plants under severe stress grow slower than normal, and insects prefer them over healthy plants. For example, trees damaged by a lawn mower are more susceptible to borer problems than uninjured trees. Cultural controls are essentially practices which discourage pest infestations and prevent problems.

This begins with tree selection in the event of a new tree installation. Inspect plant material when sourcing plants and before placing them in the landscape. New plant material can be a source of future problems. Some nurseries and garden centers may have unsanitary conditions or pest issues which can be taken to the landscape if not reviewed by the arborist or landscape specialist. Right tree,

FIGURE 10.10 Proper mulching and organic amendments improve tree health and may reduce pest issues.

right place means assigning plants in an appropriate location, considering sun exposure, soil drainage, and maintenance requirements. Be sure to install trees at the proper depth to prevent circling and stem girdling roots. This situation leads to stressful conditions making plants more vulnerable to pests. Excessive planting depth contributes largely to tree decline and failure. Fertilization and irrigation are maintenance considerations that can help or harm. Also, overwatering is just as harmful as underwatering, creating prolonged wetness which leads to disease. Always follow soil test recommendations for supplemental nutrition, using slow-release fertilizers prescribed for trees.

Current research and best practices advocate for the application of organic solutions which may prove beneficial when soil tests are possible or traditional nutrition recommendations aren't working as expected. Soil health improvement can be a better option and more efficient with resources than simply applying fertilizer. Utilizing complete organic fertilizer products and adding organic matter and amendments to urban soils determined to be very poor can lead to long-term improvements in soil health. This approach results in a more effective application and use of supplemental nutritional products.

An example of an organic amendment includes the use of biochar products. Biochar is a black carbon material proven to improve fertility and health. It is a soil amendment that can hold carbon and increase soil biodiversity. This natural product is very porous, with a high pH and carbon content. Biochar is mixed into the soil and acts to store nutrients, making them more readily available to plant roots. It also increases soil moisture as well. When mixed into soils, biochar can increase soil moisture and pH values. Research indicates it can better utilize nutrients in the soil due to its charged surface and high surface area (from its porous structure), allowing it to better adsorb nutrients like nitrogen and phosphorus. Research reveals when biochar is used in combination with compost, it can retain the nutrients provided by the compost and can help decrease the need for commercial fertilizers. It is important to source quality biochar products from a reputable company and make sure that the pH and formulation meet the requirements needed for the site or situation.

Current practice is moving towards a different perspective regarding the sustainability of trees and that management of soils in the urban forest. There isn't a "cookbook" recipe for tree pests or conditions, as evidenced by the need for appropriate soil and tissue testing. When it comes to enhancing soil and fertility, the approach should be more prescriptive. Often this begins with improving the organic content in the edaphic environment and diversifying soil management strategies. Arborists and horticulturists should recognize it takes more than just fertilizers to grow trees, but rather improving soil conditions so the soil can heal itself with better microorganism populations and biotic diversity.

Additional cultural considerations include mulching trees to an appropriate depth, typically around 2 to 3 inches, while staying away from the root collar. A good organic mulch will reduce the evaporation of water from the root zone and protect roots from temperature extremes. Another advantage of developed mulch rings is reducing the mechanical damage to the bark and cambium layer from lawn maintenance equipment. Prune plants using current standards and best practices to promote good branch structure and a dominant leader, applying good pruning cuts. Proper cuts facilitate fast, effective wound healing. When necessary, prune in the dormant season to prevent pathogen transfer, such as oak wilt and bacterial wilts.

Host plant resistance is a great way to reduce pests. Replace the more vulnerable, pest-prone plants with resistant plant varieties or species whenever possible. Finally, design for diversity, planning and selecting a wide range of genus and species where appropriate. Increased diversity with more genera in the planting scheme makes it more difficult for pests to spread between trees, adverting a disaster such as the emerald ash borer infestation on *Fraxinus* species or oak wilt on *Quercus* species.

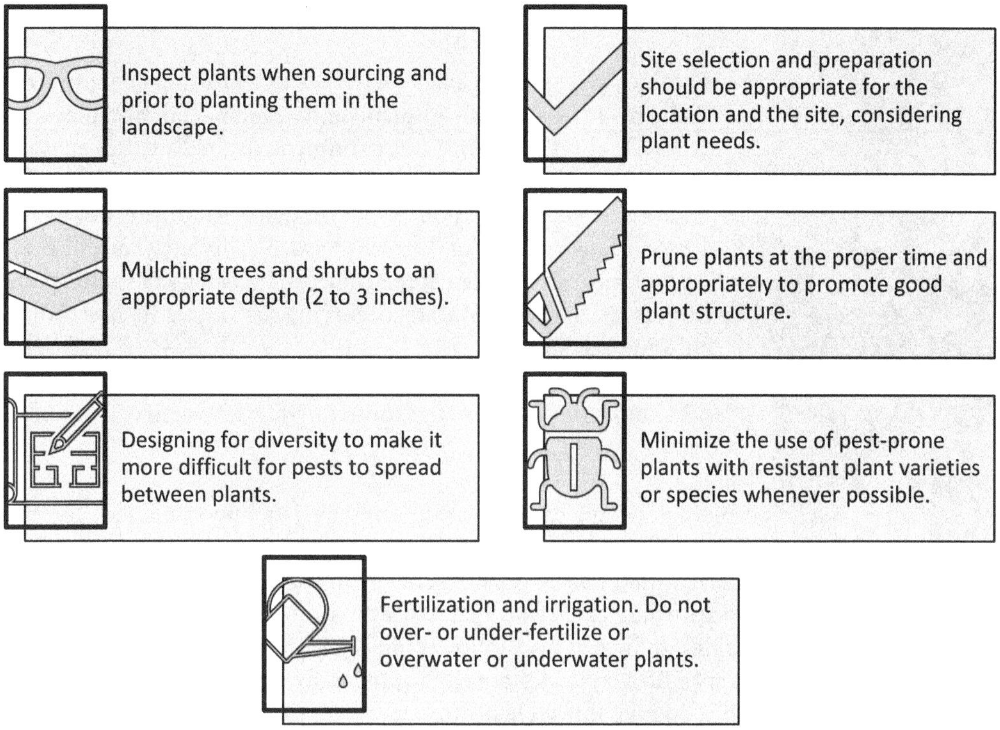

FIGURE 10.11 Cultural practices to improve sustainability.

Mechanical Control

IPM strategies include physically removing a pest and destroying it is an effective and economical strategy to manage many pest problems, when feasible. Pruning infested limbs such as tent caterpillars or removing bagworms by hand minimizes damage to vigorously growing plants. Also, this can eliminate the need for pesticide applications.

FIGURE 10.12 Physical removal of insects like bagworms on small trees is effective.

Biological Control

Biological controls are the use of living organisms to reduce or prevent plant damage as part of the IPM program. Some natural enemies feed on many pest species, while others are specialists, feeding on one or only a few species. IPM works under the realization that not all pests, especially insects and diseases, are harmful to plants; many are predators, parasites, or pathogens of pests themselves. Predators attack, kill, and eat insect pests. Parasites lay eggs in or on other insects. Pathogens are microscopic organisms, such as bacteria and fungi, that cause disease in the pest, populating the body and killing it.

There are also biorational controls for tree pests. Biorational controls are products that are biologically derived. They include non-pesticidal products that have multiple uses, including stress management, enhanced plant physiology benefits, and root growth management. They can also serve as an alternative control agent to traditional pesticides.

FIGURE 10.13 Beneficial insects are important biological controls.

The major categories of biorationals include:

- Botanicals: These are plant-derived materials and include pyrethrin, neem oil, garlic, capsaicin, and vegetable oil. Botanicals are generally short-lived in the environment, as they are broken down rapidly after application.
- Microbials: These are products that contain microorganisms, such as bacteria, fungi, viruses, or their by-products. These may include Bacillus thuringiensis (Bt) for natural control of Lepidoptera species pests.
- Minerals: These are naturally occurring substances such as sodium chloride and sulfur.
- Synthetics: These are structurally similar and functionally identical to biologically occurring controls. Bt, Trichoderma, and pheromones can be included in this group.

Biological and biorational controls are an important tool for pest control in an economically and ecologically responsible way. They have limited or no adverse effects on the environment or beneficial organisms. However, their efficacy is often much lower than pesticides. Informed decision-making is necessary for responsible control with nature in mind. Be sure IPM strategies are environmentally responsible, yet meet client needs in consideration of economic thresholds.

NUTRIENT DEFICIENCY

Many trees, shrubs, and other ornamental plantings suffer from nutrient deficiencies. Typically, nutrient deficiencies are caused by high pH (alkaline) soil. Soil pH affects plant growth directly and indirectly by affecting the availability of essential elements and microbial activity. Some of these elements include zinc (Zn), manganese (Mn), magnesium (Mg), and sometimes iron (Fe). These are essential elements required to produce chlorophyll and other important plant processes. Chlorophyll is necessary for photosynthesis and gives plants their green color. Iron (and manganese) deficiency results in leaf yellowing (chlorosis), and over time, it can lead to leaf scorch, dieback, and even the death of the tree.

FIGURE 10.14 Chlorosis is a symptom of many possible plant problems.

SYMPTOMS

Chlorosis is one of the earliest and most commonly observed symptoms of a deficiency, and it typically appears on the current years' growth first. Leaves reveal a network of darker green veins that are easily seen due to the missing green color in the rest of the leaf, leaving it light-green or yellow. If the deficiency is severe, the entire leaf may become cream-colored, while the tips and margins of leaves will turn brown.

It is important to know that not all trees exhibit similar signs of chlorosis. In advanced stages, there may be a situation where the leaves die and drop very suddenly. As an example, red oaks may reveal signs of chlorosis in their own distinctive way, such as with interveinal browning without any traditional yellowing, while still having dark green tissue wherever there isn't brown discoloration. Eventually, left unmanaged, branches may begin to die back when leaves are no longer capable of producing food by photosynthesis. Trees and plants severely affected will appear stunted, have poor root development, and may eventually die.

This condition is especially a problem with trees planted anywhere with diagnosed poor soil conditions. Typical urban soils are found as subsoils, often lacking suitable topsoil with good texture and health. The lime content, consisting of calcium carbonate, which raises the soil pH, is high in many of these subsoils. The pH range can be over 8 and as much as 10 on the scale. The application of hard water during irrigation and supplemental watering provides additional calcium content, intensifying the problem. In many regions in the U.S., pin oak, holly, river birch, red maple, sweet gum, tupelo, white oak, and white pine frequently present chlorotic symptoms.

THE CHLOROTIC CAUSE

Chlorosis is a problem resulting from multiple reasons, including (but not limited to) host genetics, site pH, compaction, competition, soil quality, and maintenance practices. Affected trees presenting chlorotic symptoms are usually deficient of one or more micronutrients, most notably iron, manganese, and occasionally magnesium. It is important to consider that some soils may be super-adequate in these elements, creating a toxic effect. The yellowing symptoms are expressing not because the elements are not present in the right quantities in the soil but rather because they may be insoluble and unavailable to the plant due to high-pH values in the soil. At these higher soil pH levels (7 and above) many trees and shrubs are incapable of absorbing adequate amounts of iron or manganese. Though iron deficiency is the most

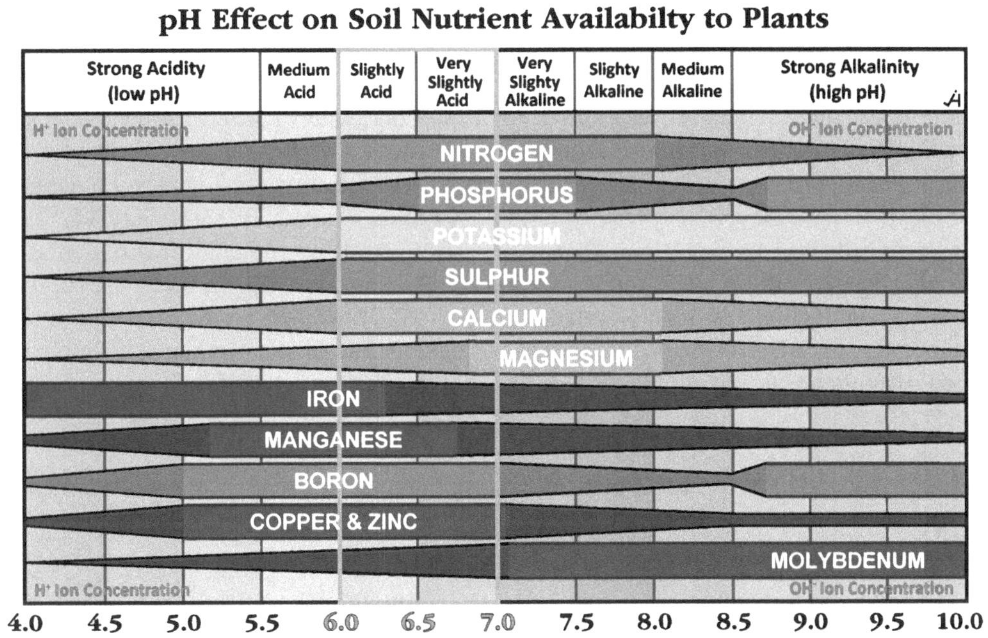

FIGURE 10.15 The recommended pH value for most trees is slightly acidic.

common suspect, manganese, magnesium, and zinc deficiencies occur equally as often and mimic the symptoms of iron deficiency. Most maple trees, especially red maples, are especially sensitive to manganese deficiency. Basically, any deficiency can create a chlorotic response. Visual identification of elemental deficiencies is unreliable. The cause for chlorosis can be virtually anything. Deficiency keys are never accurate due to the variability of symptoms. Elemental tuning to reduce deficiencies is based on proper testing, specifically tissue tests. Additionally, macro-elements such as nitrogen, phosphorus, and potassium, can drive changes in availability of other elements. Research and testing have demonstrated excesses in the major elements create deficiencies in the minor elements.

OTHER CAUSES OF CHLOROSIS

Elemental deficiency is not the only cause of leaf yellowing. Herbicide damage and other toxins can create chlorotic symptoms. Basically, visual identification of elemental deficiency is not a clear diagnostic tool. A tissue analysis will provide the best results for determining elemental deficiencies. In most cases, a foliar application of the deficient mineral early in the summer will cause a temporary greening of the leaves. If application of the mineral thought to be deficient does not cause a temporary correction of the chlorosis, some other mineral deficiency is causing the problem. Other possible causes should be investigated, including root dysfunction and damage.

MANAGING CHLOROSIS IN THE LANDSCAPE

Advanced stages of chlorosis resulting in the dieback and decline of mature trees is rarely reversible. Before treatment, a more maintenance-minded consideration is that as trees grow, they become more costly. If rescue treatments are not effective or feasible, removal is often most economical solution and replant with a suitable species tolerant to the existing conditions of the soil. An effective way to minimize pH concerns is to test the soil for pH values and to select species that can do well in existing conditions without treatment. For example, northern red oak is very similar to pin oak but tends to be less prone to chlorosis in soil with pH values that limit particular elements, such as manganese and magnesium. The following treatments are useful for cases of moderate imbalances, and depending on the treatment selected, may need to be repeated annually or biannually. For those highly valued trees, four commonly used methods to treat for iron chlorosis in plants are: (1) altering soil acidity; (2) applying appropriate micronutrients directly to the soil; (3) spraying the foliage with a solution of iron (ferrous) sulfate or iron chelate; or (4) introducing micronutrients into the main stem or trunk of affected plants by implantation, infusion, or injection.

SOIL ACIDITY MODIFICATION

Sulfur

Before attempting to modify the soil pH, determine existing soil pH with a soil test. A tissue test may be necessary as well for fine tuning the method to implement. Soil acidification with elemental sulfur is practical for annuals and perennials, along with shrubs. However, this method for large scale treatment in the root zones of trees is highly impractical due to the quantity required and large spaces to treat. If turf is present, it is nearly impossible. A topical application of powdered sulfur will severely damage turf. However, pelletized forms of sulfur can be applied with minimal damage. Moderation in urban landscape soils require between 35 and 56 lbs. sulfur/1000 sq. ft. to lower a current soil pH by 0.5. It is advised not use more than 20 lbs. of sulfur per 1000 sq. ft. at any single application. If greater modification is required, use split applications of 20 lbs. and apply in succeeding years. Always check the soil pH before making subsequent applications to determine the level of change. The amount of material required per unit area will depend upon soil texture, pH, calcium levels, and the presence of other bases in the soil. Topdressing sulfur is not as effective as incorporating sulfur into the soil. Tree response will take time and is dependent upon soil moisture, timing of application, and tree age. Several applications may be necessary over several years to effectively modify soil pH. Sulfur treatments will increase the availability of nearly all micronutrients, not just iron, and may provide additional benefits to plants by freeing up additionally unavailable nutrients.

The chemical reaction between soil and the addition of elemental sulfur produces hydrogen causing soil pH to decrease. Soil pH is, after all, a measure of the hydrogen ion concentration in soil solution, and the higher the concentration of hydrogen, the lower the soil pH values in the soil. Soil temperatures should be above 55°F when treating the soil and avoid treating saturated or flooded soil. This prevents soil microbes from converting sulfur to hydrogen sulfide (rotten egg smell) under flooded (anaerobic) conditions. At extremely high pH, elemental sulfur has limited impact in acidifying high pH (calcareous) soils because the sulfur cannot be converted to sulfuric acid by soil microbes.

TABLE 10.2 Sulfur Requirements for pH Modification

Desired Change in pH	Application Rate Based on Soil Texture		
	Sand	Silt loam	Clay
8.5 to 6.5	8.5	16.8	33.5
8.0 to 6.5	7.8	15.4	30.8
7.5 to 6.5	6.9	13.8	27.5
7.0 to 6.5	4.1	8.3	16.5
8.5 to 5.5	19.1	38.1	76.0
8.0 to 5.5	18.4	36.7	73.2
7.5 to 5.5	17.4	35.1	70.0
7.0 to 5.5	14.7	29.6	59.2

Note. These numbers assume the cation exchange capacity (CEC) of the sandy loam, silt loam, and clay soil are 5, 10, and 20 meq/100 g (which assumes soils are not calcareous). These numbers serve only as estimates and starting points to begin the acidification process. Pounds of elemental sulfur required per 1000 square feet to decrease soil pH to a depth of 6 inches.

Iron Chelate and Sulfate

Where it is not feasible or possible to affect pH by soil modifications, iron chelate and iron sulfate can be applied to the soil as a liquid solution to be taken up by the tree. These materials are less affected by soil pH, and work within a pH range of 4–10. The application requires mixing and applying the chelate into the top 1–3 inches of soil around the base of the tree or shrub and watering it into the upper soil level. It has been noted by many practitioners that iron sulfate products reduce pH much more effectively and rapidly.

FIGURE 10.16 Deep root soil injections can be applied to improve tree health.

Foliar Sprays

Symptoms may also be corrected by spraying foliage with an iron chelate or ferrous sulfate solution. Foliar sprays result in a quicker, but temporary, response. The application is short-lived and several sprays may be required throughout the growing season to maintain a desired foliage color. The application of sprays can reveal a greening response sometimes in hours, but it is often gone within a few weeks, with the treated plant returning to its chlorotic appearance.

Follow the manufacturer's directions for the iron chelate formulation used.

FIGURE 10.17 Chlorotic symptoms can be temporarily relieved with foliar sprays.

Trunk Injections

When trees are growing in well-buffered alkaline soils (with a pH above 7), trunk injection offers one of the most effective treatments available to correct chlorosis in trees with elemental deficiencies. All these treatments require invasive drilling into the tree to administer the treatment.

There are various microinjection systems for use on tree species and palms growing in alkaline soils.

Prescriptive treatments can be made to correct specific or broad elemental deficiencies in trees. Treatment rate is determined by the diameter at breast height (DBH) as well as the size of the tree and the overall crown. After drilling, a pressurized capsule (microinjector) introduces the liquid remedy into the tree. The microinfusion devices available include low pressure hydraulic systems for moving the nutrient solution into the tree's vascular system. These systems are very effective devices for improving symptoms presented from nutrient deficiencies.

FIGURE 10.18 Tree injections can administer remedies for nutrient deficiencies and pest control.

LIEBIG'S LAW OF MINIMUMS AND NUTRIENT DEFICIENCY

Healthy tree growth is regulated by its greatest need or the element in the shortest supply, regardless of the abundance of other nutrients. Some soils may have the necessary essential elements and availability, but enrichment may be necessary to overcome undersupplied soils or enhance growth.

Nutrient availability or, more accurately, essential elements and their abundance can be demonstrated by applying Justus von Liebig's Law of the Minimum. This law states yield is proportional to the amount of the most limiting element, whichever element it may be. From this, it may be inferred if the deficient nutrient is supplied, yields may be improved to the point that some other nutrient may be needed in greater quantity than the soil can provide, and the Law of the Minimum would apply in turn to that element. This translates to trees in the landscape and urban forest as well. Yield would include foliage in direct comparisons. Also, health, vitality, growth, flowering, and other characteristics are dependent on the availability and abundance of all nutrients and elements required for all plant processes. The soil can be completely balanced for the elemental requirements of the tree; however, if one is limited, health is affected with some response, typically beginning with chlorosis. For example, in the image below, many of the major elements are adequate; however, cobalt is the limiting element and growth can be improved if the soil is supplemented.

It is not clear exactly how many elements can be found in a tree. Trees are made up of many elements including hydrogen, oxygen,

FIGURE 10.19 Liebig's barrel.

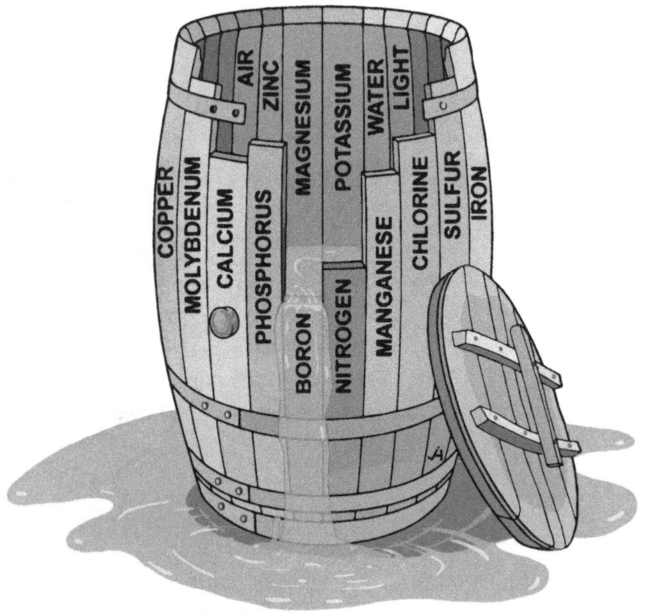

FIGURE 10.20 Plant health is proportional to the amount of the most limiting essential nutrient.

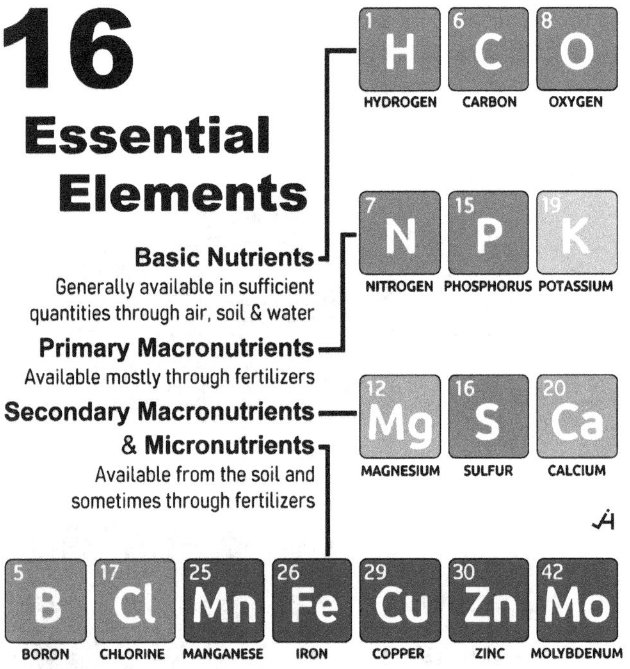

FIGURE 10.21 All elements are essential and necessary in adequate quantities for healthy trees.

carbon, nitrogen, phosphorus, potassium, calcium, magnesium, and sulfur, and micro nutrients such as zinc, molybdenum, copper, manganese, cobalt, iron, and boron. Essential elements are considered as required for normal tree growth and reproduction and these functions cannot be completely substituted for by another element. An element can be considered essential if without it the tree dies, shows severe deficiency symptoms restricting growth, survival, and defense or presents a stunted or abnormal appearance. This may be especially true in the harsh urban environment or a subdivision where soils have been displaced or replaced.

As mentioned earlier, various factors affect nutrient availability, including soil type, pH, CEC, and WHC. So, it's not just deficiencies that need to be considered, but also availability. The tree care professional is required to understand soil properties for a better diagnosis and directed enrichment.

Soil analysis is essential in formulating and improving a fertilization program. The analysis reports key information on organic matter content, pH, and extractable nutrients. Soil analysis is particularly useful when conducted for several consecutive years because trends can be observed. However, plant health care professionals should not rely on soil analysis alone to formulate a fertilizer program or to diagnose a nutritional issue.

Quantifying values of soil nutrients eliminates guesswork when adjusting a fertilizer program. Elemental deficiency or surplus causes trees to grow poorly or affects them negatively in other ways. Soil testing and leaf tissue analysis do not assess all the same factors, so care must be taken to choose the correct test when diagnosing nutritional concerns. If feasible, both testing methods should be used for prescriptive nutrition programs.

Leaf tissue analysis is a very useful tool to detect problems more specifically and adjust fertilizer programs because leaf nutrient concentrations are the most accurate indicator of nutritional status. Considerable research involving leaf testing has established its reliability as a management tool, but sampling guidelines should be followed precisely to ensure that analytical results are meaningful.

Leaf tissue analysis tests many of the factors that influence nutrient availability and uptake. Tissue analysis shows the relationship of nutrients to each other as well. For example, potassium (K) deficiency may stem from a simple lack of K in the soil or it can result from excessive calcium (Ca), magnesium (Mg), and/or sodium (Na). Similarly, adding nitrogen (N) when potassium is low may result in further potassium deficiency since the increased growth caused by nitrogen requires more potassium. There are several nutrient interactions when it comes to supplementing nutrition for the tree. This could be considered similar to drug interactions with people and their prescriptions. Effective PHC requires knowledge of these interactions to prevent complications.

EXCESS N: < K and Cu
P: < Fe and Zn
K: < Ca, Mg and B

Soil Tests: ***nutrient availability or toxicity.***
Tissue Tests: ***nutrient or elemental deficiencies.***

Soil tests are quite useful for pH, organic matter content, and other qualities. Relative to nutrition, soil tests primarily measure extractable elements present in the sample. However, it doesn't measure the total amount of nutrients present nor the quantity available to the tree. Comparatively, leaf analysis is a useful tool to detect problems because leaf nutrient concentrations are the most accurate indicator of crop nutritional status. Leaves reflect nutrient accumulation and redistribution throughout the plant, so the deficiency or excess of an element in the soil is often reflected in the leaf.

Finally, a sampling or testing program can be most effective if conducted annually, especially on problem plantings. Leaf tissue testing is valuable for all elements. Soil testing is most useful for consideration of pH and nutrient values. Interpretation of leaf and soil tests should be used to make fertilizer or supplement decisions. A broad-spectrum or "shotgun" approach is never accurate, less effective, and often not economical either. Informed use of the test results allows for optimal health and growth, minimizes environmental impacts, and reduces fertilizer waste with inaccurate applications.

DIAGNOSTIC TOOLS

Plant health care (PHC) is the science and practice of understanding and overcoming the impacts of biotic and abiotic factors limiting plants from achieving their full genetic potential in our landscapes and urban forest. Plant health care has been practiced as long as modern arboriculture itself, and as a science-based concept is an important component in overall integrated pest management (IPM).

Advanced PHC tactics means recognizing the technology and science-based tools available to improve the IPM program. These can include analyzing common host-pest interactions and applying concepts like phenology and growing degree days (GDD) in the diagnostic process. This provides better informed decisions on questions such as, "When is the optimal time to apply insecticides for control of bagworms or emerald ash borer?" "When should we expect an outbreak of scale insects on shade trees at the commercial park?" and "When is the best time to control mites downtown?" These are common questions that can be perplexing for PHC professionals at times. A good way to start the diagnostic process is by recognizing plant and pest development thresholds. This helps deliver a more effective and timely approach to managing pests, when they are most vulnerable to treatments.

GROWING DEGREE DAYS

Pests and plants alike are dependent on specific temperatures for development and growth. For example, some insects can emerge earlier in warmer years than cooler years. Also, plant development and bloom times may be seen sooner with those "early spring" seasons. Monitoring plant events, such as bloom times, can be employed as a biological calendar to track growing degree-day (GDD) accumulation and predict pest activity accurately. GGDs, while not perfect, are a reliable method for predicting plant and insect development. This approach is more reliable than relying on a calendar for outbreaks and allows for targeting the most vulnerable stage of the insect's life. An important benefit is allowing for better control with reduced amounts of active ingredients in pesticides.

PHENOLOGY

Phenology is the study of periodic events in biological life cycles and how they are influenced by seasonal variations in temperatures. It is the study of naturally recurring events in plant and insect life cycles. These include examples such as bud expansion and bloom times or when scale crawlers appear on trees. Regardless of the time of year, bloom sequence is reliably consistent with pest emergence. Growing degree day (GDD) is a tool in phenology that measures heat accumulation to estimate the growth stages of plants and life stages of insect and insect-like pests. GDD tracking and calculations combined are effective tools but require diligence in collecting and assembling data and information. There are at least a couple of ways to determine this temperature-based function; however, there are tools available that compute this information and provide the necessary daily metrics. Phenological episode tracking is tedious and time-consuming but quite effective in combination with other tools for scheduling applications.

The information found in phenological models originate from published lists usually from nearby land-grant universities or pesticide manufacturers and distributors. Typically, they are verified by pathologists and entomologists for accuracy. PHC technicians can assemble their own pest calendar model to include as part of their exclusive IPM resources for even greater local or regional accuracy. However, confidence in any model should be field tested. This simply involves comparing the predicted generation time (degree-days) for an insect using a phenology model to the actual degree-days for a generation using a sampling method such as a pheromone trap.

TABLE 10.3 Phenology and GDD Example

Pest Name	Plant Event	GDDs	Vulnerable Stage	Average Date
Bagworm	Catalpa Full Bloom	600–900	Larvae	Late May–Early June
Emerald Ash Borer	Black Locust Full Bloom	780–1040	Adult	Early May–Late July
Eastern Tent Caterpillar	Forsythia First Bloom	90–350	Larvae	Early April–Mid May
Japanese Beetle	Little Leaf Linden Full Bloom	950–2000	Adult	Early June–Late July

Now, once the GDD value has been determined, pest stage of development and plant phenology can be used to determine pest appearance and vulnerability for best pesticide application times. Utilizing GDD in conjunction with indicator plants from phenology charts allows for accurate applications. These examples are from North-Central Indiana and will vary in other areas.

PHEROMONES

Insect pheromones are neurotransmitters that work as a chemical communicator between similar insect species. They are produced in special glands and released as a wind-borne signal for a variety of needs, including finding a sexual partner.

Pheromone traps are a type of insect trap using pheromone analogs to lure insects. The traps usually consist of a small glue trap impregnated with the specific sex pheromone. These hormone scents, usually emitted by the female insect, are recognized by the male as a cue for mating. Male pests are drawn to the trap for the purpose of mating and are captured. These traps assist with determining the mating activity of the insect pest by counting insects trapped over a specific period. This information enables better decision-making regarding the need for control measure based on population dynamics.

Scouting is a strategic tool for monitoring trees as the growing season progresses. This tool relies heavily on phenological events and GDD to be timely. Pheromone traps are helpful for determining pest populations in a region and are an excellent scouting tool. The PHC technician is searching for outbreaks associated with tree and pest development. Scouting may begin too early or too late if using just the calendar itself, resulting in wasted time and overlooking damage or injury on trees or missing the most effective control windows for the pest. GDD and the application of phenology signals when to start looking for pest problems and ideal times to plan applications.

DISEASE PREDICTION

So far, the focus has been on insect pests, which can be perceived as a bit less challenging than identifying and controlling disease dilemmas. A major complication with diseases is often the pathogen has arrived and infected the plant long before any symptoms are expressed by the tree or shrub. The damage and the disease are already in progress and taking a firm hold on its host. It's important to recognize as well that some diseases and insects encountered in trees are merely an aesthetic concern, and are not always lethal to the plant and may not warrant control measures.

Somewhat similar to insect development, timing is an important factor in when to expect certain diseases relative to spring and summer infectious agents. Some diseases are considered cool-weather problems and can be ruled out in the warmer summer months, such as *Anthracnose*. This is a common type of leaf spot and canker disease affecting many trees, including maple, sycamore, birch, elm, and dogwood. Symptoms are especially severe in years with cool, wet spring weather. Whereas bacterial leaf scorch, a serious vascular disease in oaks, for example, is typically a summer weather disease concern.

Recognizing how these science-based tools work and assist us in our quest to control pests is important; however, uncovering the impact of control products such as fungicides, insecticides, fertilizer, biorationals, organics, and the list goes on, well, that is another consideration.

Specifically, determining how such products impact a plant's ability to survive and thrive is a key area of knowledge necessary in plant health care. The stronger and healthier the tree, the better its ability to withstand increasing pest pressures to grow and survive.

If a tree is stressed by diseases, insects, or the environment, its energy is diverted from thriving to surviving. Understanding the environment has a major significance. Sunlight. Air. Water. Soil. Nutrition. These fundamental components are brought together through a complex web of plant processes to create a sustainable, healthy tree with an effective defense system.

When tree defenses are compromised by drought stress or other issues, the disease or insect can gain an advantage within the tree, resulting in serious damage which can spiral out of control to end up in death for the tree. Often it is a matter of a compromised immune system which can predispose the tree to pest issues. Some of those predisposing and inciting factors can be:

- environmental and site stressors, such as drought, planting trees too closely together, and mechanical damage to the stem or root system;
- trunk damage, which invites infection by creating points of entry for pathogens, causing disease and decay;
- root damage, which creates points of entry for pathogens and reduces the tree's ability to transport water, increasing stress and susceptibility to pests; and
- drought, whether resulting from a lack of rainfall or root damage, which reduces the tree's ability to procure and process resources needed to compartmentalize infections and prevent spread through the tree.

Good cultural practices are the cornerstone of an integrated pest management approach. The PHC technician should be capable of identifying, analyzing, and evaluating trees in the urban forest for biotic and abiotic issues. Probably the greatest collection of success stories for plant health care is the number of insect and diseases managed by good cultural practices, aboveground and belowground with minimal impact to our environment.

TREE SUPPORT AND LIGHTNING PROTECTION

CHAPTER 11

LEARNING OBJECTIVES

- Identify when trees support is necessary and the protocols for newly established trees.
- Describe tree support systems on mature trees how to determine which trees are candidates and the appropriate strategies.
- Explain the different ways trees can be supported with cables, braces, including the benefits and limitations of cabling and bracing systems.
- Recognize tree characteristics affecting the support strategy decision.
- Recognize situations where lightning protection may be beneficial for trees.
- Identify the basic components of a lightning protection system and the installation steps.
- Explain the maintenance protocols for lightning protection systems.

Tree support systems are often necessary for various reasons such as risk mitigation and stabilization on exposed sites and recently installed trees with limited root systems. The practice of adding support systems to newly installed trees should be determined based on nursery stock type, tree size, and site conditions. Many transplanted trees establish themselves in a stable, vertical position, under normal situations. However, on exposed sites or vulnerable locations, support is required. Typically, support systems, such as staking and guying are unnecessary and can be detrimental to tree development. In fact, movement caused by the wind is crucial in helping saplings and young, newly established trees develop into strong, structurally balanced trees. When necessary, a support system must be installed properly and removed at the appropriate time to prevent damage.

WHEN TO STAKE TREES

When supports are needed on newly planted trees, the decision of when to stake trees can be made based upon site conditions and the nursery production method of the tree. Bare-root trees and container-grown trees with lightweight root balls often require support until lateral roots develop outside the planting pit to better anchor the tree. Another factor to consider is crown size, as a larger crown represents more leafy mass for the wind to move,

FIGURE 11.1 Newly established trees, especially evergreens are subject to wind throw if not supported properly.

FIGURE 11.2 Tree staking and guying provides support and protection from mower damage as well.

transferring increased stress to the stem and root system. Large evergreen trees have a high wind resistance (drag) and often need to be supported to prevent tipping under heavy wind loads. Open sites exposed to strong winds, such as new housing development sites, sites with sandy soils, or elevated, berm plantings, present conditions where support of newly planted trees is necessary.

Taller trees with undersized root balls may need to be staked as well. Without support, trees will become tilted and movement of the root mass within the planting hole damages the fine, absorbing roots trying to establish themselves into the new soil. Also, trees in areas with high rates of vandalism should be considered candidates for support systems. Staking or guying has helped deter theft and damage in many instances.

An additional advantage of tree supports is that the staking materials provide physical barriers against mechanical damage of tree trunks. This is particularly true regarding damage from lawn mowers and other lawn maintenance equipment. The support hardware acts as a visual cue to stay away from the tree because equipment may be damaged. Leaving the stakes as tree guards after the supporting guys are removed from a tree may be useful if they don't pose a hazard. Staking and guying trees in areas of heavy public use such as city parks and playgrounds should include mulch rings outside the support system area and high-visibility tape as an extra precaution to prevent tripping or other accidents.

There can be distinct disadvantages to support systems. Improperly staked trees suffer from poor development, such as poor trunk taper and smaller root systems, and may be unable to stay upright after supports are removed. A tree supported by stakes or guys too long will often grow taller than an unstaked tree, but the staked tree has a smaller stem diameter near the ground, producing a trunk with little or no taper. This tall, thinner tree is more likely to fail after the support materials are removed. This may also be the case when tree supports have been left in place for too long. A tree free to move in the wind has increased taper because of the increased growth at its base (response growth) and is more capable of remaining upright in the wind.

Often trunk tissue suffers cambium damage from rubbing on poorly installed hardware and may even be girdled by support equipment. Also, due to poor development and taper, previously supported trunks are more likely to break off in high winds or blow over after stakes are removed. Support wires left too long can girdle the tree, either killing the tree or making it more prone to snapping in the wind. Stakes left in the ground create a hazard and damage tree structure as well.

HOW WIND AFFECTS TREES

FIGURE 11.3 staking and guying equipment left on too long will girdle the tree.

Most of the time we don't think too much about the effect of wind on trees until it rages, littering the ground with broken branches or uprooting trees. However, wind, under normal conditions, is necessary for strong tree development. Gentle breezes have profound influences on trees, and stronger prevailing winds affect the way trees develop their strength and shape. Wind sway creates a biomechanical reaction within the tree, stimulating added growth of wood fibers in the lower trunk. Internal responses in the tree create response growth in form of flexure wood. Ethylene and other hormones collectively affect tree growth in response to the movement caused by wind. A simple explanation of a tree response to swaying in the wind is the bending stress increases the production of the hormone ethylene, which increases growth in the stressed region of the stem. Research has shown when branches bend, the production of ethylene increases markedly from 30% to 300%, causing increased diameter in the flexing region. Virtually and bending stress stimulates localized ethylene production and is associated with the formation of response wood (wood that forms in trees as a reaction to external stimuli). The mechanical bending and pulling of the branches, main stem, and roots affect their growth in both length and diameter. Movement by wind causes shortened, stronger stems, increased trunk diameter, and enhanced root development. The result is a better balance between the crown, trunk, and root system. This increase of response wood helps resist failure in unfavorable weather conditions.

PROPER METHODS AND MATERIALS FOR GUYING AND STAKING

Staking and guying a tree to keep it upright is often necessary as a temporary support system. However, it does not compensate for poor root development and long-term establishment. Refer to the current standards and ISA best management practices included in the references for details on tree support installation.

Trees that are guyed with three points of attachment should have these attachments placed at approximately two-thirds the height of the tree, using flexible webbing materials.

Staking allows for support using upright poles or stakes with flexible attachments. This method connects the trunk to a nearby steel or wooden post and is a common approach for smaller trees and containerized tree stock. These posts are usually from 2 to 6 feet long and are driven vertically into the soil near the trunk. Typically, two to three stakes with separate flexible ties are used for support. Installing three stakes provides better support. The materials used, the thickness of the stake, and the depth driven into solid ground contribute to the level of support provided. Lower staking can keep the newly installed tree sturdy and stable in the ground while allowing for trunk and crown movement. This is better for improving taper and development. Proper tension is important in staking; allow some movement for trunk and root development. If a single post is used, it should be placed on the windward side of the tree.

FIGURE 11.4 Leftover stakes embedded in a tree.

Guying is another mechanical method of tree support for trees of various sizes. It is typically used on larger trees that are transplanted balled-and-burlapped and containerized trees. Larger bare-root transplanted trees require guying equipment as well. Three points of attachment provide the best support for equipment installations. This is done by selecting wide, smooth, flexible tie-materials, such as soft webbing weave, that can help dissipate pressure exerted on the tree at the contact point. This material minimizes abrasion and damage to the cambial tissue just beneath the bark on the trunk and stems. The material should resist breaking and allow the tree trunk to flex or move slightly in wind. When guying, attach the guy-wires on the lower half of the trunk, less than two-thirds of the height of the tree to reduce the chance of the tree breaking at the attachment point. Proper tension is important for support as well, but some movement should be allowed to stimulate trunk taper and root anchorage. Excessive tension in the guying material can cause the trunk to break where it attaches as the tree bends in the wind. Poor tension provides no support for the tree. To prevent injury to the bark, the ties should be examined at least once during the growing season and adjusted as necessary.

Prior to current standards, the process involved attaching a wire to an anchor and slipping it through hose material around the tree. However, the hose material and wire cause damage from abrasion and compression of the cambium at the attachment point. This compression and physical damage can compromise long-term tree health. Advanced stages of compression progress to girdling the stem, which is usually lethal for the tree.

FIGURE 11.5 Guying trees provides greater support in exposed conditions.

FIGURE 11.6 Close up of guying materials attached to a tree.

FIGURE 11.7 Wire and hose systems are not recommended or considered best practices for support.

FIGURE 11.8 Tree death can occur if equipment is left on too long.

Wood or metal stakes should be driven into existing soil for the best anchorage. Be sure to get well past any backfill material encountered and into solid ground. There are specialized, ground anchors available to attach the guywires and provide easier installation. These small, metal augers can be reused and easily adjusted for depth. Space stakes or anchor points evenly around the tree in the ground so there is equal space between each anchor point. Guys should be attached between the anchor point and the attachment point at a 45-degree angle. Where possible, make allowances to place all support equipment inside the mulch ring. This prevents damage from lawn equipment and helps eliminate trip hazards. Mulching out to the anchor points is a recommended practice.

Belowground stabilizing systems are effective and economical for stabilizing the root balls on larger balled-and-burlapped trees. This is especially helpful to eliminate aboveground hardware that can be a hazard to people while keeping visible equipment to a minimum. There are several commercial anchor systems available. However, a simple untreated wood frame held in place with wooden stakes can provide an aesthetic option which degrades naturally over

FIGURE 11.9 Belowground strap kits can provide support where aboveground equipment is unacceptable.

FIGURE 11.10 Proper staking and guying should be necessary only for the first year of establishment.

time. Vertical stakes are attached to the ends of horizontal boards over the ball. Stakes should be driven up to 4 feet deep, or as deep as possible in the existing soil to keep them from getting loose or pulling out. The system can be easily covered with mulch after installation.

Tree support systems, such as staking and guying, should only be considered when conditions and exposure to the wind make it necessary to keep a tree upright. Improperly installed support systems cause stem abrasions and trunk girdling. Review the anchor, attachment point, and tension on a regular basis, adjusting as needed to make certain the supports are effective and not damaging the tree. If a tree is supported, the ties and guys should be removed as soon as feasible, usually no later than after one growing season or one year. Trees prevented from moving for longer periods usually grow taller than trees free to move in the wind, but they grow less in stem diameter, have a smaller root system, and are often unable to support themselves or break easily in the wind after the supports are removed. Staking trees too high, too tightly, and for too long are the most common causes of tree damage. Consult industry best management practices listed in references for details on installation and inspections.

FIGURE 11.11 Improper tree staking.

Staking trees too high, too tightly, and for too long are the most common causes of tree damage.

LARGE TREE CABLING AND BRACING

Trees can grow in ways that aren't always the most stable or sustainable. This happens for several reasons such as species characteristics, location, or lack of proper pruning in early life stages. Cabling, bracing, guying, or installing props may help prevent branch or whole-tree failure. Tree support systems act to reinforce weaker locations of the tree by limiting the movement of their branches and leaders. Properly selected and installed equipment can reduce the likelihood of failure, which also works to reduce consequences of injury to humans and damage to property by providing supplemental support for structurally weak areas of the tree. However, it is important to understand that not all risk can be mitigated by tree support installations. Also, not all trees can be reinforced with supplemental support. Close examination of the tree and its parts are necessary to determine a tree's eligibility for a support system and will be discussed.

CODOMINANT STEMS AND OVEREXTENDED BRANCHES

FIGURE 11.12 Cabling systems are mitigation options for trees to reduce likelihood of failure.

The most common risk of tree failure is the presence of codominant stems. Codominant stems and "V-crotch" branch unions are structurally weak compared to single stems or those with proper branch aspect ratios and spacing. The strength loss is due primarily to the lack of connective tissue anchoring the stem to the tree trunk. The presence of included bark between the stems creates even more strength loss in the union. The longer or taller the codominant stems, the greater the likelihood of failure, especially on larger trees. The best solution for problems associated with codominant stems is to avoid them altogether by selecting plants with good branch structure. For a tree with poor branch structure, an alternative is to remove or reduce one of the codominant stems as early as possible in its growth, allowing for the development of a dominant leader. For older, mature trees with this weak structure, a support system installation may be the only option for strengthening the weak area of the tree in efforts to retain it.

Overextension can also create potential problems at the position of the branch attachment. This condition is the presence of unusually long, heavy, or far-outstretched limbs. The condition can be more problematic where most foliage mass is concentrated toward the end of the branch. Overextended limbs often break at the connection point of the branch and stem. Alternatively, a branch may crack due to tension and compression forces near the middle of the beam (branch). These failures usually occur when a branch is under a heavy load of wind, snow, or ice. Installation of cables or propping may be used to avoid making large removal cuts; however, early corrective pruning is the best course of action to prevent this condition. Long, extended branches are often a natural growth habit of certain tree species, such as *Quercus virginiana* and capable of managing the loading forces through creation of response wood. A key component to consider in branch formation is adequate taper to support the branch and the branch attachment point.

FIGURE 11.13 Codominant stems are often candidates for tree support systems.

FIGURE 11.14 Overextended branches may require cabling to reduce loading on the branch attachment.

TREE SUPPORT DEVICES FOR ESTABLISHED TREES

Cables and brace rods are supplemental structural supports intended to reduce the risk of failure of weak stems, branch unions, and multiple leaders. A static cabling system consists of flexible, extra-high-strength steel strand cables attached to bolts installed in the upper crown of a tree by drilling through the wood. Cabling may reduce failure from high wind or the weight of ice or snow. The system transfers weight from a weak branch union to a stronger one, prevents breakage, and restricts the distance a branch can move in relation to the rest of the tree. Cables may be installed on overextended branches for additional support. Another alternative is dynamic or soft cabling systems. This involves the use of synthetic ropes attached to the tree without the need for invasive drilling. These systems can be less damaging to the tree and can be used where hardware installation isn't possible. All cabling systems are intended to limit the movement of the supported branches so they are less likely to fail during storms.

The dynamic systems allow for bracing large branches and stems, while allowing for shock absorption and stretch without damaging the tree. The installation requires adjusting the length of the poly rope-like material by splicing and can be done without special tools. The system can also be installed without the shock absorber for higher-risk limbs that need a more static bracing solution. The dynamic, low-tension support is designed to work with the tree's natural growth and movement. This is much different than static cables which have little stretch or give and are more rigid. Some trees will require static, limiting support systems. However, many trees do not qualify for invasive installation of metal hardware. This is due to lack of solid wood fiber because of limited branch size or perhaps the presence of decay. The non-invasive nature of dynamic systems makes them an excellent option for trees with poor compartmentalization or those trees with existing internal structural issues or decay. No drilling holes are required and the cable around the stems can be adjusted much easier than static systems as the tree grows.

Props are another option especially for long, extended branches. This is another non-invasive option for reducing loads on the branch attachment and the need for large pruning cuts.

FIGURE 11.15 Dynamic cable system.

FIGURE 11.16 Propping is another option for heavy, overextended branches.

Before installing any system, the tree should be inspected carefully and objectives of the installation should be clearly defined. Pruning is an integral part of the process and should be completed prior to any installation. Maintenance following tree support installation is minimal, but important. Ground inspections should be conducted annually for loose hardware or adjustments. Usually, every 3 years the support system should be inspected by an arborist who ascends the tree for a closer look and adjusts where necessary. Consult best management practices for details on installation and inspection.

TREE SUPPORT CONSIDERATIONS

Typically, support systems are used to reduce the risk of tree failure, such as when cables and/or braces are utilized to strengthen structural weaknesses (e.g., codominant trunks) on an otherwise healthy tree. Installing tree support systems can help reinforce weak branches, mend split trunks, and reduce the possibility of failure. Often, significant branches and even entire trees are needlessly removed where supplemental support systems may mitigate the issue and reduce the risk.

An important consideration for the arborist to remember is that not all trees are candidates for support systems. Some trees cannot be cabled or braced due to poor health, decayed wood, or location. The primary deliberation is the question, "Can the risk be reduced to an acceptable level for the tree manager or owner through support installation?" Also, consider that the condition of concern still remains, regardless of the support strategy implemented. Cabling and bracing do not eliminate the risk of tree failure, but when done correctly, they reduce the risk and can extend the life of a tree.

Some considerations and questions important in the decision-making process include:

- Is this a reasonable treatment for the tree, or am I just trying to preserve a tree which ultimately should be removed?
- Is this expenditure reasonable or is a new tree a better long-term investment?
- Is the tree healthy enough to allow support equipment to be installed?
- Do I understand that support systems are no guarantee against failure?
- Will the client be prepared to have the system inspected regularly, then perhaps replaced after 7 to 10 years?

INSTALLING CABLE SYSTEMS

There are various cabling systems to apply based on the situation presented by the tree. Typically, single branches, codominant stems, and multiple leaders are the common scenarios that require support from cables. A single branch requires cabling only when a branch is overextended and the attachment is determined to be weak.

Codominant stems considered weak would receive a single cable called the simple direct system. Direct cabling between the two stems can support each other when installed correctly. This will reduce the potential for failure or prevent further failure should the codominant already be splitting. A brace rod system would be required to complement the cabling system.

Three cables attached to three stems or leaders in a triangle system provides direct support to each of the stems and reduces loading on the often-clustered leader system. This will reduce loading forces on the attachments and minimize the branch movement.

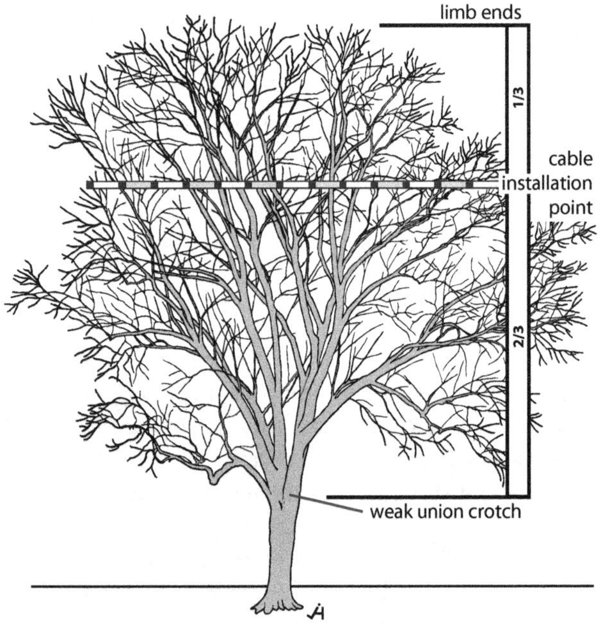

FIGURE 11.17 Cable position diagram.

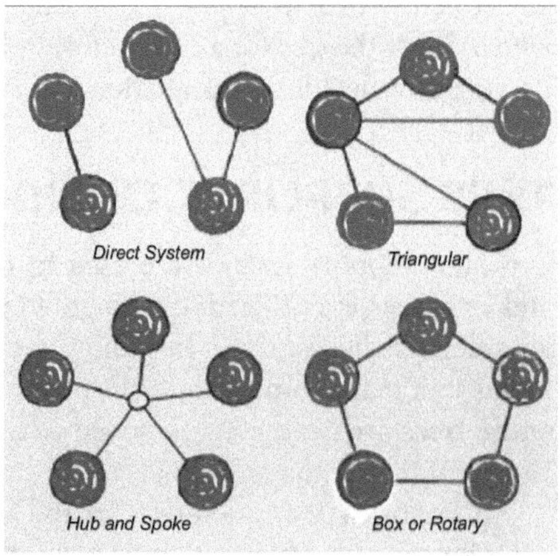

FIGURE 11.18 Recommended cabling configurations.

A box or rotary system connects cables to multiple stems considered the main leaders in the crown. Box systems can be used to connect four or more branches. This system provides minimal direct support, thus allowing more branch movement than with direct or triangular systems. This system allows for more movement than the previous systems, but still provides support under loading. Increased support can be applied if additional interior cables are placed to create a group of triangles within the system.

Hub-and-spoke systems consist of a centrally attached steel-ring hub with cables radiating from it to three or more branches. These systems are installed when there is no usable central leader or when there is a need for many cables. The central anchor point may allow for a reduced number of cables. Hub-and-spoke systems tend to be more difficult to install because the tension of each cable must be adjusted independently.

Cable position is critical for support and mitigation of the condition of concern. Whether using dynamic or static cabling, cables should be attached to branches or stems approximately two-thirds of their length from the attachment or junction point. When using multiple cables, each attachment should be approximately the same distance from the junction or crotch. It may be necessary to make adjustments to these recommendations if branches vary in length or if the diameter doesn't support the cables according to the standards.

- Direct is the most common method when there are two limbs needing support.
- Triangular is the preferred method when maximum support is required.
- Hub and spoke should only be used when other installation techniques cannot be installed.
- Box should only be used when minimal direct support is needed.

CABLING TOOLS AND HARDWARE

The tools used for installing tree support systems are relatively common and easy to obtain and determined by the type of system and dynamic or static cables. These include:

Cable puller and cable aid	Come-along / block and tackle
Cutting tools	Drill and drill bits
Hammers	Various wrenches
Tool bags	Safety equipment

Typical hardware used in cabling includes many components. For static systems, they are steel forged and resistant to environmental exposure. If using steel hardware, it should be stainless, galvanized, or treated to reduce rust and corrosion. In dynamic systems, synthetic ropes are treated for UV resistance.

Hardware may include:

Cable	Amon eyebolts
Terminations	Forged eyebolts and hooks
Cable clamps	Threaded lag hooks, lag eyes
Thimbles	Washers
Dead-end grips	Cable stops

Dynamic systems include synthetic rope specially manufactured for tree cable systems. This type of rope is typically a hollow-braid polypropylene rope. Simple splices, straps, slings, and compression springs or shock absorbing inserts may also be used. The equipment is determined by the manufacturer.

BRACING TREES

Braces are threaded rods installed through unions of weak codominant stems. They are installed by drilling through the tree trunks and fastening them together to provide more rigid support. Brace rods reduce the risk of stems spreading apart or moving sideways in relation to each other. In many instances, bracing can be used to repair a splitting trunk crotch. Often a combination of cabling and bracing is implemented to obtain the most stable support for the defective conditions.

Weak codominant stems already split are usually braced with a threaded rod to complement the selected cabling system in the crown. Proactive bracing may be installed

FIGURE 11.19 Brace rods are employed to support stems that are at risk of splitting

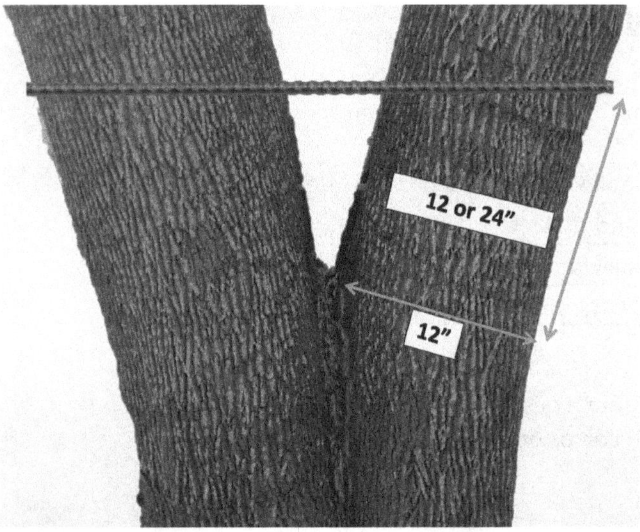

FIGURE 11.20 Single brace rod location.

on unsplit codominant stems. Typically, a single rod is sufficient to hold smaller diameter stems that are unsplit to reduce the potential for failure. Larger diameter stems that are split or weak can require two or more threaded rods to hold the two sections together and minimize the risk of failing under loads. Standards provide the recommended numbers and sizes to apply to the various situations of tree conditions.

Brace rod installations may pass completely through the tree or dead-end braces. Rods used for through-bracing are threaded with heavy-duty washers and nuts that fasten each end of the rod. If a lag-threaded rod is used, special nuts with lag threads are required. If internal decay is present in a tree, only through-braces can be used, and there must be a minimum thickness of sound wood of no less than 30% of the diameter of the tree at the point of installation. It is important to determine the extent and location of decay to determine the best strategy.

The placement of the rods depends on the condition, whether it's a split or unsplit union. The recommended location for a single brace rod to support a union that has not split is above the union at a distance of 1 to 2 times the diameter of the larger branch, measured above the union. For example, if the larger leader is 12 inches in diameter above the union, the brace rod should be installed 12 to 24 inches above the union. For larger trees with a DBH of 20 to 40 inches and with a single union and a split below the union, an alternating rod system should be used. This system includes one rod above the union and two or more rods below the union in a staggered pattern from each other. When practical, the vertical distance between rods should be equal to or greater than the diameter of the tree at the rod, or 12 inches apart, whichever is less. Extremely large trees require even more bracing to support the trunk.

- Educate the tree's owner/manager of the need for periodic inspection.
- Client is responsible for scheduling inspections.
- Periodically inspect for tension, condition, and position.
- Annually inspect from the ground.
- Aerial inspection every 5 years.
- Pruning and maintenance are critical for support systems.

Cable and bracing installation can be a complicated process and the tree should be reviewed carefully and objectives of the installation clearly defined. Maintenance following tree support installation is minimal but also important. Annual ground inspections should be conducted to check for loose hardware or to make any necessary adjustments. Usually every three years, the support system should be inspected by an arborist who will ascend the tree for a closer look.

Additional consideration for tree support systems:

- Is this a reasonable treatment for the tree, or am I just trying to preserve a tree which should be removed?
- Is this expenditure reasonable, or is a new tree a better long-term investment?
- Is the tree healthy enough to allow support equipment to be installed?
- Do I understand that support systems are no guarantee against failure?
- Am I prepared to have the system inspected regularly, then perhaps replaced after 7–10 years?

LIGHTNING PROTECTION FOR TREES

Lightning protection systems installed in trees are effective in reducing the risk of damage to trees from lightning strikes. These systems protect trees by providing an alternate path for the energy. Instead of lightning passing down a tree, a conductor is used to take the lightning from the top of the tree to the ground. Properly installed and maintained systems are often effective at preventing serious damage to trees. Refer to the standards and ISA best management practices for detailed information on installation and maintenance.

A tree lightning protection system (TLPS) works by providing a path of least resistance for the electrical charge of a lightning strike to follow. The system typically consists of a copper conductor that runs from the top of the tree down to the ground and out to a ground rod or plate. This conductor provides an alternative path for the electricity to follow, rather than traveling on the tree itself. When lightning strikes, the electrical charge is conducted down the copper conductor and into the ground, where the energy is dispersed. This helps prevent damage to the tree and reduces the risk of a side flash to adjacent structures.

It's important to note while lightning protection systems can significantly reduce the risk of damage to trees, they cannot guarantee complete protection. Lightning strikes are unpredictable and can still cause damage even with a protection system in place. Trees equipped with a system that has been installed in accordance with ANSI A300 standards can minimize damaged by lightning strikes, but protection cannot be considered absolute.

There are a few factors to consider when deciding to install a lightning protection system on a tree. Trees within 10 feet of a structure, or have branches extending to heights above the structure, should be considered as candidates for a lightning protection system. This can reduce the danger of side flash, fire, or superheating of the moisture in the tree, which could result in severe tree damage and destruction. Additionally, trees that are tall, growing alone in open areas such as on hills, in pastures,

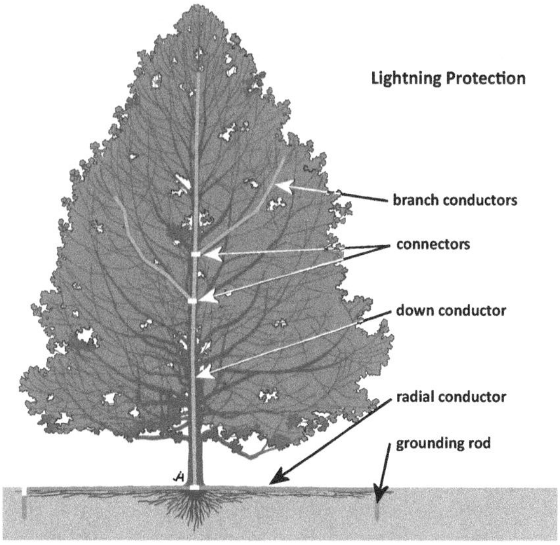

FIGURE 11.21 Tree lightning protection systems can reduce damage to trees from the effects of lightning.

or near water, as well as certain species including tulip tree, oak, pine, and maple are more susceptible to lightning strikes. Also, it may be prudent to consider installing a lightning protection system on trees that have historical significance or high value in the landscape.

FIGURE 11.22 Lightning strikes leave varying types of damage.

Damage and Injury from Lightning

- Acoustic wave
- Electrical disruption
- Extreme heat
- Root damage from grounding process
- Water loss due to tissue damage
- Secondary infestation/infection

DAMAGE FROM LIGHTNING

The amount of damage can vary tremendously upon species and conditions. In some cases, no noticeable damage may be found. In other cases, back scarring and other injury to the bark and cambium is found. In worst case, complete destruction is possible. The area just under the bark layer contains moisture in the form of sap and water. Since water is a better electrical conductor than wood, lightning striking a tree tends to travel just underneath the bark. If the tree is wet with rain, the energy may travel down the outside of the tree. This is a better path which can occur with potentially no damage. The explosive expansion of the lightning's return stroke will literally blast off the bark, and sometimes some of the wood, along the length of the lightning channel. The result is a visible scar along the trunk of the tree. Often, the core of the lightning channel will carve a half-inch wide groove or larger in the wood. If the lightning flash consisted of more than one return stroke, there may be two or more of these grooves in the tree.

PROTECTION FACTORS

Trees close to houses or other buildings and much taller than the building should be considered for protection. Trees that are more prone to strikes and are located within more open recreational areas, parks, golf courses, or other areas where people congregate should be considered for lightning protection. A significant factor in the probability of a tree being struck is its location and height. Lightning is more likely to strike the tallest tree in a group, trees growing in the open, tall trees that border woods or line a street, trees near large bodies of water, or trees on hilltops or in geographic regions where lightning is common.

SYSTEM COMPONENTS

A tree lightning protection system typically consists of several key components, including:

- Air terminals: installed at the highest point of the tree to intercept lightning. An air terminal should be installed in 3 inches or smaller branches as close to the top of the tree as possible.
- Fasteners ("stand-offs"): separate the conductor from the surface of the tree. The fasteners are sized to match the conductor and are used to attach the cable to the tree.

- Conductor cables: 14- or 32-strand copper cables providing the connection between the air terminals and the ground, directing the lightning strike into the earth where it can safely dissipate.
- Ground rods: metal rods or plates buried at least 8 feet in the ground, providing a path for the electrical charge to be dispersed. Multiple rods or plates may be used depending on the soil and application. The rod location should be at least 10 feet from the trunk.

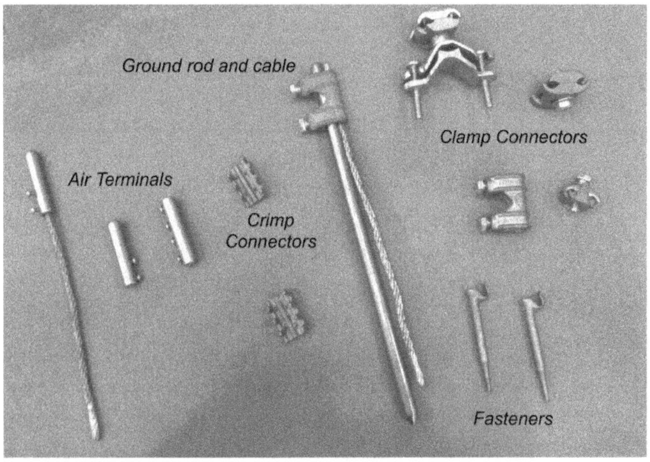

FIGURE 11.23 TLPS parts.

SYSTEM INSTALLATION

The most common lightning protection systems for trees consist of heavy-gauge copper wires running the length of the tree. These wires are connected to the ground using a series of copper fasteners and terminate with the air terminal. Most trees receive between one and three conductors. For trees with a crown spread radius of 35 feet or less, one centrally located conductor is sufficient. For trees with a crown spread radius greater than 35 feet, one or more branch conductors are required.

Installing a tree lightning protection system includes the following steps:

- Assessing the tree: The first step is to assess the tree for any risk issues requiring mitigation prior to equipment installation. Then determine its height, structure, and the best location for the air terminal(s).
- Installing the air terminal: The air terminal or lightning rod is installed at the highest point of the tree. The terminal should be as high as possible and no other branches should exceed the height of the terminal.
- Running the conductor cable: A copper conductor cable is then run from the air terminal along the branch, down the trunk of the tree and out to a ground rod or plate. The route of the primary conductor should follow the natural shape of the tree and branches down the trunk. Additional connectors are required when multiple conductors are applied.
- Installing the ground rod: A ground rod or plate is installed in the ground at least 10 feet away from the base of the tree. The conductor cable is then connected to a ground rod, providing a path for the electrical charge to be dispersed. Multiple rods or grounding plates may be necessary, depending on the system and the soil type.
- Testing and inspecting: Once installed, test and inspect the system with a continuity test. This will help determine if there are any defects in the system, reducing protection.

FIGURE 11.24 Copper cable attached with a fastener.

TREE LIGHTNING PROTECTION SYSTEM MAINTENANCE

An effective TLPS should be maintained and inspected as determined by the ANSI standards. However, scheduling this work is the responsibility of the tree owner or manager. As with tree support systems, lightning protection requires periodic inspections. The frequency varies with growth rates. For example, annually for fast-growing trees, and every 2 or 3 years for slow-growing trees. This inspection should include a check of all splices and connections to ensure continuity.

Every 5 years or less, the tree should receive an aerial inspection and all cables and connectors should be examined and tightened as needed. If the tree has overgrown the wire, an electrical continuity or resistance check should be made. The grounding system's electrical resistance should also be tested to make sure the cable has not been severed. Conductors that have been overgrown by the tree are more difficult to inspect but likely do not present a problem to the integrity of the system. It is best to address them before overgrowth occurs, and often more easily accomplished during the dormant season, where visibility is best.

MAINTENANCE INSPECTION PROCESS

- Locate air terminals ensuring near the top of the tree and in an upright position.
- Check fasteners to make sure they are securely attached to the tree but not overgrown by the tree.
- Inspect primary, branch, and bonding connectors to ensure they are tight and free of corrosion.
- Inspect conductors to ensure they are free of breaks, nicks, pinches, kinks, and serious corrosion.
- Determine if the ground conductor is intact with cables. Investigate the site for any evidence of excavation or trenching that may have severed the ground conductor or recent landscape activities that may have affected the grounding electrode.

CHAPTER

12

DILEMMAS, REMEDIES, AND REPARATIONS

LEARNING OBJECTIVES

- Identify the various above and below ground challenges for trees in the built environment and the possible mitigation options.
- Recognize dysfunctional roots systems and possible remedies to improve tree health.
- Explain the various types of severe weather damage trees experience and the recommendations for restoration.
- Identify the effects of drought on trees and their legacy effects. Explain how arborists can prevent drought injury and improve tree health during drought conditions.
- Identify how to prepare trees for dormancy for a healthy recovery the following growing season.

DYSFUNCTIONAL ROOTS, ROOT FLARES, AND COMPACTION

Trees in the built environment experience many challenges that imperil their survival, both aboveground and belowground. The below-ground issues are the most difficult problems arborists must diagnose. The origin of dieback and decline symptoms in landscape trees can be very difficult to diagnose due to the many cultural, environmental, and biological factors that disrupt tree health. The common culprits belowground include excessive planting depth, buried root collars, and compaction. The most consistent tree symptoms from these issues are flagging, decline, or chlorosis in the crown of the tree. Typically, this indicates soil and root disturbances that should be examined more closely. Exploration of root dysfunctions

FIGURE 12.1 A mini-mattock can be used to carefully excavate soil around the root flare.

requires the use of specialized tools by trained arborists. The most common and least invasive are simple tools such as trowels and garden shovels. Advanced methods include the use of high-speed forced air aimed at the target areas in the root zone, displacing the soil around the root flare and root system for closer examination.

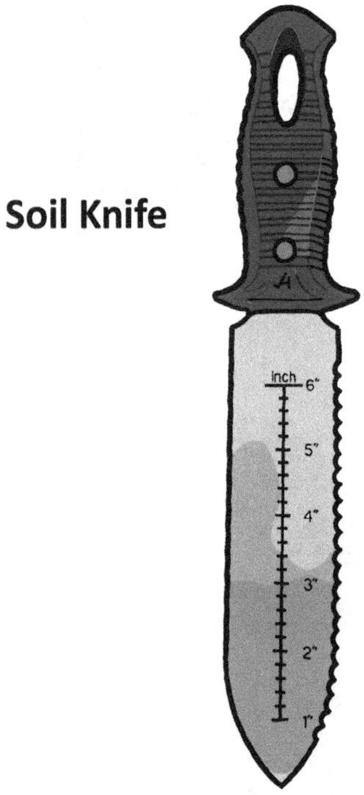

FIGURE 12.2 Soil knife.

Every arborist should have basic diagnostic tools available to more closely examine belowground concerns. Two very useful tools include a hori-hori, sometimes referred to as a "soil knife," and garden trowel or cultivator. A soil knife is a heavy serrated blade that is sharp on both sides and comes to a semi-sharp point at the end. These tools can easily excavate small areas of soil around the root flare to examine soil conditions, root health, and planting depth.

Soil excavation using compressed air, also known as air-spading, has been prevalent in the tree care industry since the mid-1990s. This practice allows tree roots to be exposed by displacing soil efficiently and with minimal damage to the roots. This tool makes it possible to better care for a tree's soil environment and root system, especially in highly trafficked landscapes where soil becomes compacted or otherwise unsuitable for natural tree growth, or in scenarios involving construction projects like underground utilities, pavements, or building foundations. In such cases, trees might otherwise be severely damaged or removed. The application of forced air tools can be dangerous. The safety of the operator, nearby people, and property is important when using compressed air powered tools. Always be sure to reference the tool manufacturer for their most updated safety procedures and operational materials.

FIGURE 12.3 Tools such as soil knives can remove soil without damaging roots.

STRATEGIES FOR IMPROVED TREE HEALTH

Stem girdling roots (SGRs) are a common causes of tree decline. These misguided roots begin with circling roots from smooth-walled containers. Often, they result in dysfunctional root systems encircling the stem or trunk by growing around it rather than laterally away from the tree. Often roots circling close to the trunk can be embedded or fused in the trunk or cross over main roots at the base of the trunk at advanced stages. The condition worsens as the tree grows and expands. Root and trunk diameter increases, which creates compression where the root circles around the trunk, just at or below ground level. The effects are the roots begin to strangulate the tree, disrupting water conductance and nutrient flow from the roots. Additionally, the compression slows or stops carbohydrates and other assimilates from moving downward from the leaves.

A common cause of SGRs is when the finish grade is set too high above the root flare or root collar. Usually, excessive planting depth or when trees settle into the planting pit due to excessive planting pit depth below the root ball contributes to the problem. Alternatively, root collars can be buried over time due to the accumulation of mulch or the addition of soil around the tree, creating problems.

FIGURE 12.4 Stem girdling roots can be surgically corrected after soil is excavated.

A useful strategy to examine dysfunction roots is with an air excavation tool. Look for the following symptoms and signs:

- Root rot
- Root flare compression
- Roots circling and crossing
- Roots girdling trunk or fused into trunk
- Roots deflected at sharp angles
- Only three or four main roots extending from the stem
- Excessive planting depth
- Excessive mulch depth

The symptoms of girdling roots are varied depending on the severity. In some instances, trees may not present typical decline indicators, but rather begin leaning in the landscape or fail during heavy winds due to the SGR's completely girdling the trunk, creating acute compression at one point, leading to structural instability.

FIGURE 12.5 Girdled roots belowground leads to failure.

The tree response may manifest as a one-sided trunk flare, no trunk flare (especially if the SGR is below the soil line), or a thickened area at the root-trunk interface. SGRs tend to develop because of adverse site conditions, improper planting procedure, or poor root formation in containerized nursery trees. If the soil is severely compacted or the roots encounter an impermeable surface, such as a sidewalk, curb, or foundation, the roots will change their growing direction. In compacted planting pits, particularly with heavy clay soils, it is quite common to encounter situations where the roots circle back towards the tree and begin to encircle it. These redirected roots can eventually become SGRs.

Even if a tree is planted at the correct depth, with the root flare exposed, SGRs can easily develop due to excessive mulching (volcano mulching). Tree roots grow in soils and other media with the least resistance and the most moisture and oxygen levels. Deep mulch rings create a misperceived elevated soil level. This promotes the development of adventitious roots into this "new soil medium." Then, the tree potentially develops SGRs just below the mulch surface. Another common problem leading to SGRs is pot-bound nursery stock. When trees are held in smooth-walled containers for too long, roots will extend to the sidewall of the container and begin to follow the outer edge in a circling pattern. If not corrected during the planting process, these circling roots will eventually create girdling roots. Removing circling roots as a pre-plant treatment on containerized trees can rescue the tree from certain SGRs if the circling is not too severe. On established trees, the opportunity for mitigation may be less effective or even lethal if there are very large roots in proportion to the tree's size or if it has fused with the stem.

FIGURE 12.6 Severe volcano mulch.

FIGURE 12.7 Circling roots on container-grown trees.

FIGURE 12.8 Root surgery removes girdling roots.

Root removal or severance is a critical practice that can help the tree or harm the tree. Air excavation around tree root collars and girdling roots exposes the roots, making it possible to identify the SGRs and surgically remove them. Most recommendations are based on anecdotal evidence and practitioner experience. For example, if the offending root can be easily accessed and removed without damaging the trunk, proceed with complete removal. However, if root removal is difficult to accomplish without collateral damage, consider dissecting them without complete removal. A general guideline for root removal is that root mass area can be safely removed until the total cross-sectional area is at or exceeds about 25% of the trunk's cross-sectional area at a height of 4.5 feet from the ground. When removing roots, the final cut should be made tangent to the trunk because new roots often grow back in the direction of the removed roots.

Tree health, environmental conditions, and post-surgery maintenance play a role in the decision-making process. More aggressive surgery can be performed if conditions are favorable and post-surgery maintenance is rigid. This means pre-surgery preparation with improved cultural practices such as watering to reduce drought impacts. After root treatment, soil is then returned to the excavated root area and the grade is reestablished at the appropriate grade level. Finally, composted hardwood mulch is applied to the surface in a layer 2–3 inches deep and supplemental irrigation provided. Additional organic treatments may be applied, depending on the tree and the arborist's preference for treatment.

RADIAL TRENCHING

FIGURE 12.9 Radial trenching improves gas exchange to the roots.

Radial trenching is another mitigation option for modifying soil structure, improving aeration, and encouraging root growth to additional depths (10–12 inches deep). The extent of trenching should occur to the dripline or beyond, if possible. After the trench pattern is determined (it may be helpful to paint guides on the ground), the arborist can begin the process. The operator will continuously move the tool while holding it at a 30–45-degree angle to achieve the specified depth and work side to side to control the trench width. Recommendations are to layout the radial trenches in a pattern like a wagon wheel, with the spokes at approximately 45-degree angles and a 4-inch-wide trench. The distance between the trenches should not exceed 8–10 feet. While excavating a trench, it is recommended to cover nearby trenches to prevent refilling. The soil can be collected for replacement or amended with various organic materials. There are several organic materials and combinations that can be used in the areas of displaced soils.

VERTICAL MULCHING

FIGURE 12.10 Vertical mulching allows for deeper root growth.

This soil modification technique is used to help alleviate compaction and promote gas exchange with the roots as well. This application is useful in areas with poor drainage and shallow compact soils. Vertical mulching encourages roots to grow deeper into the soil profile, improving the tree's stability as well. Like radial trenching, vertical mulching is implemented throughout the tree's dripline or beyond. Create the layout, depth, and size of the holes and then proceed with the forced-air tool perpendicular to the ground with the nozzle near the surface, slowly pushing the tool into the soil. When at the desired depth, slowly withdraw the tool, clearing the hole of loosened soil, and then reinsert. For best results, depths of 18–24 inches are preferred with 3–5-inch diameter, 2 feet apart and if possible. Often, severely compacted soils are challenging to get much below 12 inches deep. Excavated soil can be removed, replaced, or amended before the vertical holes are replenished, similar to radial trenching.

ROOT INVIGORATION (AIR TILLING)

Trees with shallow, compacted root zones or lacking good root growth from poor soil quality can benefit from root invigoration as another modification option. Since most of a tree's roots are near the surface where the greater amounts of oxygen and nutrients are found, air tilling is an excellent strategy for promoting improved tree health by improving root growth and anchorage.

Root invigoration should be implemented to the dripline, at least. The larger the area, generally, the more effective the treatment. The process starts with the removal of any turf, mulch, or other groundcover within the specified treatment area. Then the soil is loosened using an

FIGURE 12.11 Air tilling loosens soils and mixes amendments to invigorate soils and roots.

air excavation tool working in circular or linear patterns, moving slowly until the soil is visibly loosened. The soil should be displaced at least 6 inches deep or more, if possible. The speed and action depend on soil conditions and the extent of compaction and texture. The tool is held nearly vertically, directing the airflow downward. After the initial "digging", the prescribed soil amendments are applied evenly over the disturbed soil area. Proceed to blend the mixture into existing soil in the same way as it was disturbed and at a similar depth as the original the soil grade. Next, add composted hardwood mulch to the treated area 2–3 inches deep. Monitor soil moisture and provide supplemental irrigation to promote the best results. The newly treated and disturbed soil can dry out quickly and should be placed on a regular watering cycle.

FIGURE 12.12 Surface roots under tree.

Surface Root Syndrome

Tree roots growing at or slightly above the soil surface are called surface roots. They are often problematic for tree owners and managers whether it be in the lawn or landscape or in urban planting spaces.

In the developed landscape, roots grow predominantly near the soil surface with nearly all roots located in the upper 24 inches of the soil. In fact, a tree's root system can be surprisingly shallow, dominated by long and relatively small lateral roots spreading out close to the soil surface, rather than being characterized by a deeply penetrating taproot system found in the natural system. Primarily, genetics determine the characteristics of root growth in a tree. However, soil conditions can manipulate their direction. Roots growing horizontally from the upper part of the taproot are called lateral roots. These roots, which include surface and shallow roots, are often the most vigorously growing and form the main framework of support and anchorage for the tree. Lateral roots near the soil surface will thicken with age as well, creating a large woody root system framework. The shallower roots rise to the upper, nutrient-rich layers at the surface of the soil. Then, as traffic occurs—such as mowers and foot traffic—the soil is eroded away and compacted, exposing the upper surface roots.

This type of naturally shallow root growth is found to be especially common in trees growing in the riparian zone, which is an area where streams, rivers, and wetlands are located nearby. Examples of trees with shallow root growth include maples, sycamores, willows, and some ash species. Their roots naturally grow close to the soil surface as a strategy to escape saturated or prolonged wet soils more quickly. However, any tree can develop surface roots in heavy, compacted soil or wet soil, or because of erosion from pedestrian and mechanical traffic. Disturbed soils, such as those found in the urban forest, are usually compacted, making desirable root development more difficult if not impossible.

Surface roots present challenges for property owners when it comes to turf management and maintenance activities. They can cause damage to equipment and become trip hazards in the landscape as well. When trees and grass compete for moisture and nutrients, tree roots usually win due to the prolific network of roots. Also, roots are usually denser near the trunk, so grass growing over or near these surface roots may become thin and lose its dark green color because of the competition for nutrients and lower light intensity.

There are few viable options available to remove problematic surface roots that won't adversely affect the tree. Damage to the root system creates an increase likelihood of failure and health concerns for the tree. The following are examples of *negative* impacts to long-term tree health and safety.

Removing Visible Surface Roots by Cutting or Grinding. Cutting or grinding out surface roots is the worst decision. Physical damaging and removing roots reduces the ability of the tree to absorb nutrients and water, causing decline and leading to death. Also, since those large roots serve as an anchor to keep the tree strong and upright, damaging or removing these roots can severely diminish stability, which can cause the tree to fall over in heavy winds (windthrow). Instead, careful root pruning or root surgery may be a possible alternative.

Adding a Layer of Topsoil to Cover Exposed Roots. This approach is commonly used; however, adding topsoil will likely create long-term problems with the tree and possibly lead to its death. Roots need oxygen because, like the crown, roots require active respiration to survive. The established soil grade allows for gas exchange, allowing roots to obtain oxygen in addition to absorbing water and nutrients. If additional soil is added or any covering with a high bulk density, the roots are basically covered and smothered. This restricts the roots from obtaining oxygen from the air above, causing root cell death. In some cases, adding a light layer of one-half compost and one-half soil mixture spread less than 2 inches deep can create a temporary remedy to restore the turf, but the roots will most likely migrate back to the surface again.

Planting a Raised Flower Garden or Installing Groundcover Over Surface Roots. This is another commonly used approach, but the effect is the same as adding soil. Any additional soil coverage will reduce the gas exchange between the tree roots and soil surface, causing a decline in tree health and potentially the death of the tree. Also, added soil and vegetation near or against the trunk buries the root collar and can cause decay, further compromising tree health. The planting process of tilling and preparing the soil is damaging to the roots as well, regardless of whether annuals or perennial groundcovers are planted.

FIGURE 12.13 Enlarged mulched ring.

The Reasonable Solution

The most reasonable solution for problematic surface roots is simply developing larger mulch beds. The application of wood mulch to form enlarged mulch rings or create larger mulch beds is the best way to minimize the challenge of surface tree roots. Adding 2 to 3 inches of wood mulch reduces the need to mow over the roots, also which creates compacted conditions on the roots. A mulched root zone provides an improved growing environment for the tree and less damage to the roots. Mulch beds provide an alternative for a cleaner landscape appearance where it is otherwise impossible to grow thick turf. For best results, create a mulch bed extending out to the dripline, if possible, or at least to where the surface roots have dissipated enough for healthy turf growth and easier lawn maintenance.

Gravel or stone mulch products can be utilized in special situations, but there are negatives to their use. Using rock or stone mulch can increase the levels of compacted soil, while using more organic mulch can improve aeration in the root zone. Also, stone mulch can increase soil temperatures where applied by absorbing radiant heat from the sun. Any mulch product should be advantageous for the tree without creating additional challenges.

TREES AND STORMS

Trees are damaged and destroyed every year from storm-related events. Major damage to parts of the tree or even the entire tree can result from high winds, snow, and ice. Seasonal storms can cause extensive damage depending on timing and intensity and increase risk situations as well.

FIGURE 12.14 Tree damage is more common in trees with poor structure.

Generally, loads are forces on the tree such as wind. As trees grow, they develop additional wood fiber in response to wind loading conditions. If conditions are consistent, trees will initiate modifications in their trunk and root structure to compensate for these loading patterns. This is done by adding additional wood fiber to compensate for the loads presented.

Also, ice and snow can cause damage from excessive weight or force on branches. Normal weather events are typically not an issue for trees. They are engineered to withstand normal conditions. However, some storms can exert extreme forces such as tornadoes and hurricanes, resulting in severe damage.

The response to storm destruction should be based primarily on the level of risk presented by the damage. Tree risk assessment is the systematic process of identifying, analyzing, and evaluating the tree for conditions of concern and defects. If the tree presents a high likelihood of failure, creating an unacceptable risk to people, property, or the disruption of activities, the decision becomes more critical. Often this situation requires removal especially if whole tree failure is the condition of concern. Broken or damaged branches with an imminent likelihood of failure and high likelihood of impacting a target may be mitigated to a safer level with only pruning.

Types of Storm Damage

There are several types of tree damage that occur from violent weather. Trees will fail at their weakest point. Typically, failure is caused by the combination of a defect and external forces being applied to the tree. All parts of the tree should be inspected during a post-storm assessment. Some common damage found includes:

Wind Throw. This situation refers to when an entire tree is "pushed over" by high winds or partially dislodged from the soil. The loading forces on the aboveground tree parts are greater than the ability of the root system to hold. Previous damage from activities, such as construction and attacks from pests, predisposes the tree with poor anchorage and health. Inspect the root plate of each tree on a regular basis to determine if there are any issues. Look for fungal fruiting bodies, exposed roots, lifting, or other damage.

FIGURE 12.15 Saturated soils increase the likelihood of wind throw.

Crown Twist. A tree contains an array of leaves, twigs, and branches, making up the crown portion of the tree. Many times, the crown can have an asymmetrical form either from competition or poor pruning, leaving an unbalanced form. The result of this situation in high winds can be troublesome. Uneven wind loading on the lopsided crown produces a damaging twist on the stem. This rotation creates torsional stress, much like twisting a rope backwards, resulting in splits and cracking. This is especially damaging around old wounds or other defects, which will lead to failure. Close investigation is necessary to spot these often-subtle cracks.

Branch Failure. Tree branches are easy victims of loading forces. Whether it is axial loading from snow or ice accretion or lateral side loading from wind, damage is frequent. Because branches can sometimes be poorly attached to the main stems, they can easily experience injury. However, the strategic and structural arrangement trees produce allows branches to be flexible and sometimes disposable, easily shedding small branches

FIGURE 12.16 Parts of the tree to inspect during an assessment.

FIGURE 12.17 Unbalanced crowns are subject to torsional cracks.

and leaves to help dampen the impact of wind, minimizing damage to the larger, structural branches. However, poor architecture and development, such as codominant stems and included bark in branch junctions, predisposes branches to failure and often are the first to experience damage.

Branch failure can often be easily repaired and is not necessarily lethal to the tree unless a large portion of the crown has been damaged. However, branch failure can create high-risk situations. Often twisted and cracked branches are difficult to see and may need closer inspection by an arborist to identify and mitigate. Be sure to use pruning best practices when correcting damage in the crown, discussed later in this chapter.

Stem Failure. Old injury sites or wounds are common on tree trunks, and these damaged areas can lead to tree failure under excessive loads. Also, trees newly exposed to prevailing winds from development or removal of neighboring trees can be at greater risk. Trunks can snap or buckle if the tree is unable to withstand the newly experienced wind forces. Trees with major defects on the trunk such as large cavities, should be examined carefully to determine the degree of risk for the situation. Look for decayed areas, cavities, nesting holes, and fungal fruiting bodies for indicators which lead to increased likelihood of failure.

Root Failure. Stability and anchorage are important roles of the root system for keeping a tree upright. Both the structural, woody roots and the fine absorbing roots create an effective network of strength to withstand loading forces. As the tree becomes larger, increased stress is experienced by the roots and they must be able to withstand these loads.

FIGURE 12.18 Branches with poor spacing are more likely to fail in high winds.

FIGURE 12.19 Codominant stems often fail during high winds and storms.

However, roots that are restricted, diseased, or damaged can cause trees to lean and fall. Careful examination of the root plate is needed to determine if anchorage is compromised in any way. Look for heaving soil, exposed roots and roots potentially girdling the tree trunk, which can present high-risk situations. The presence of pulled or broken roots, as well as any fruiting structures present such as mushroom development around the root system, also can be indicators of decay in the roots.

FIGURE 12.20 Diseased and damaged roots fail under wind loads.

Storm Damage Prevention

It is impossible to protect a tree from storms or prevent damage from weather events. However, there are some actions which can minimize injury to trees. Structural pruning benefits trees with a healthy, more wind-resistant crown. Reduce or remove vulnerable codominant stems. Select a strong dominant leader on trees of any developmental age, except mature and veteran trees. Implement structural pruning practices by selecting branches with good spacing, radially around the tree for balance and symmetry. Under no circumstances should a tree be "topped" or thinned to prevent wind or ice damage. Topped trees produce many small, poorly attached sprouts and as they grow, the canopy weight and density are actually increased. In addition, topped branches often develop decay and diseased stems, which further weakens the tree and makes the branches more susceptible to breakage. Topping may seem to make a tree safer, but it increases the potential for storm damage. Trees that have been thinned create even more stress on the tree structure by reducing the mass damping effect provided by crown biomass.

POST-STORM RESPONSE

Safety first! Stay clear and look for dangerous hanging limbs, broken branches, and other failures before beginning cleanup or inspections. Keep others clear of the areas beneath and around damaged trees. Be alert for utility lines involved with damaged trees. All utility lines should be considered energized and dangerous. Damage assessment is the next step. Review the affected tree or trees to determine the extent of damage. Some trees may not require action and can recover on their own. The challenging part of the assessment is deciding which trees can be pruned properly to a healthy state and which trees should be removed. Many factors are involved in the decision, and damage may be hidden or difficult to determine. Often, post-storm inspections require an advanced assessment to help determine the level of risk presented and proper mitigation options.

STORM RESTORATION

Every tree will fail at some time in its life. Trees fail when they experience loads or forces that exceed their capacity to manage those loads. Trees are strategic and economical with their resources and will only produce enough strength (wood fiber) to withstand loads under normal

FIGURE 12.21 Crown restoration is possible with time and a light-handed approach to pruning.

conditions. They cannot anticipate loads and are incapable of over-engineering their structure and form. Trees are thrifty when it comes to energy use and growth as well. They react to mechanical and physiological stresses internally by growing more vigorously to reenforce these weak areas, while depriving less stressed parts. The salient point is that trees do all they can in a way that will achieve an even distribution of stressors, provided they have the vigor and vitality. This is the action that helps minimize potential failure points and maximizes their chances of survival.

Trees damaged in storms and those that have been deformed by topping can sometimes be restructured through a sequence of pruning treatments. Crown restoration efforts improve structure, which increases strength within the crown. Many damaged trees can never be restored to their previous structure; however, they can remain as functional elements in the landscape, if the investment is feasible and the risk level is acceptable. In some cases, such as failure of a large codominant stem that tears down one side of the trunk, the damage may be so severe that restoration is impractical. The wound is likely too large for the tree to seal and recover, or the tree may remain structurally compromised, leaving unacceptable risk levels or residual risk.

In most cases, damaged trees will develop sprouts in the weeks and months following damage. Epicormic and dormant sprouts will develop at or near the ends of broken or cut branches. This is a response by the tree to reiterate itself and restore loss of crown mass. This responsive sprouting is essential for recovery because it provides a mechanism to replace lost leaves and restore biological processes. In most cases, sprouts should be allowed to grow for a

Guidelines for Pruning Restoration
• Do nothing except mitigate risk issues, leaving new sprouts for at least one growing season.
• Select and allow development of one main branch at the location of each damaged branch end that will replace the broken branch, spacing the sprouts as far apart as is practical to prevent crowding.
• Remove or reduce unwanted sprouts and branches. Ideally, the retained sprouts should have plenty of lateral branches. A rule of thumb is to remove one-third of the other sprouts and reduce one-third to encourage the retained sprouts to develop lateral branches and taper.
• Some of the retained sprouts may have to be reduced or removed. This prevents all the sprouts from growing too long, and it also makes room for the growth and development of lateral branches from the retained sprouts.
• A few years later (perhaps sooner if growth rate is rapid), retained sprouts may need to be managed with reduction cuts to help develop a new leader and good structure for the branch and overall crown.

year or more before any pruning begins so that energy reserves can be restored and wounds can begin to heal. The length of time needed before restoration pruning begins depends primarily on the growth rate of the sprouts. Each tree will respond differently, depending on its pre-casualty health, vigor, and environmental conditions before and immediately after the damage.

Retaining and managing sprouts provides other benefits as well. A robust flush of leaves at or near the damaged area can help prevent decay and facilitate wound recovery. After waiting the appropriate time for maintenance to begin and as sprouts begin competing for the same space, some can be reduced and some can be removed. It is important to retain some spouts in desirable positions on the branch, ensuring they are free of defects, while culling others. Even though sprouting proceeds, decay will develop on certain (but not all) species behind broken tips and headed branches. Current recommendations are to remove one-third, reduction prune one-third, and retain one-third of the sprouts. Select a new branch leader to develop with supporting branches to improve the crown formation. Sprouts can be viable branches as overlapping layers and branch collars begin to develop. Larger sprouts reduce the amount of dysfunctional wood which helps reduce the spread of decay.

Reparations Lion's Tailing
Allow new sprouts to develop along the branches for a few years. The goal is to eventually develop a few sprouts into permanent branches.
Remove some of the sprouts and shorten others 2 or 3 years after the tree was mispruned, making sure that the ones retained are more or less evenly distributed along the branches.
Be sure to retain sprouts toward the base of the main branches closer to the trunk to develop taper on the main branches.
Sprouts selected as permanent branches should be positioned so that future growth can be pruned clear of other branches in the crown. Sprouts can be pruned to direct them into a desired position in the crown.

The combination of aggressive sprout growth and decayed wood can be a cause for concern for many arborists. Epicormic sprouts growing from damaged branch ends are considered poorly attached, at least for a period of time, and can crowd each other. Managing regrowth is crucial to redeveloping improved branch structure. Sprouts on species such as poplars (*Populus spp.*) and most maples (*Acer spp.*) that are weak compartmentalizers may need to be continually reduced by regular pruning sessions. If left to extend without management, there is a lack of branch diameter to support the weight, predisposing it to failure.

It is acceptable by standards, to apply some heading cuts on sprouts and broken branches especially in the early sessions of the restoration process. Heading cuts may be necessary occasionally when sprouts have grown too long without lateral branches, so there are no options for reducing to a lateral. Protect the trunk or larger branch by not removing too many sprouts. The restoration pruning process requires patience and likely will have to be repeated for a period of years.

FLOOD DAMAGE TO TREES

Annual flooding from excessive precipitation in the form or rain or melting snow can create challenges for trees. The extent of damage from prolonged exposure to wet and waterlogged soils depends on the season and can have a legacy effect. Tree damage generally occurs from prolonged saturated soils and physical damage from the flood waters. Most urban trees are not adaptable to lengthy flooded conditions and each species can have varying degrees of tolerance. Some species such as *Taxodium distichum, Platanus occidentalis, Populus spp.*, and *Acer sacharrinum* or more tolerant to extended wet conditions. These trees, known as riparian species, are better suited for wetter conditions, while some upland species like *Quercus, Carya*, and *Fagus* species prefer drier, upland conditions.

Plants differ in their ability to survive and grow in hypoxic conditions. Oxygen is a necessary component of respiration, which is the process that converts stored sugars into the energy needed to keep all living organisms alive. In the case of trees, oxygen is a necessary component in the roots as well. Unlike animals, plants do not have an active transport system for oxygen. Oxygen moves slowly and passively within a plant through diffusion. When oxygen is not present at the root-soil interface, soil-absorbed nutrients and oxygen become limited. In the absence of oxygen, root growth stops and over time, the begin to die. When flooded for a prolonged period of time, root dieback can be significant. Similar to drought stress conditions, the stomata start to close and photosynthesis is reduced. Respiration continues as photosynthesis slows and the tree operates at a deficit. A prolonged period of operating at a deficit leads to root and crown dieback, greater susceptibility to disease and pests, and eventually death.

FIGURE 12.22 Prolonged flooding can cause serious tree damage and increased failure risk.

Flooding has various physiological impacts, including poor aeration, where water moves into the pore spaces replacing gases like oxygen. This leads to an accumulation of CO_2, methane, and other gases, instead of a suitable oxygen exchange aboveground. This creates an anaerobic condition unsuitable for growth. Anaerobic conditions refer to soil that has little to no available oxygen in the soil. When saturated conditions persist, the microbial, fungal and plant activities quickly use up any remaining oxygen. These conditions occur most frequently in winter but can also be a result of over-irrigation and soil compaction. Anaerobic soil conditions are common in heavy clay soils where pore space is already reduced.

Flood damaged trees express symptoms such as reduced or no growth, poor leaf expansion, smaller leaves, chlorosis, and premature leaf drop with decline in growth being the primary issue. This may be attributed to denitrification, where bacteria remove the nitrates and nitrites from the soil, leading to a reduction of available micronutrients such as manganese, iron, and sulfur. Disease issues can be expected as well. *Phytophthora* is a root attacking fungi that tolerates low soil oxygen levels and is stimulated by tree decline leading to root loss. This root mass loss from disease and decay leaves the tree prone to drought and windthrow with post-flooding conditions.

The extent of flood damage on trees depends on factors such as the season of occurrence, water quality and temperatures, as well as the condition of the trees when they experience protracted flood conditions. Flooding during the growing season is far more harmful than winter flooding. Warm-season flooding magnifies the injury because of plant respiration and continued water loss through the leaves. The higher the temperatures, the faster the tree will experience oxygen deficiency, leading to decline. Just a little over 2 weeks of submersion can lead to the decline and death of some trees. Some species observed can tolerate up to 6 to 8 weeks of flooded soils with minimal damage.

Siltation is another concern as water displaced soil from flooded waters can collect and deposit around tree trunks and over roots systems. If affected trees don't respond with the formation of new root systems in these elevated locations, they will eventually decline and dieback. Conversely, soil erosion can be a concern as some trees will be left with bare roots and exposed roots plates. This creates a poor anchorage system and leaves trees more susceptible to windthro.

Secondary pest problems are a concern as well. Due to the reduced tree vigor and changing growing conditions, opportunistic pests can become a problem for several years after the flooding conditions. These might include boring pests and soil-borne pathogens causing disease.

Post-flooding management practices are important to initiate quickly to reduce the impacts of flooding on trees, especially if these conditions occur during the growing season. Consider the following steps:

- Conduct an assessment to investigate tree health and structural conditions that could present unacceptable risk levels. Expect trees to sometimes die suddenly, decline in the current year, or decline over the next few years.
- Perform a soil test to determine if elemental deficiencies have developed from the prolonged wet conditions. Provide supplemental nutrition as indicated by the soil tests.
- After flooding recedes, apply a light dose of nitrogen, approximately 0.5–2.0 pounds of nitrogen per 1,000 square feet. Very small amounts of calcium sulfate can be added to return the sulfur to the soil, if indicated by the tests.
- After this application, do not fertilize with nitrogen again until after full leaf expansion in the next growing season. In each of the next 3 years, it is recommended to apply 2–4 pounds of nitrogen per 1,000 square feet.

- Reduce or refrain from any pruning that removes living tissue. Leaves are needed for carbohydrate production and food supplies.
- Monitor for fungal and insect attacks on stressed trees and take informed action. Both the trees under attack and the surrounding trees may need rescue or preventative treatments.
- Continue to monitor until tree health has returned to pre-casualty conditions.

It is important to recognize the legacy effect of flooding long after the floodwaters have receded, as trees can still be at risk from a health perspective and a risk consideration.

TREES AND DROUGHT

Water scarcity is a critical limitation for any plant, especially trees. Drought is an extended period of time without natural precipitation. This means moisture in the ground becomes less available to the tree with diminished resources. If extreme drought occurs, any water at all in the soil profile cannot be extracted for biological processes. There may be some moisture in the ground; however, it is held so tightly by the soil that the tree cannot pull it off the soil particles.

Water is absorbed by tree roots from the soil moves through the xylem vessels through transpiration pull and finally exits as a vapor through the pores or stomata on the surface of leaves. Transpiration is a physiological process responding to soil and atmospheric factors. It is a passive movement of water through the tree system that allows columns of water to move great heights. Water movement through a tree is controlled by the tug-of-war between water availability and water movement in soil versus water loss from leaves. The rate at which water travels can be measured and approximated. For example, water movement in a ring-porous vascular system tree, like a red oak, is about 90 ft/hr, while in a diffuse-porous system tree like a basswood, it is about 10 ft/hr. For pine trees, this rate is often less than 6 ft/hr. Trees can absorb water at rates between 10 and 150 gallons of water daily, yet of all the water absorbed by plants, less than 5% remains in the tree. Trees rely on available water in the soil to "rehydrate" especially during the nighttime hours, attempting to replace the water lost during the daytime hours.

The second process for water gain is the interception of water by the leaf surface and branches during rainfall, and its following evaporation. Together, these two processes are often referred to as evapotranspiration, which is a combination of evaporation and transpiration. Evaporation is a physical process that transforms liquid water into vapor by absorbing heat. Transpiration is the biological process of trees and plants releasing water vapor through their leaves. This water was previously pulled up by the roots from the soil, creating the transpirational pull necessary to move water through the tree. Both transpiration and evaporation are strongly affected by the amount of sunlight, the temperature and humidity of the air, as well as wind speed. Trees release nearly 95% of the water they absorb into the atmosphere as vapor.

Nearly every plant process, including photosynthesis, respiration, and transpiration, rely on water to function properly. Water is an essential element in the tree. It is as important, if not more important, than the other elements because it is required in order to put all other nutrients and elements into a form usable by the plant.

EFFECTS OF DROUGHT

Water is the most limited resource for a tree, and without adequate moisture, decline and death are imminent. Drought can have a major impact on tree health and survival by:

- slowing and reducing growth,
- reducing carbohydrate production,
- lowering energy reserves for survival through winter,
- inhibiting the production of defense chemicals against pests, and/or
- causing death to a portion or even the entire tree.

TREE SURVIVAL STRATEGIES

Trees have developed drought survival strategies for reacting to this interruption of life processes, prioritizing their response from the least to most damaging. However, roots are the first to experience drought stress.

The growth of the primary root system is not severely affected by drought stress. However, the growth of lateral roots is significantly reduced, mainly due to the suppression of lateral root meristems activation. In addition to lateral roots, the growth of small roots, such as adventitious roots develops which is considered as an adaptive strategy to increase water uptake by providing more absorptive surface.

Drought stress is known to reduce photosynthesis by decreasing both leaf area and photosynthetic area. The immediate response of some plants being exposed to drought stress is stomatal closure. The major phytohormone, abscisic acid, is synthesized in the roots and then translocated to the leaves, where it initiates the adaptation of plants to drought stress through stomatal closure. This protective process not only diminishes water loss through transpiration but also reduces CO_2 and nutrient uptake. As a result, it alters metabolic pathways such as photosynthesis.

Changes in cell conditions are initiated as well during drought conditions through osmotic adjustments. Osmotic adjustment (OA) is solute accumulation in dividing cells when water potential is reduced, involved in maintaining cell turgor. Under conditions of drought stress, OA has been connected to maintaining stomatal conductance, photosynthesis, leaf water volume, and growth. The accumulation of compatible solutes helps in protecting the plants from the detrimental effects of drought stress.

Another adaptation to counter drought stress involves leaf-level morphological and structural changes aimed at reducing water loss. This adaptation, known as sclerophylly, results in plants forming hard leaves that resist permanent damage from wilting and can fully recover their functionality when normal conditions resume. This is accomplished by building parenchyma tissues and increasing protective structures like the upper cuticle and both the upper and lower epidermis.

WATER MANAGEMENT BEHAVIOR

How trees manage their internal water resources is important for their survival in short-term and long-term drought conditions. Water management strategies which have evolved in plants are linked directly to the transpiration process. Research indicates changes in moisture availability have stronger near-term impacts on forest dynamics than changes in temperature.

Urban planting sites, with their limited rooting space and harsh environments, create situations which make trees more vulnerable to decline and death. However, trees have developed strategies to manage water deficits and have evolved to manage challenging situations, especially in more natural environments. Examples include deep rooting characteristics to access moisture deep in the soil profile, stomatal closure during the early stages of drought to reduce water loss from their leaves, and the development of marginal leaf scorch and early leaf abscission to decrease water demands in the crown. These strategies help trees avoid water deficits. These tree characteristics are considered isohydric traits, aimed at reducing water loss by avoiding drought stress. Isohydric plants, such as *Ulmus parviflora*, regulate xylem water potentials under levels of moderate soil moisture by progressively closing their stomata.

In contrast, some trees have an ability to tolerate lower water availability or drought stress. They have the capacity to maintain cell turgor pressure during dehydration. Anisohydric species such as *Quercus virginiana* maintain high transpiration rates and are generally independent of soil moisture content until there is little water left to withdraw. These tree types tend to keep their stomata open and can maintain hydraulic conductivity in their stems for longer during the drought cycle. This strategy could be a bit risky for anisohydric trees. It might be considered beneficial when water is abundant, as well as under moderately stressful conditions. However, under conditions of intense or prolonged drought, this trait might endanger the plant. Each type of response has some advantages and disadvantages, impacting the plant's capacity to withstand environmental stress.

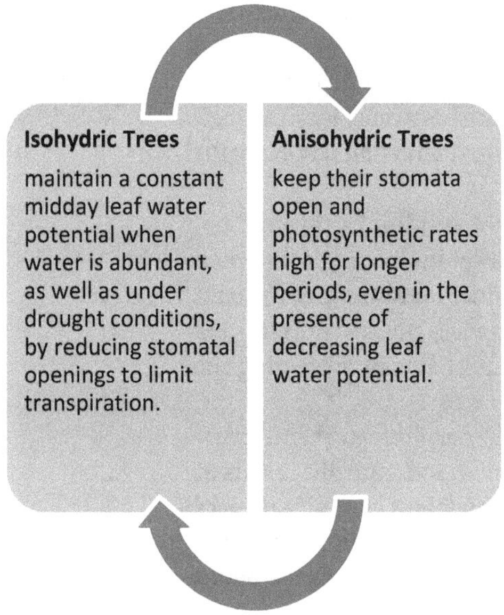

FIGURE 12.23 Tree traits for water management.

Species which seem to avoid water deficits tend to decline more readily when they have limited access to usable soil and water volumes. They are more susceptible to carbon starvation, and often lose their leaves early in the growing season from water deficits. These internal tree strategies may allow trees to survive periodic drought stress, but none of these avoidance qualities really improves tree performance or sustainability in the long term. For us, this translates to reduced returns from ecosystem services from these trees. It is much better to select species with an inherent tolerance to water deficits, as they can maintain normal plant functions longer during drought conditions. Regardless of a species' drought tolerance ranking, the full expression of drought tolerance will only come about in well-established trees. However, the selection of drought tolerant trees should never be seen as a substitute for good post-planting procedures, such as mulching and supplemental irrigation when required by the tree.

Although drought affects tree physiology and growth directly, the impact of secondary factors (e.g., insect pests, pathogens, and fire) is often greater than the impact of the drought

Transpiration is controlled by the atmospheric demand for moisture and is limited by stomatal conductance, which is regulated to a certain extent by plant water status and water availability.

stress and can lead to tree mortality. Prolonged drought conditions weaken trees, making them more susceptible to pests that normally may not invade a healthy tree. Trees in decline are known to produce certain volatile chemicals (VOCs) that can attract insects. For example, woodborers are positively influenced by prolonged water stress and the decline of host tree resistance. In contrast, leaf defoliators benefit from increased nitrogen in plant tissues linked to moderate or intermittent water stress.

More severe drought stress creates physical injury to tissues through drying; damaged tissues are then easily invaded by otherwise "weak" pathogens. During extended dry periods, it is common to see an increase in borer activity, infestations of scale insects, and diseases such as wilt, canker, and root decay. At this point in the environmentally induced decline, expect serious, and often permanent, damage to trees especially in the built environment, where the trees are most likely already under stress.

Drought Strategy for Trees

Low water resources available
Cell alteration and osmotic adjustment
Stomatal closure duration is longer
Increase in fine root production
Energy reserves exhausted
Root suberization and recession
Root and foliage senescence
Leaf abscission

Drought stress incites disease and insects through several different mechanisms:
• It alters the plant's physiology, making trees more susceptible.
• It reduces the ability of the plant to produce defensive chemicals.

LONG-TERM EFFECTS

How a tree responds to subsequent years of drought depends largely on its health and vitality before the prolonged dry conditions and the maintenance inputs during these drier periods. The best defense against any environmental or biological challenge is good cultural practices and plant health care. A physiologically stressed tree is a compromised tree, vulnerable to a myriad of maladies.

A biological lag effect is common in trees where environmental conditions are at their extreme and prolonged. Most noticeable will be decreases in the number of new leaves formed and shorter stem segments (internodes) present. Drought influences the number of leaves, the surface area of leaves, and the extension of twigs as buds expand in the following year.

The results of prolonged dry conditions might not inhibit the first growth flush but may decrease the number of stem units formed in the new bud that will expand during the second (or third, etc.) flush of growth. If the drought continues, all growth flushes will be affected. Thus, tree growth the following year will be atypical and, again, create predisposed conditions to diseases and insects if not monitored and managed.

DEALING WITH DROUGHT

Obviously, drought cannot be prevented, however, there are measures to take which make trees more drought-tolerant and may reduce the long-term effects of prolonged dry conditions.

Always protect tree trunks, especially those of young trees, from mechanical damage, such as that caused by string trimmers, lawn mowers, and other equipment. Preventing damage to the bark and wood at the base of the tree allows a continuous pipeline of water- and food-transporting vessels. Reduce competition for available moisture from other plant materials such as turf, shrubs, and groundcovers, where feasible. Remove those "expendable" plants and extend the mulched area. They are much more readily replaced than established trees.

Mulch trees! Never has there been a more compelling reason to maintain adequate levels of mulch than during periods of drought. Use organic mulch, such as clean woodchips, and

FIGURE 12.24 Proper mulch levels can reduce evaporation from the soil.

apply it to a depth of 2 to 3 inches around the plant root zone. Be sure to replenish the mulch throughout the year as the mulch decomposes or is displaced. This will help reduce moisture loss through evaporation and moderate soil temperatures. Extend the mulch beds and tree rings to the edge of the tree canopy (drip line) to protect those fine "feeder" roots from drying out.

It is important to water trees, especially young ones, whenever rainfall is insufficient for extended periods. A proven recommendation is the 5+ rule, which says to provide at least 5 gallons of water for every diameter-inch of tree trunk. This should provide adequate water to help the tree during times of inadequate moisture. In landscape plantings, the equivalent of one-half gallon of water per square foot of mulched area should be applied at ground level at slow rates to reduce evaporation and runoff. Be sure to check soil moisture for adequate drainage. It is possible to overwater in poorly drained soils, which is equally as harmful. Symptoms of overwatering are often similar in expression to underwatering, which may include chlorosis and dieback. This prolonged wetness and saturation of soils can lead to disease and is also harmful. Soil type is important and has an impact on moisture levels as well. Sandy soils usually require more frequent irrigation than the heavier clay content soils. Understanding your soils helps in deciding how much to water them. In landscape plantings, the equivalent of ½ gallon of water per square foot of mulched area should be applied at ground level at slow rates to reduce evaporation and runoff.

For mature and well-established trees, provide at least 1 inch of supplemental water every week or so to maintain adequate soil moisture. To determine the amount of irrigation needed, place a catch device at least 1 inch deep in the irrigated area to measure the amount of irrigation provided to the root zone of the tree-planting space. The critical root zone is located within the drip line of the tree. It is advisable to water plants though the fall until the ground is frozen, so that trees have adequate moisture to survive the winter months and are ready for spring growth. Nighttime irrigation is optimal, as this is when the tree is physiologically catching up from the water loss during the hot daytime temperatures.

Decrease or cease certain maintenance activities such as pruning and fertilization. Pruning live branches is not recommended in a drought, because trees must expend energy to repair pruning wounds and loss of green mass. It is not a good practice to fertilize any tree under stress conditions, especially drought. Trees expend additional energy to process the

Watering Facts

- An inch of water per square foot is equal to 0.62 gallons.
- An inch of water will soak a clay soil about 4 inches deep.
- A cubic foot of water contains 7.5 gallons.
- To determine hose output, measure the time it takes to deliver water into a 5-gallon pail. Then, divide the 5 gallons by the time in minutes to determine the gallons-per-minute output.

nitrogen, the fertilizer can act to pull water from the roots due to the osmotic gradient which can worsen tree health. Do not disrupt the soil or areas under the drip line or canopy of the tree in drought. This will reduce the capacity of the tree to take up water due to root disturbance and damage. Every viable root is needed to absorb the already scarce water.

Trees have managed to do quite well without our interventions, growing to be incredible, biological machines providing life-sustaining benefits. However, in urban and suburban environments, trees often require assistance to overcome challenging urban conditions. Tree health and vitality stems from good cultural practice which includes adequate moisture. Regardless of the delivery system or technique, when nature isn't supplying enough water, it is essential to provide supplemental water.

WINTER DORMANCY

In temperate climates, "winter" and "dormancy" are similar terms that describe a state of rest for trees. Dormancy is not death; it is a natural state in which trees adapt to cold conditions with physiological and biological adjustments. Even though the leaves are changing colors and dropping to the ground, trees are still active with important life processes. Winter conditions can make obtaining moisture difficult for any plant. Potential sources of winter water include unfrozen soil, internal reservoirs, and the subnivean zone (the area above the ground, but just under the snow cover). If trees cannot find the water they need at the right times in these areas, expect a response the following spring. This response may be in the form of winter desiccation, fewer and smaller leaves, as well as other symptoms.

FIGURE 12.25 Snow and ice are significant loads on a tree's structure.

Summer and fall drought conditions can place trees in an overall water deficit, predisposing them to pest issues and poor health the next growing season. Likewise, these adverse weather conditions affect trees' ability to survive dormancy. Young and newly planted trees are even more vulnerable, requiring supplemental irrigation because of limited root systems capable of obtaining water in their new environment. The key to survival is giving trees adequate moisture before winter freezes the world around them.

Good cultural practices and plant health care make a difference in how much water is available to your trees in winter and how well they respond to winter conditions.

Winter desiccation injury is a common problem for needled and broadleaf evergreens. Desiccation injury may occur because transpiration is still ongoing and the plant is unable to replace the water loss due to frozen soils. Water is lost through plant tissue more quickly than it can be absorbed though the roots. Typically, this is caused by long dry periods of cold accompanied by winter winds which increase transpiration rates.

Some of the more common trees and plants that experience winter desiccation or winter burn include:

- Broadleaf evergreens such as boxwood, holly and rhododendron.
- Conifers such as arborvitae, cedar, spruce, juniper and pine.

Winter burn symptoms often become apparent as the snow melts and spring temperatures rise. Foliage starts to brown at the tips of branches with browning progressing inward toward the center of the plant. On broadleaf evergreens, leaf edges typically brown first, followed by the browning of entire leaves. Tree crowns facing south, southwest, or west is most often

TABLE 12.1 Guidelines for Improving the Winter Experience for Trees

Trunk Wraps	Smooth or thin-barked trees, like honey locust, crabapples, linden, and especially maple, are susceptible to sunscald and SW injury because of the temperature fluctuations and sun exposure in the winter. The wounds caused can leave the tree exposed to fungal organisms causing decay. Prevention includes protecting the trunks of younger and smooth-barked trees up to near the first branches using a flexible, white tree wrap. Leave the wrap on until after the last freezing temperatures. Remove tree guards in the spring to reduce potential damage from disease and insects on the trunk.
Mulch	Mulch benefits trees all year long, not just during the summer months. Refresh the mulch layer placed in the spring so that there are about 2–3 inches of wood chips, bark, or other organic mulch over the critical root zone of the tree. Start at the base of the trunk, but not against it, and extend mulch to the edge of the outer branches or dripline. This will reduce evaporation, improve water absorption, and insulate soils from temperature fluctuations. Existing mulch should be cultivated lightly to loosen and allow mixing with the new application.
Fertilization	If trees are not stressed from moisture deficits, and tests indicate, apply a complete fertilizer with micronutrients. Use a low-nitrogen formulation (5–10–10 or similar ratio/combination) applied after fall color begins until leaf drop to prevent a late, unwanted flush of new growth. Be sure to broadcast the material uniformly over the root zone followed with sufficient water, allowing the roots to absorb nutrients. Water-soluble fertilizers may also be applied during watering sessions to provide supplemental nutrition as well. Follow label directions. This practice is directed more towards newly established trees.
Irrigation	Whenever rainfall is insufficient for extended periods, supplemental water is critical, especially on newly planted and early established trees. Follow the "5-gallon rule," which is to provide 5 gallons of water for every diameter-inch of tree trunk. This should provide adequate water to help a tree during times of inadequate rainfall for most situations. For mature and well-established trees, 1 inch of supplemental water applied to the root zone every week should keep soil moisture adequate. Continue to water through the fall until the ground is frozen, so that trees have ample moisture to survive the winter months and do not go into spring under a deficit. If limited rain occurs in the fall or snowfall in winter indicates drought, it may be necessary to water in the winter. The best time for winter watering is on a warm day, when the temperature is above 40°F.
Pruning	Late fall and early winter are acceptable times for the functional pruning of most tree species when wood is not frozen. Prune trees to remove dying, diseased, or dead branches. This is also a good time to remove basal sprouts as well. However, limit the extent of green wood pruning as winter approaches to reduce the amount of energy reserves the plant must use to heal the pruning wounds.

affected due to common prevailing winds. Symptomatic foliage often begins to drop off starting in spring and continuing through mid-summer as new foliage is produced. In extreme cases, entire plants can brown and die.

One way to help avoid winter burn is by applying antidesiccants or antitranspirants. These specialized sprays add a protective waxy coating to the leaves of broadleaf evergreens to reduce water loss. If antidesiccants are going to be used, they should be applied in the late-fall or early-winter, per label directions.

The use of an antidesiccant does not guarantee foliage will not experience winter burn-type injuries. Also, these treatments do not prevent, but only reduce moisture loss through the leaves. Furthermore, antidesiccants do not protect against plant cells being damaged from the formation of ice crystals within the foliage during excessively cold temperatures. Tests indicate antitranspirants are most useful when the air temperature is above freezing, but the soils are still frozen or cold (water movement is extremely slow in soils or plant stems when temperatures are below 40°F).

One important physiological impact is on stomatal openings. Although the waxy, film-forming products do help minimize water loss through the leaves, they can reduce the amount of carbon dioxide allowed to pass through the stomatal openings. Evergreen plants continue to photosynthesize even during the winter months. The reduction of a sufficient level of carbohydrates can cause a reduction in energy for the normal growth of leaves, stems, roots, and flowers. Therefore, blocking the stomata of evergreens to protect them from winter wind can be a benefit, but it could result in a weaker plant being more susceptible to other stressors, such as drought, later.

Generally, antitranspirants are tools that help trees growing in otherwise optimal conditions endure stressful, short-term conditions. Common situations include trees exposed to high winds in open area plantings and newly established trees adapting to their new environments. It is important to remember that these applications are not likely to be successful as stand-alone protection. Good cultural practices, such as watering and proper mulching going into cold weather will increase the efficacy of antitranspirants. Finally, sometimes the use of burlap barriers may be the best remedy to protect plants in more extreme environmental sites. Although these covers are often considered unsightly looking winter protections, they will also provide the best defense against sprays of deicing salts on roads or driveways.

CHAPTER 13

ETHICS AND PROFESSIONALISM: WHAT DOES IT MEAN FOR ARBORISTS AND BUSINESS?

Professionalism and ethics are two words that are seemingly important to many individuals, and yet, in our behavior, we often, unconsciously, or perhaps deliberately, overlook these very important interpersonal attributes. From a business perspective, ethics are guidelines used when interacting inside and outside the company. However, they are often missing in many business models because they aren't typical content in the organization's handbook.

Ethics are the science of conduct and have been a topic of philosophers for over 2,500 years since Socrates and Plato. Ethics comes from the Greek word, *ethos* which is the distinguishing character, moral nature, or guiding belief of a person, group, or institution. Ethics are the standards by which society measures its own behavior and that of other people, groups of people, and institutions. We think of ethics as the principles which guide our behavior toward making the best choices that contribute to the common good of those around us. Ethics are what guide us to tell the truth, keep our promises, take responsibility for our actions, and offer mutual respect for others. There is a framework of ethics underlying our lives daily, helping us make decision that create positive impacts, and steering us away from negative outcomes. The decision to behave ethically is guided by our internal moral compass. However, a moral compass can only point someone in the right direction; it can't make them go there. Ethics are a learned quality that individuals choose to practice throughout their lives, every day.

FIGURE 13.1 ISA certified arborists project professionalism.

Business ethics are just as important as ethics in our personal lives. Business leaders have a unique role and a great responsibility in shaping the ethical culture of their businesses, and thereby influencing their broader communities as well. Ethics can be basically defined by the three B's. For us, as arborists, this translates to *being* an arborist distinguished by our character and our guiding beliefs as it relates to planting, maintaining, and caring for trees; *behaving* like a professional arborist by our decision-making principles guided by our moral nature; and *becoming* a better arborist, not just in science and practice but morally.

Simply stated, ethics involve recognizing right from wrong, but just as importantly, making the right choices, and being consistent about the principles guiding behavior in every situation. Business leaders and managers are under tremendous pressure and, at times, may face significant ethical challenges. Decisions taken within an organization may be made by individuals or groups, but whoever makes them will be influenced by the company culture. The attitudes and culture of the company transcends to the client about their perception of you and the company you represent.

When placed in a compromising situation or challenging ethical dilemma, you enter a crossroad which requires analysis of the situation and to decide. Hopefully the decision you make is a moral decision that positively contributes to the common good of all those around you. Ethics shouldn't be compromised or confused with success or achievement in business performance. Ethical business is about acting without bias in your business activities, consistently.

FIGURE 13.2 Relational equity in a company improves trust and safety.

Another important concept that factors into company culture and relationships is called relational equity. This is a behavioral theory that studies the distribution of resources between relational partners. According to this social theory, relationship equity matters because people seek relationships where there is balance. In other words, when someone receives more than they give, they often can feel guilty about it and especially the other way is true as well.

One of the strongest bonds we can have with those around us is relational equity. The greater that equity is, the greater trust when working together. It is an intentional investment in a relationship over time, allowing for trust, value, and a better appreciation of the connection. However, this equity is not built quickly. One of the advantages of long-term friendships in the personal arena is the opportunity to build the social parity over sometimes many years. The longer we have a healthy relationship with others, the greater the trust that is developed and the more we can mutually count on one another, help one another, and positively influence one another.

The success of training and learning relies upon the relationship between the trainer and the learner. We listen to those we trust and who we know has our best interests in mind. On the other hand, where such parity does not exist, we often resent or discount the advice or mentoring from others who try to push too hard into our lives. This is not surprising since we may not recognize their motivations or know whether we may not be able to trust their intentions.

Factors in building relational equity:

- It is built over time because time validates our intentions and builds trust.
- It must be clear to others that we always have their best interests in mind through our actions and service to them.

- We must not have overstepped appropriate boundaries in the relationship or taken advantage of others.
- Our interactions must be respectful and lift others up, making those around us better.
- We must be able to be relied on when needed and be faithful friends, colleagues, and peers.

All of these qualities build significant equity in our relationships. When any of these principles are violated, the social parity is diminished or even lost, and in some cases, it may be challenging to rebuild or may never be regained.

CARE FOR BOTH KINDS OF CUSTOMERS

We all recognize the importance of taking care of the external customer (the people who purchase our products and services). They are often seen as the most important part of our business and our very livelihood. However, successful organizations equally recognize the importance of taking care of the internal customer (the people who work on behalf of your organization).

Determining whether the client or "external customer" is satisfied with the work is easy. As we all know, external customers can be very emphatic at times—both about the good and the bad. But it can be challenging to determine if the "internal customer" is happy. Taking the temperature of company morale is an important administrative practice. For example, you may ask yourself, "How are the crews doing?" or even more importantly, "What is their tenure going to be here at my shop?"

I recommend, as a former manager and supervisor, taking time as the leader of the company to witness crew activity each morning or evening silently, subtly, and respectfully in the shop or in the office. This occasional look-in will provide the opportunity to listen and observe your employee's language, actions, and relational qualities. Do your observations uncover respect, courtesy, and care, or do they demonstrate carelessness, rudeness, dishonesty, harassment, and offensive behavior towards each other? These are key indicators of potential issues that may also be occurring outside the shop, which proves to be even more awkward and damaging for business.

Internal customers serve as a direct link to external customers and have a significant impact on the quality of the product or service provided. The presence of an ethical corporate culture with your internal customers is important because the delivery of a great product begins with them! The supply chain starts even with the receptionist who answers phone calls and is often seen as the face of your company, and continues to the crew, the branch manager, the sales representative, and to every person in the supply chain. If the culture is broken or less productive internally, imagine what the external customer is seeing and experiencing on the jobsite. Consider how your own behavior and the behavior of your employees affects every aspect of the business.

IDEAS FOR *CREATING* AND KEEPING AN ETHICAL WORKPLACE

Training for Ethics

When a company gets a new hire, is it too late to talk about ethics and train them about ethics? Or are ethics just an unspoken expectation? According to a few business experts

I follow, some believe ethics *can* be taught, while others say that it depends on how a person was raised to an adult. Can they distinguish between being morally right and being ethically wrong? Business psychologists today say ethics consist of knowing what we should do in sticky situations, especially at work, and that the concept can be taught in most cases.

Communicating Ethical Expectations

FIGURE 13.3 Ethical behavior is a part of company culture which can be taught.

Owners, managers, and supervisors set the tone when they talk about doing the right thing consistently with staff. Communicate clearly, consistently, and frequently the desired behaviors and why they are important to the organization. It is an easier task to train employees on what to do as opposed to telling them what not to do; by then, usually the damage has been done. When discussing ethics and behavior, be clear in your language, message, and expectations to leave less room for misinterpretation.

The company culture, whether it is ethics, safety, or any other attribute, flows downhill and begins with the boss. Subsequently, leadership often have the greatest responsibility and visibility on the job. It can't be said enough that one of the best ways to implement and ensure ethical business practices is a "top-down" approach. Those at the top of a company are often seen as role models for other employees, and if they are acting in a way that is unethical, the lower-level employees will likely follow suit.

Certainly, senior staff are important, but they often only talk to each other, not to the crew and other staff. Ethics should be a regular conversation in any training event. It should involve all ranks in the discussion, making everyone feel like an important and confident part of the team. Remember, a leader's consistency between their words and actions speaks loudest. We cannot expect staff to do what we say and not what we do, as they are watching and modeling our behavior. Whether facing minor quandaries or major problems, constantly raise the question, "What do our organizational values tell us about how to handle these decisions?" That makes ethics very real and tangible, especially if those values are in place and discussed regularly. But how do you keep the ethical ship afloat? In other words, once the concept becomes an integral component of your organization's mission and goals, how do you keep the momentum?

Atta Boy Moments

> Bottom line, employees tend to separate from their boss, not their job.

Exemplary behavior should be recognized and can be used as a teaching moment. This builds the team and the image of the company and creates loyalty within the crew and staff. Kudos are good for everyone and not just those who receive them. It creates an

FIGURE 13.4 Build company morale with recognition of exemplary performance.

aspiration in others for positive attention and recognition. This is the best way to build people and retain long-term employees. Discussions with former students and early career professionals reveal their dissatisfaction often stems from issues such as inconsistent policies, radical and unplanned changes, unclear company goals, and poor employee relations. These situations demonstrate a lack of ethical leadership.

Keeping Them Around

Retention and productivity can be increased by developing the team, personally and professionally. This is an investment in the person and the employee. Professional development is a key benefit articulated by nearly every professional. Behavioral development is a key training aspect as well. It provides the opportunity to discuss ethical concepts such as character and conduct. Those implementing ethics training have good intentions, but once introduced, it often dies on the vine because companies lack continuity in their training programs, stopping after a few months. Long-term investments in training and development help encourage employees to stay with a company for the long haul. Frequent communication of desired behaviors, such as accountability, professionalism, honesty, and respect, and why they are important to the organization, should be discussed with everyone, not just the new staff or lower-level staff. organization's goals and culture can only be clearly and consistently understood if everyone is involved in the journey.

Sustainable Growth, Ethically

It has been proven that ethics-driven companies are more productive and sustainable.

When profit is viewed and embraced as the sole purpose of the company, it distorts and corrupts the whole process. Corporate purpose built on quality relationships based on trust with customers and employees inspires the highest levels of ethical leadership and success. It is critical to hold people accountable for demonstrating ethical behaviors, raising the bar for excellence, safety, and production.

Adding ethics to our day-to-day interactions, work, and personal connections creates a better world and workplace. It is evident that maintaining a professional and ethical mindset, as individuals and organizations, is vital for our future, industry, and careers. When companies set the tone with meaningful ethical standards, the result is a positive corporate culture, where people want to go to a safe workplace, and clients keep coming back. I was once told many years ago, "What is good to know is often the most difficult to learn." This certainly gives credibility to the ethical business challenges ahead of us and the need to instill and maintain an ethical practice in the tree care industry.

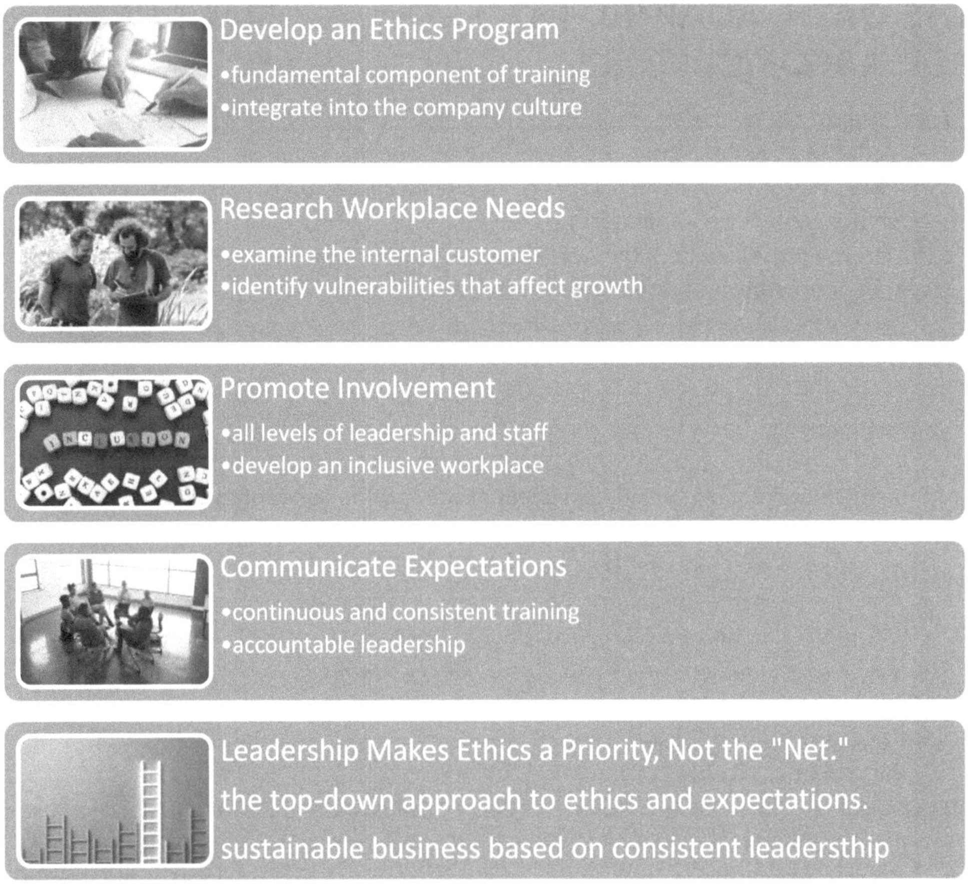

FIGURE 13.5 Ethical development in business.

Many arborists possess professional credentials or have plans to obtain them, which indicates a level of knowledge and a representative skill set, such as the ISA Certified Arborist or TCIA Certified Treecare Safety Professional certifications and other titles. These credentials represent a consistent body of information across the industry and provide guidance for interacting with a client. However, these credentials and accomplishments do not express the need for a responsibility of care or ethical treatment of our patient, which are the trees in our urban forest.

This led to the consideration of a long-standing, historical set of guidelines that are science-based, yet interactive with the patient (tree), the practitioner (arborist) and the client. The Hippocratic oath is one of the oldest binding ethical codes in history and is an oath historically taken by physicians and other healthcare professionals. It involves a commitment to practicing medicine honestly and fostering strict loyalty to peers.

The tenets of the medical profession, when adapted to arboriculture, connect us to the science with an ethical directive. This is the foundation for an ethical practice that includes accountability to the patient (tree). The parallels between humans and trees within this doctrine are strikingly similar. Both are dynamic living organisms that respond to external stimuli in various degrees in which the principles can be applied.

THE OATH, ADAPTED INTO ETHICAL PRACTICES FOR ARBORICULTURE

- *I will respect the hard-won scientific gains of those physicians in whose steps I walk, and gladly share such knowledge as is mine with those who are to follow.*
    - I will respect the scientific gains of others, keeping current with industry research, while supporting and mentoring others.
- I will apply, for the benefit of the sick, all measures which are required, avoiding those twin traps of overtreatment and therapeutic nihilism.
    - I will apply all curative measures required, avoiding overtreatment, or empty assurances of a cure.
- *I will remember there is art to medicine as well as science, and that warmth, sympathy, and understanding may outweigh the surgeon's knife or the chemist's drug.*
    - I will remember arboriculture is an art as well as a science, requiring mind and muscle in its practice, being sensitive to client needs and the needs of the tree.
- *I will not be ashamed to say "I know not," nor will I fail to call in my colleagues when the skills of another are needed for a patient's recovery.*
    - There is no shame in saying, "I don't know." Be resourceful, call on your peers, never letting the situation exceed your skills.
- *Tread with care in matters of life and death. If it is given me to save a life . . . but it may also be within my power to take a life.*
    - I recognize and understand not all trees can be saved, nor should they, mitigating for the greater good of the client.
- *I will remember that I do not treat a fever chart, a cancerous growth, but a sick human being.*
    - I will consider the impact of any intervention on the tree in whole, not causing further damage. No intervention should contribute to making the condition of the tree worse.
- *I will prevent disease whenever I can, for prevention is preferable to cure.*
    - I recognize prevention is preferable to cure wherever possible, applying proactive cultural care whenever possible.
- *I will remember that I remain a member of society, with special obligations to all my fellow human beings.*
    - I understand that I belong to a society with an obligation of respect to my colleagues and peers with representative responsibility.
- *May I always act so as to preserve the finest traditions of my calling and may I long experience the joy of healing those who seek my help.*
    - I will act to preserve the science and practice and enjoy the experience of this discipline.

This philosophy could be considered as a guide for the practitioner (arborist) of modern arboriculture. It delivers basic criteria for evaluating the appropriateness of arboricultural interventions and deciding upon the ethical choice of care for short- and long-term objectives. As a society and professionals, we are accountable for our decisions and actions as well as the consequence of those actions. We must consider our responsibility of care for the tree and the urban forest and abide by the physiological demands of the tree. The first consideration for ethical decision-making with the treatment of trees should be based on biology, not business.

People are concerned for the planet and we, as a society, want to help the Earth. As air and water quality are challenged, people are taking a more serious look at this perfect biological machine—the tree. This patriarch of our landscape is essential for our survival on the planet. As professional arborists, our primary duty is to maintain safe and healthy trees to improve our quality of life. With such an important charge to us as tree care professionals, we should abide by an ethos that includes ethics and a sensible duty of care for our patients, the trees.

APPENDIX 1

PRUNING PRESCRIPTIONS

Adapted from the Florida ISA Prescription Pruning Qualification

Tree pruning can be complicated by the many variables involved, including species, growth habits, previous pruning implemented, and various other factors. Therefore, prescriptions should be prepared by professional arborists with proven and demonstrated knowledge in creating and implementing clearly written pruning prescriptions easily interpreted by the crew. These prescriptions created by the prescriber are essential for communicating what is to be accomplished by the pruning intervention to the producer (climbing arborist).

If prepared properly, pruning prescriptions establish tree owner objectives and describe how the arborist can meet them with pruning. Prescriptions guide and protect both the tree owner and the tree care professional by ensuring all involved clearly understand the objectives and scope of the work. Municipalities, utilities, commercial arborists, and tree owners benefit from writing and using pruning prescriptions. The language and practices contained in national standards (ANSI A300 Part 1) should be applied when writing pruning prescriptions.

Arborists, horticulturalists, or any tree care professional drafting prescriptions should have working knowledge and experience with modern arboricultural pruning techniques. This includes being familiar with the applicable safety standards (ANSI Z133) and best management practices and being up-to-date with current research. Additionally, anyone pruning should have demonstrated success and experience with proper pruning techniques. The most important qualification for a prescription writer is an understanding of pruning objectives and the ability to describe them in a clear, concise, and consistent way. This allows for efficient and productive collaboration between prescribers and practitioners. Sample forms are available in the pruning best management practices resource.

A pruning prescription should include the following information.

1. **Trees to prune** (plant names, common and/or scientific)
2. **Location in the landscape** (street address and position in the landscape e.g., all trees lining the driveway)
3. **Pruning system** to be applied (e.g., natural, pollard, topiary); the natural pruning system is by far the most common and will be the only system described in this document
4. **Purpose** or reason for pruning: (e.g., improve sight lines, provide clearance, improve a view, reduce risk)

5. **Pruning objective(s)** (e.g., improve branch structure, clearance, reduce risk, manage size, improve health)
6. **Type of parts to remove** (e.g., live branches, dead branches, dead fronds, fruit)
7. **Size range of branches to remove** (e.g., diameter range, maximum or minimum diameter, length, e.g. cut only 3-4" branches)
8. **Location within the crown** of the parts to remove (e.g., largest-diameter first order branches, in the periphery of the crown, top half of the crown, over roadway, largest aspect ratio branches, branches on the travel lane side of tree, over the building)
9. **Type of pruning cuts** to utilize (e.g., branch removal cut, reduction cut, heading cut)
10. **Amount to remove** (e.g., all dead branches over 2 inches in diameter; number of pruning cuts of a certain diameter; 5 to 7 feet of length; a percentage of length or height; a percentage of foliage such as not to exceed 30% of live foliage)
11. **Limitations** (e.g., no removal cuts on the trunk)
12. **Plan for disposal** of debris (e.g., remove all debris, chip brush, leave wood chips, leave firewood rounds for client)
13. **Time frame** for completion of the work (e.g., certain date, over the winter, July)
14. **Reinspection** or pruning interval (e.g., 3 years, annually during the growing season, upon review of tree health and recovery)

The prescriptions should be based on tree owner's objectives. These parameters should be adjusted according to the tree owner's tolerance of severe pruning voids in the crown, the proximity of the tree to pedestrians and structures, and the severity of any defects. Details, especially drawings or images, should be incorporated as needed and are often essential to communicate what is to be performed by the crew. Prescriptions can be used for single trees or several trees. Single-tree prescriptions can be very specific, detailing what is to be done on the tree; prescriptions written for a large grouping or population of trees are likely to be less specific and broader in language to account for the variation among trees in the population to be pruned.

prescribing arborist: The arborist preparing the pruning prescription with details on the pruning strategy.
production arborist (practitioner): The arborist responsible for carrying out the pruning prescription by pruning in the tree in accordance to the prescription.

Prescriptions should be short, concise, and simple to write, ensuring the contractor can easily understand the instructions for executing the work. An experienced, skilled arborist with a history of meeting objectives and providing quality service may execute the work exactly as written, whereas a newer, less experienced arborist or one unfamiliar with this methodology may require more detail or training and practice in advance.

Calibration between the prescriber and practitioner is often the best way to collaborate on pruning projects successfully. This requires the team to develop a written prescription on paper and to go to the field and apply it to real trees. As the arborist prunes the trees according to the prescription, the prescriber may realize that the prescription requires an adjustment to meet the defined objectives. The prescribing arborist will become better at preparing clear prescriptions after some practice sessions on a variety of tree species, sizes, and scenarios. Items that may need adjusting include the type of cuts applied, the diameter of pruning cuts, the diameter or aspect ratio of primary branches to be pruned, the number of pruning cuts to apply to primary branches, and the location of pruning cuts. This is a major paradigm shift from decades of pruning practices and will require a commitment to employ this methodology accurately. However, in the end, a detailed prescription provides a more reliable system of operating and is a better way to see that work is being performed in a manner that meets the needs of the arborist, the client, and the tree.

BRANCH ORDER AND LABELING

The application of written prescriptions requires consistency to be effective. This includes a clear understanding of the standards and best practices and identifying branches specifically as well as describing which branches are to receive pruning and the amount.

Lateral branches grow from larger branches, called parent branches, which originate from the main stem or trunk. In prescriptive terms, this would be the 1st order branches. Ideally, each proceeding branch should be at least half the size of the previous branch order, 2nd order branch and so on. Prescribers should use the concept of branch order. First-order branches, regardless of their size, grow from the trunk or main stem.

Branches are considered first order up to the point where bifurcate into (equal diameter) branches. This bifurcation can occur close to the trunk or out at the edge of the crown. The two or more branches beyond the dividing point are considered second-order branches. Most pruning operations are performed on the periphery of the largest first-order branches. The smaller first-order branches can be left alone. The better option for the trunk is to retain first-order branches on mature trees because removal cuts are larger and lead to decay. This can be managed by stating in the prescriptions first-order branches remain on the trunk or no removal cuts on the trunk are permitted.

Arborists often reduce a first-order branch back to a second-order branch to accomplish objectives. This reduction cut is described in the ANSI A300 standards as pruning to a one-third lateral cut. The cut should be made just beyond the lateral branch and the remaining branch should be one-third to one-half the size of the branch removed. This remaining branch will then assume the terminal role for support and survival. Second-order branches can be reduced back to first-order branches or removed where doing so would meet objectives. Third-order branches are lateral branches that grow from second order.

PRESCRIPTION EXAMPLES

Sample Pruning Prescription for Single Tree

Tree to prune: Large oak (*Quercus*) in the front yard.
Pruning system: Natural.
Purpose: Improve tree structure
Objectives:

- Reduce the risk of live and dead branch failure; increase light penetration to the ground.

Branches to prune:

- Remove dead, dying, and broken branches greater than 1 inch in diameter at the point of attachment locating a healthy branch (a branch with foliage) or on the trunk.

- Reduce the live branch on the north side (about 8-inch diameter) with a crack and dead twig using two 3-inch reduction cuts.
- Reduce the five largest live upright-growing first-order branches using two 2.5–3-inch-diameter reduction cuts on each branch for a total of 10 reduction cuts on the crown periphery.

Limitations: No removal cuts should be used to eliminate live branches.
Completion date: Early spring of this year.
Reinspection: 2 years.

Sample Pruning Prescription for Pruning Trees at Planting

(Note to prescriber: The structure on most newly or recently planted trees can be improved significantly with structural pruning at planting or within the first year or two after planting. This prescription could be applied to new trees with good vitality planted in any location. Pruning can be conducted to improve branch structure without compromising health. The goal is to create one dominant leader near the center of the crown by reducing the growth rate of the largest competing branches. Growth suppression on these largest branches is accomplished by substantially reducing their length, or by removing them, or by a combination of reduction and removal cuts.)

Trees to prune: 45 trees (a mix of medium- and large-maturing tree species) with a caliper size ranging from 3.5 to 4 inches.
Pruning system: Natural.

Purpose: Create strong branch structure; limit the need for major branch removal as tree matures. (*Note to prescriber: Removing large branches later can negatively impact tree health with large wounds vulnerable to decay.)*

Objectives:

- Improve crown structure with a combination of reducing and removing branches and codominant leaders with an aspect ratio larger than one-half. Encourage growth in one dominant leader. *(Note to prescriber: Aspect ratio is calculated as the diameter of the branch compared to the diameter of the trunk measured directly above the branch union. Two-thirds can be adjusted to one-half for a more aggressive long lasting pruning application.)*

Branches to prune:

- Reduce length by two-thirds on all primary branches with an aspect ratio larger than one-half by cutting back to a lateral branch that is larger (where possible) than one-third the diameter of the pruning cut; do not retain any pruning cuts on the main stem as the leader.

Limitations:

- Less than 20% of cuts on live branches shall be branch removal cuts. *(Note to prescriber: Including this statement in the prescription requires the contractor to use mostly reduction cuts to prune the trees.)*
- Small branches on the lower 3 feet of the trunk shall not be removed unless they have a diameter larger than 1 inch.
- $<$ 50% of the foliage shall be removed from an individual tree.

Debris disposal: All debris will be removed from the work area and recycled as mulch or disposed of at the green waste recycling center.

Completion date: TBD

Sample Pruning Prescription for Pruning Young Medium Trees

Typical location would be a median, interchange, street tree, or other location within about 25 feet of the roadway.

Trees to prune: 122 oak (*Quercus*) street trees, ranging from 14 to 120 inches in DBH and 30 to 40 feet tall.
Pruning system: Natural.
Purpose: Improve branch structure; create clearance for pedestrians and vehicles under the crown; prevent large branch removal from the lower trunk later; reduce likelihood of tree failure.
Objectives:

- Improve crown structure by reducing branches and leaders with an aspect ratio larger than two-thirds. Encourage growth in one dominant leader. *(Note to prescriber: Aspect ratio is calculated as the diameter of the branch divided by the diameter of the trunk measured directly above the branch union.)*
- Provide clearance for adjacent sidewalk and street. *(Note to prescriber: Adjust language for the situation; for example, some communities require 8 feet on the pedestrian side and 14 feet on the clearance street side).*

Branches to prune:

- Remove at the trunk all primary branches and sprouts less than 4 inches in diameter. *(Note to prescriber: Adjust the diameter of the branch to meet conditions)* that originate from the lower 10–14 feet of trunk. Reduce all larger branches growing from the lower 16 feet of trunk using 3 to 4 reduction cuts of approximately 3 inches in diameter to remove about 50% of foliage on each. Remove additional secondary branches from the lower crown to create 8 feet of clearance under the crown on the sidewalk side and 16 feet of clearance on the street side. *(Note to prescriber: Adjust clearance height for the current and expected form of the species and the situation. You will notice that clearance increases on slightly older and larger trees in*

the prescription that follows. This amount of clearance is governed by the location of the tree, growth rate, species crown form, and when trees are expected to be pruned again. Be sure to mind the recommended 60% LCR).

- Reduce the three branches with the largest aspect ratio in the upper half of the tree using one 3-inch reduction cut on each.
- Remove dead, dying, diseased, and broken branches greater than 1 inch in diameter at the point of attachment. *(Note to prescriber: Adjust diameter of branches to remove based on species attributes and site conditions).*
- Reduce (preferred) or remove (less preferred) branches which are in conflict with adjacent trees.

Limitations:

- Less than 35% of cuts on live branches shall be branch removal cuts. *(Note to prescriber: Including this statement in the prescription forces the contractor to use reduction cuts to prune the trees).*
- No more than 40% of the foliage shall be removed from an individual tree.
- Trees shall not be lion's tailed during the pruning process.

Debris disposal: All debris will be removed from the work area and recycled as mulch or disposed of at the city green waste recycling center.
Completion date: TBD
Reinspection: 3 years.

Pruning Mature Trees

Typical locations might include street trees, open lawn areas, parks, or highway interchanges.

Trees to prune: 12 trees with a DBH greater than 30 inches in the picnic area of the park.
Pruning system: Natural.
Purpose: Preservation of mature trees.
Objectives:

- Improve aesthetics.
- Reduce risk of branch, trunk, and whole tree failure.
- Enhance visibility under tree crowns.

Branches to remove:

- Remove dead, dying, and broken branches that measure 2.5 inches and larger in diameter at the point of attachment through the entire crown.
- Remove branches less than 6 inches in diameter from the lower 10 feet of the trunk.

Branches to reduce in length by one-quarter to one-third:

- Reduce branches that extend outside of the shape of the main crown. Reduction in length shall be accomplished using reduction cuts on branches 2 to 4 inches in diameter. Some trees may not require this depending on crown shape and where no branches extend beyond the main crown.
- Reduce the largest four branches (four is an average per tree but could range from one to six depending on the tree) measured at the trunk that originate 10 feet or more from the ground with four 3–4-inch diameter reduction cuts on each for a total of 16 reduction cuts per tree. *(Note to prescriber: You could write instead "reduce the four largest aspect ratio branches." Specify the number of branches that make the most sense for the situation considering the compartmentalization ability of the species, the pruning cycle, and the condition of the trees.)*
- Reduce length by one-third on primary and secondary branches with bark inclusions or visible cracks.

Limitations:

- No heading or shearing cuts shall be made.
- No more than one-fourth of the foliage shall be removed from an individual tree. *(Note to prescriber: Use your own number here; 25% is only a suggestion. More may have to be removed in certain circumstances to meet one or more of the objectives.)*
- Trees shall not be lion's tailed in the pruning process.

Completion date: TBD.
Reinspection: 5 years.

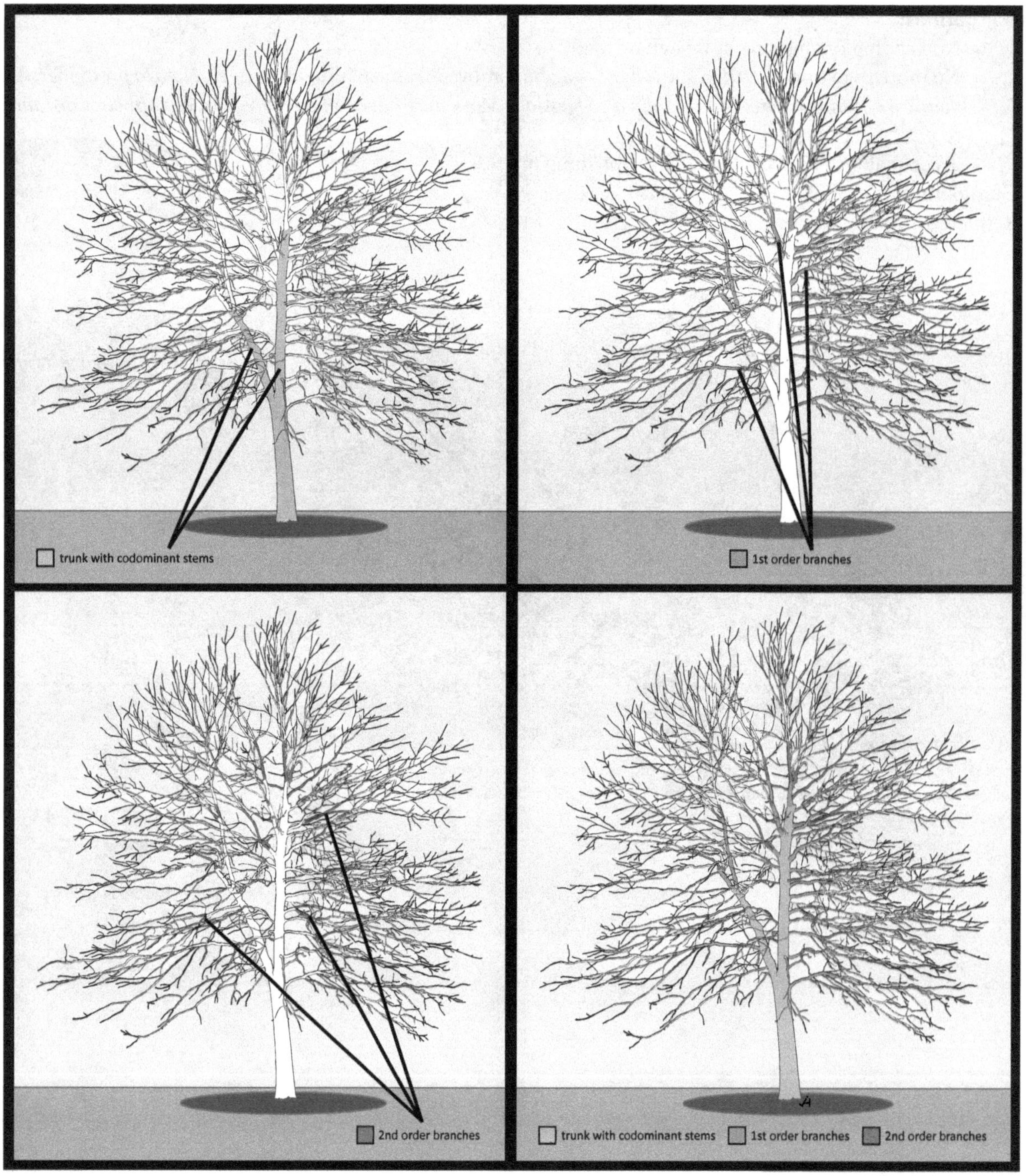
trunk with codominant stems
1st order branches
2nd order branches
trunk with codominant stems
1st order branches
2nd order branches

APPENDIX 2

MODERN ORGANIC SOIL PRODUCTS

Modern organic soil amendments are materials derived from plant and animal sources added to soils to improve their structure, fertility, and overall health. They are organic substances that have been modified in some way by humans for practical and specific uses.

Some examples of common organic soil products include:

Compost: organic material made from decomposed plant and animal matter that improves soil structure and provides nutrients.

Manure: animal waste products often used as a soil amendment for organic forms of nutrients, capable of improving soil fertility.

Mulch: made from a variety of organic materials, including wood chips and pine straw, helps to retain moisture in the soil and modify temperatures.

Biosolids: treated sewage sludge used as a soil amendment such as Milorganite. They are rich in nutrients and can help to improve soil fertility.

Biochar: a soil amendment made from biomass that has been heated in a low-oxygen environment which can improve nutrient and water retention in soils.

These amendments are a dynamic collection of compounds, nutrients, microbes, and other organic materials. The choice of amendment depends on the specific needs of the soil and the PHC program.

BIOCHAR

Research showed biochar increased soil performance whenever included with any remediation metric. It would not be a stretch to assume properly made biochar, abiding by the standards, would aid in soil improvement. Multiple studies have shown positive improvement in urban or degraded soils, and increased performance when used in conjunction with compost, fertilizer, air tillage, and/or mulch chips. Recent studies also reference a modern study where fertilizer was used more effectively when combined with biochar. Based on applications in the field, biochar generally has a compounding effect when used in conjunction with organic matter such as humates, compost, and mulch. The studies applied 5% by volume mixed into soil and when combined with appropriate amounts of compost or humates, using large amounts of biochar over 5% have yielded no negative effects. This information supports the use of biochar, however rates are based on the most current research, using a 5% biochar incorporation by volume as recommended.

FIGURE A2.1 Application of soil amendments to deficient soils.

HUMATES

Humates are also organic substances created because of the decomposition of plant and animal tissues found in the earth's soil. They are a natural result of the process of humification, in which organic matter is formed from the remains of plant and animal life over a period of thousands of years. Humates are a critical component of soil fertility and plant nutrition. Plants grown in soil rich with humic substances produce healthier trees, provide better nutrition, and are less subject to stress, compared with trees grown in most urban soils. They're a rich source of energy and beneficial bacteria which aid overall plant health. These bacteria also protect against plant pests, thereby providing a natural form of protection.

Two commonly used forms of humates include soil humus and humates (Leonardite). *Soil humus* is mature organic matter that resists further decomposition. Soil humates are very old, compressed organic materials similar in form to coal that have been oxidized (brown coal). Apparently, Leonardite is primarily formed from prehistoric sea/lake beds, resulting in concentrated and very diverse humic substances, rich in organic acids (humic, fulvic, ulmic, etc).

The industry typically provides whole products, such as Leonardite, in varying compositions. Other products include humic acid or fulvic acid. These are further processed and then added as a single humic substance or group of substances. For example, 3-3-3 fertilizer with fish hydrolysates and humic acid—sold as a fertilizer or possibly as a biostimulants. This should be considered as a complex of natural products for improved nutritional value.

These humic substances improve soil performance by generally improving the basic functions of physical, chemical, and biological soil properties. They increase nutrient availability through cation exchange and nutrient chelation, increase pore space and WHC by forming connections to clay particles, creating larger aggregates, and creating a "bridge" for improved CEC and pore space. Additionally, they may provide salt and heavy metal mitigation, plant hormone stimulation, and increased microbial diversity.

FIGURE A2.2 Air excavation tools opens the soil to add necessary soil supplements.

BIOSTIMULANTS

Soil biostimulants are a group of additives found to increase nutrient and water uptake, tolerance to stress, and improve root growth. These substances give a boost to a plant's natural processes, growth, and defense mechanisms to help the plant more effectively use what is already present in its environment. Biostimulants are products consisting of microorganisms or substances that stimulate plant metabolism, enhance growth, and increase plant resistance to biotic and abiotic stresses. Humic acids, seaweed, and symbiotic bacteria are just a few types of biostimulants available for use. As the term indicates, this group of organic amendments can mean the stimulation of microbial and plant activity. Typically, they are quick acting and can be used as foliar sprays or soil additives. There is much research needed to interpret their feasible use on trees in the urban forest.

The application of data-driven research supporting the use of these products are growing. More information is being provided which does show promise for improving soil health. Often these amendments are strong companions to interventions, such as correcting compaction, root collar air tillage and other modification on generally poor soils. It is possible they can be a viable approach in an IPM program for reducing chemical inputs and increasing soil health. Hopefully, this results in longer lived trees.

GLOSSARY OF ARBORICULTURAL TERMS

As adopted from the International Society of Arboriculture

A

abiotic: nonliving.

abiotic disorder: plant ailment caused by nonliving, environmental, or man-made agents (contrast with *biotic disorder*).

abscisic acid: plant hormone that stimulates stomatal closure and promotes seed and bud dormancy. Inhibits shoot growth but may promote root growth. Plays a role in wounding responses and in abscission of leaves and fruit.

abscission: leaf or fruit drop induced by hormonal changes.

abscission zone: area at the base of a petiole, small branch, or flower where cellular breakdown leads to leaf, flower, or fruit drop.

absorbing roots: fine roots with functional root hairs that are responsible for the uptake of water and minerals.

absorption: taking up (contrast with *adsorption*).

acceptable level of risk: degree of risk that is within the tolerance or threshold of the owner, manager, or controlling authority.

acclimation: physiological adaptation process of plants and other living organisms to a climate or environment different from their prior growing conditions.

acid: having a pH less than 7.0 (contrast with *alkaline*).

acidic: state or quality of being acid (see *pH*; contrast with *alkalinity*).

acute: 1) disorder or disease that occurs suddenly or over a short period of time (contrast with *chronic*). 2) leaves with straight sides tapering to a pointed apex.

adaptability: genetic ability of plants and other living organisms to adjust or acclimate to different environments.

adaptive growth: see *response growth*.

adhesion: force of attraction between two different materials; in soils, it is the attraction of water molecules to soil particles.

adsorption: adhesion on contact of the molecules of gases, dissolved substances, or liquids to the surface of solids or liquids with which they are in contact (contrast with *absorption*).

advanced assessment: assessment performed to provide detailed information about specific tree parts, defects, targets, or site conditions. Specialized equipment, data collection and analysis, and/or expertise are usually required.

adventitious: arising peripherally from parts of the root or stem and having no connection to meristems or existing buds.

adventitious branch: branch arising from a stem or parent branch and having no connection to apical meristems.

adventitious bud: bud arising peripherally from a place other than a leaf axil or shoot tip, usually because of hormonal triggers.

adventitious roots: root-like growth arising from roots or stems and having no connection to apical meristems.

adventitious shoot: shoot that develops from an adventitious bud.

aeration: provision of air to the soil to allow root and microbial respiration.

aerial inspection: inspection of parts of a tree not visible from the ground, including the trunk, stems, and branches; typically done by climbing or from an aerial lift. Aerial inspections may include evaluation of internal decay.

aesthetic: pleasing to the senses, visually or otherwise.

air excavation tool: device that directs a jet of highly compressed air to excavate soil. Used to avoid or minimize damage to tree roots or underground structures such as pipes and wires. May also reduce hazards associated with excavation near pipes or wires.

air knife/spade: see *air excavation tool.*

alkaline: having a pH greater than 7.0 (contrast with *acid*).

alkalinity: state or quality of being alkaline (see *pH*; contrast with *acidity*).

allelochemicals: substances produced naturally by plants as part of a defense against pests and other plants. May adversely affect the growth and development of other plants.

allelopathy: the influence, usually detrimental, of one plant on another, by the release of chemical substances.

American National Standards Institute: a private, nonprofit organization that oversees the development of voluntary consensus standards by accredited representatives of government agencies, industry, and other stakeholders (see *ANSI A300*, *ANSI Z133*, and *ANSI Z60.1*).

anaerobic: without—or with a restricted supply of—air. Process that occurs in the absence of oxygen.

angiosperm: plant with seeds borne in an ovary. Consists of two large groups: monocotyledons (grasses, palms, and related plants) and dicotyledons (most woody trees, shrubs, herbaceous plants, and related plants) (contrast with *gymnosperm*).

anion: ion that carries a negative charge (contrast with *cation*).

anisohydric: trees and plants which continue the process of transpiration and water loss even when soil water is limited.

annual: 1) a plant living only one year (contrast with *biennial* and *perennial*). 2) anything that recurs yearly, such as an assessment, inspection, program, or treatment.

annual rings: see *growth rings.*

ANSI: acronym for American National Standards Institute (see *American National Standards Institute*).

ANSI A300: in the United States, industry-developed, national consensus standards of practice for tree care.

ANSI Z133: in the United States, industry-developed, national consensus safety standards of practice for tree care.

ANSI Z60.1: in the United States, industry-developed, national consensus standards for nursery stock.

anthocyanin: red or purple pigment responsible for those colors in some parts of plants.

antigibberellin: plant growth regulator that inhibits the action of the plant hormone gibberellin, which, among other things, regulates cell elongation.

antitranspirant: substance applied to the foliage of plants to reduce water loss (transpiration).

anthropogenic: environmental change caused or influenced by people, either directly or indirectly.

anvil-type pruning tool: hand pruning tool with a sharp, straight blade that cuts against a flat metal surface (anvil) (contrast with *bypass pruner* and *secateurs*).

apical bud: bud at the tip of a twig or shoot (see *terminal bud*).

apical control: inhibition of lateral buds, decreasing from the top down, by apical buds over many seasons, resulting in trees with an excurrent growth form.

apical dominance: condition in which the terminal bud inhibits the growth and development of the lateral buds on the same stem formed during the same season (see *crown hierarchy*).

apical meristems: growing point in buds, and at the tips of shoots and roots.

appraisal (valuation): 1) act or process of developing an opinion of value, cost, or some other specified assignment result. 2) a report stating an opinion of appraised value. 3) particularly outside the United States, an evaluation of nonmonetary landscape or plant characteristics.

arborestry: a term created by the author, combining urban forestry management with arboricultural practices and science.

arboriculture: practice and study of the care of trees and other woody plants in the landscape.

arborist: professional who possesses the technical competence, through experience and related training, to provide for or supervise the management of trees and other woody plants in residential, commercial, and public landscapes.

aspect ratio: the ratio between branch diameter and trunk diameter measured just beyond the union.

asymptomatic: without symptoms.

auxin: plant hormone that promotes or regulates the growth and development of plants. Produced at sites where cells are dividing, primarily in the shoot tips. Auxin-like compounds may be synthetically produced.

available water: water remaining in the soil after gravitational water has drained and before the permanent wilting point has been reached (compare to *field capacity, gravitational water, permanent wilting point,* and *saturation point*).

water holding capacity: the soil moisture (water) that can be extracted and used by plants and often referred to as capillary water.

axil: point of attachment of a leaf petiole to a stem.

axillary bud: bud in the axil of a leaf; lateral bud.

axillary shoot: shoot rising from the axil of a leaf.

B

B&B: acronym for balled and burlapped (see *balled and burlapped*).

backfill: 1) soil or amended soil used to fill the hole when planting a tree. 2) soil, common fill, aggregates, or contaminants in various combinations put back after an excavation. May not be suitable for tree root growth and function.

balled and burlapped: tree or other plant dug and removed from the ground for transplanting, with the roots and soil wrapped in burlap or a burlap-like fabric (contrast with *bare root, container grown, containerized,* and *in-ground fabric-bag grown*).

bare root: 1) tree or other plant removed from the ground for replanting without soil around the roots (contrast with *balled and burlapped, container grown, containerized,* and *in-ground fabric-bag grown*). 2) the harvesting or transplanting of a tree or other plant without soil around the roots.

bark: protective outer covering of branches and stems that arises from the cork cambium.

bark tracing: cutting away torn or injured bark to leave a cleanly cut edge.

barrier zone: chemical and anatomical barrier formed by the cambium in response to wounding. Inhibits the spread of decay into xylem tissue formed after the time of wounding. Wall 4 in the CODIT model (contrast with *reaction zone*).

basal rot: decay of the lower trunk, trunk flare, or buttress roots; also called *basal rot* or butt rot.

basal swelling: increased wood growth in the area near or where the trunk and roots come together.

basic assessment (Level 2): detailed visual inspection of a tree and surrounding site that may include the use of simple tools. It requires that a tree risk assessor inspect completely around the tree trunk looking at the visible aboveground roots, trunk, branches, and site.

bending moment: the turning, bending, or twisting load that occurs when a force acts on a lever arm, defined as the force (acting perpendicular to the lever) multiplied by the length of the lever.

bending strength: maximum fiber stress that wood incurs just before it cracks or breaks when force is applied laterally.

beneficial organisms: insects and other organisms that promote plant health or assist in the control of pest populations.

berm: wall or mound of dirt that directs, diverts, or holds water; screens objectionable views or reduces objectionable noise; or provides additional rooting volume for trees or other plants.

best management practices: best-available, industry-recognized courses of action, in consideration of the benefits and limitations, based on scientific research and current knowledge and standards.

bifurcation: natural division of a branch, stem, or root into two parts.

biodiversity: biological diversity in an environment as indicated by number of different species of plants and animals.

biological control: 1) method of managing plant pests or weeds using natural predators, parasites, or pathogens. 2) biological methods—management of vegetation by establishment and conservation of compatible, stable plant communities using plant competition, allelopathy, animals, insects, or pathogens. Cover-type conversion is a type of biological control.

biomechanics: application of physical or mechanical principles to biological organisms.

biorational: product or pesticide formulated from naturally occurring plant extracts, microbes, or microbial by-products that poses very low risk to nontarget organisms and has limited environmental persistence.

biorational pesticide: 1) pesticide formulated from naturally occurring plant extracts, microbes, or microbial by-products that poses very low risk to nontarget organisms. 2) pesticide that has limited environmental persistence and poses very low risk to nontarget organisms.

biotic: pertaining to living organisms.

biotic agent: living organism capable of causing disease.

biotic disorder: disorder caused by a living organism (contrast with *abiotic disorder*).

bleeding: 1) flow of sap from plant wounds, injuries, or pathogen invasion. 2) flow of blood from a human or animal wound. 3) process of removing air from a fluid system (such as hydraulic brakes on a vehicle).

botanical pesticide: pesticide derived from plants.

bow: lean characterized by the top of the tree bending over more than the lower trunk, creating a curve.

box cable system: tree support cabling system that forms closed polygons. Used to join more than three stems (contrast with *direct cable system* and *triangular cable system*).

bracing: installation of lag-threaded screws or threaded-steel rods in limbs, leaders, or trunks to provide supplemental support to weak branch unions.

bracing rod (rigid brace): metal rod used to support weak sections or crotches of a tree.

branch: secondary shoot or stem in a woody plant; generally smaller than the parent.

branch angle: angle between a branch and stem (or parent branch).

branch architecture: orientation and distribution of primary branches on the trunk.

branch arrangement: orientation and distribution of branches along a larger stem or main trunk.

branch bark ridge: raised strip of bark at the top of a branch union, where the growth and expansion of the trunk or parent stem and adjoining branch push the bark into a ridge.

branch collar: swollen area where a branch joins the trunk or another branch that is created by the overlapping vascular tissues from both the branch and the trunk.

branch orders:

first-order branch: a branch (regardless of its diameter) attached directly to the trunk or leader.

second-order branch: a branch growing from a first-order branch that is smaller in diameter than the first-order branch from where it originates.

third-order branch: a branch growing from a second-order branch that is smaller in diameter than the second-order branch from where it originates.

branch protection zone: chemically and physically modified tissue within the trunk or parent branch at the base of a smaller, subordinate branch that retards the spread of discoloration and decay from the subordinate stem into the trunk or parent branch.

branch removal cut (removal cut): pruning cut that removes a branch back to a larger stem such as the trunk; pruning cut that removes a smaller branch from a larger branch.

branch taper: natural, gradual reduction in the diameter of a branch from its point of origin to the tip (see *taper*).

branch union: point where a branch originates from the trunk or another branch (see *crotch*).

breach of duty (of care): failure to take reasonable care to avoid injury or damage to a person or property in a situation where the law imposes a duty of care.

broadcast fertilization: application of granular fertilizer over the soil surface (contrast with *drill-hole fertilization* and *liquid fertilization*).

broad-spectrum pesticide: pesticide that kills a large number of unrelated species.

brown rot: fungal wood rot characterized by the breakdown of cellulose (contrast with *soft rot* and *white rot*).

buckling: failure mode characteristic of collapsing under compressive stress.

bud: undeveloped flower or shoot containing a meristematic growing point. Small lateral or terminal protuberance on the stem of a plant that may develop into a flower or shoot.

bud trace: vascular connection extending from the base of latent buds inward to near the pith. Often accompanied by altered wood grain. Grows in length with each annual increment and appears as a thin, continuous line when viewed in longitudinal section.

buffer yard: an intentional landscaped area intended to separate and partially obstruct the view of two adjacent land uses or properties.

buffering capacity: ability of a soil to maintain (i.e., resist change in) its pH.

bulge: swellings on branches, trunks, or root flares; often caused by new tissue formed as a response to movement and that reinforces the wood structure at the weak area.

bulk density: mass of soil per unit volume; used as a measure of soil compaction. Often written as grams/cubic centimeter.

burl: abnormal swelling of a tree trunk characterized by swirling wood grain and meristematic tissue. Wood with these structures is prized for woodworking.

burlap: coarsely woven cloth made from fibers of jute, flax, hemp, or synthetic materials. Also referred to as hessian.

buttress roots: roots at the trunk base that help support the tree and equalize mechanical stress.

bypass pruner: tool that pushes a sharpened blade through a small-diameter stem, past an unsharpened hooked or curved blade. Also called secateurs, pruning shears, or hand pruners (contrast with *anvil-type pruning tool*).

C

cabling: installation of steel or synthetic cable in a tree to provide supplemental support to weak branches or branch unions.

caliper: thickness or diameter of the trunk of a tree taken at a standard height, used in specifying nursery stock.

calipers: instrument used to measure the diameter of the trunk of a tree.

callus: undifferentiated tissue formed by the cambium, usually as the result of wounding.

cambium: thin layer(s) of meristematic cells that give rise (outward) to the phloem and (inward) to the xylem, which results in secondary growth (increasing diameter) of stems and roots.

canker: discrete, localized, usually necrotic area on stems, roots, and branches. Often sunken and discolored. Most canker diseases require laboratory isolation and microscopic examination to positively identify.

canopy: collective branches and foliage of a tree or group of trees' crowns. Aggregate or collective tree crowns.

capillary water: water held in the capillary pores of the soil; much of this water can move in any direction and is readily available to plant roots.

carbohydrate: chemical compound, combining carbon, hydrogen, and oxygen in a proportion of C:2H:O (CH_2O), that is produced by plants because of photosynthesis (sugars) or derived from assimilates (e.g., starches, cellulose, hemicellulose, and lignin).

carbon dioxide: colorless gas, soluble in water, used by green plants to make carbohydrates during photosynthesis; CO_2.

carbon sequestration: capturing and storage of carbon. Most often used in reference to the capturing and retention of atmospheric carbon dioxide through biological, chemical, or physical processes.

carotenoid: yellow, orange, or red pigment often responsible for those colors in some parts of trees and other plants.

cation: positively charged ion. In soils, the most abundant cations are calcium (Ca), magnesium (Mg), potassium (K), sodium (Na), and aluminum (Al) (contrast with *anion*).

cation exchange capacity: ability of a soil to adsorb and hold cations. Affected by soil pH and particle size. A measure of soil fertility and clay composition.

causal agent: biotic or abiotic agent that induces a disease or disorder.

causation in fact: in determining legal liability, the test of the proximate cause and foreseeability of the consequences.

cavitation: the formation of gas bubbles in xylem which interrupts the continuous stream of water.

cavity: open or closed hollow within a tree stem, branch, or root, usually associated with decay.

CEC: acronym for cation exchange capacity (see *cation exchange capacity*).

cell: basic structural and functional unit of living organisms.

cellulose: long-chain, insoluble glucose polymer found in the cell walls of the majority of plants.

central leader: main stem of a tree, particularly of an excurrent specimen; dominant leader.

center of force: the point toward or from which a force acts; the resultant point of aggregate forces.

chain saw: portable power saw with teeth linked in a loop of chain.

chelates: chemical compounds that keep plant nutrients—usually iron (Fe)—soluble and available for plant absorption over a broad range of pH.

chemical control: 1) control of pests using conventional, manufactured pesticides. 2) management of incompatible vegetation through the use of herbicides or growth regulators.

chemical injury: plant tissue damage caused by pesticides (e.g., insecticides, fungicides, herbicides, plant growth regulators) or other chemicals (see *phytotoxicity*).

chlorophyll: green pigment of plants found in chloroplasts. Captures the energy of the sun and is essential in photosynthesis.

chloroplast: specialized organelle found in some cells. Site of photosynthesis.

chlorosis: whitish or yellowish leaf discoloration caused by lack of chlorophyll. Often caused by nutrient deficiency.

chronic: disorder or disease occurring or recurring over a very long period of time, typically multiple growing seasons (contrast with *acute*).

circling roots: roots that grow in a circular pattern when a plant is confined in the same container longer than necessary. If not corrected, root pattern will prevent a tree from establishing well when transplanted and will create future problems, such as blow over, girdling roots, and strangulation.

city forester: individual specializing in the fields of arboriculture and urban forestry and having responsibility for the management of all or part of planted and naturally occurring greenspaces on public land in communities.

clay: 1) soil particles with a typical grain size less than 0.004 mm. 2) a soil predominantly composed of such particles (contrast with *sand* and *silt*).

clearance: amount of open space between a tree or tree part and another object, the ground, or pedestrian or vehicle traffic on the ground.

client: the person or organization for whom professional services are rendered under a formal agreement. Usually, the owner or manager responsible for the trees specified in the agreement.

clinometer: instrument used for measuring the height of a tree or other structure.

CODIT: acronym for compartmentalization of decay in trees (see *compartmentalization*).

codominant branches: forked branches or stems of nearly the same diameter, arising from a common union and lacking a branch collar; may have included bark.

codominant stem: a bifurcation of the main trunk with stems growing about the same rate, and with nearly the same diameter, from the same union; often the piths are connected in the union.

column of decay: wood decay inside a tree that extends longitudinally up and down a stem or through a branch.

compaction: closely or firmly packed or joined together (see *soil compaction*).

compartmentalization: natural defense process in trees by which chemical and physical boundaries are created that act to limit the spread of disease and decay organisms (see *CODIT*).

complete fertilizer: fertilizer containing the three primary elements: nitrogen (N), phosphorus (P), and potassium (K).

complex : combination of factors contributing to the stress or decline of a tree. See *disease complex*.

compost: 1) *n.* organic matter that has been intentionally subjected to decay processes and is more or less decomposed. 2) *v.* to subject organic matter to decay and decomposition processes.

compression: in mechanics, the action of forces to squeeze, crush, or push together any material(s) or substance(s) (contrast with *tension*).

compression crack: fracture caused by compressive stress.

compression wood: reaction wood in gymnosperms, and some angiosperms, that develops on the underside of branches or leaning trunks and is important in load bearing (contrast with *tension wood*).

conifer: cone-bearing tree or other plant that has its seeds in a structure called a cone.

consequences: effects or outcome of an event. In tree risk assessment, consequences include personal injury, property damage, or disruption of activities due to the event.

consequences of failure: personal injury, property damage, or disruption of activities due to the failure of a tree or tree part.

constant occupancy: present at nearly all times, 24 hours a day, 7 days a week.

container grown: tree or other plant that has been grown and marketed in a container (contrast with *balled and burlapped*, *bare root*, *containerized*, and *in-ground fabric-bag grown*).

containerized: field-grown plant placed into a container for a time and then sold as a potted plant. Term does not include a plant initially grown in a container (contrast with *balled and burlapped*, *bare root*, *container grown*, and *in-ground fabric-bag grown*).

cork cambium: lateral meristem from which the corky, protective outer layer of bark is formed. Also known as phellogen.

corrected lean: tree lean characterized by a leaning lower trunk and a top that is more upright as a result of self-correcting growth; sweep.

Council of Tree and Landscape Appraisers (CTLA): group of representatives of several tree care and landscape associations that works to research and author the *Guide for Plant Appraisal*.

cracks: separation in wood fibers; narrow breaks or fissures in stems or branches. If severe, may result in tree or branch failure.

critical root zone: area of soil around a tree where the minimum amount of roots considered critical to the structural stability or health of the tree are located. CRZ determination is based on the drip line or a multiple of the DBH, but because root growth can be asymmetric due to site conditions, onsite investigation is necessary.

crotch: branch unions; the angle formed by two branches in the tree.

crown: upper part of a tree, consisting of the branches and foliage; its length is measured from the lowest branch to the top of the tree.

crown cleaning: in pruning, the selective removal of dead, dying, diseased, and broken branches from the tree crown.

crown raising: in pruning, the selective removal of lower branches from a tree crown to provide clearance (see *raising* and *lifting*).

crown reduction: pruning to reduce the height and/or spread of a tree crown; reduction.

crown restoration: pruning to restore the natural growth habit of a tree that has been topped or damaged; restoration pruning.

crown rot (root crown rot): disease or decay at the base of a tree or root crown (see *basal rot* and *butt rot*).

crown thinning: an unacceptable practice in pruning involving the selective removal of live branches to reduce crown density.

CRZ: acronym for critical root zone (see *critical root zone*).

CU Structural Soil: a patented soil mix from Cornell University that consists of crushed stone and clay loam. It is designed to provide adequate rooting space and load-bearing capacity for urban trees and pavement.

cultivar: cultivated variety of a species that cannot be reproduced without human assistance. Usually propagated asexually (cloned) (contrast with *variety*).

cultural control: 1) method of controlling plant pests by providing a growing environment favorable to the host plant and/or unfavorable to the pest (see *Plant Health Care* and *integrated pest management*). 2) management of vegetation through alternative use of the right-of-way that precludes growth of incompatible vegetation through establishment of crops, pastures, prairies, parks, successful cover-type conversion, or other managed landscapes.

cultural problems: too little or too much sunlight, water, fertilizer, air, pests, or other factors resulting in poor health or growth.

cuticle: waxy layer outside the epidermis of a leaf that retards water loss and helps protect the leaf from insects and diseases.

cytokinins: plant hormones involved in cell division, leaf expansion, and other physiological processes. Compounds with cytokinin-like activity may be synthetically produced.

D

dead wooding: objectively pruning to remove dead or dying tree parts. Removing dead and dying branches from a tree. Outdated term for *crown cleaning*, which is more comprehensive.

decay: 1) *n.* the process of decomposition. 2) *v.* the process of degradation by microorganisms.

decay detection device: any device used to assess the presence (and often extent) of internal decay in trees.

deciduous: tree or other plant that sheds all of its foliage annually (contrast with *evergreen*).

decline: 1) gradually diminishing health or condition of a tree. 2) a pattern of decline conditions in a particular species not necessarily associated with a single cause (contrast with *decline disease*).

decline disease: conceptual model explaining development of specific, recognizable disorders resulting from interaction of predisposing, inciting, and contributing factors that act in succession over a prolonged period of time, resulting in progressive deterioration in tree health.

decline spiral: interaction of causal agents that accelerates decline (see *mortality spiral*).

decomposition: breakdown or separation of a substance into simpler substances.

decurrent: rounded or spreading growth habit of the tree crown (contrast with *excurrent*).

defect: imperfection, weakness, or lack of something necessary. In trees, defects are injuries, growth patterns, decay, or other conditions that reduce the load-bearing capacity.

deficiency: pertaining to plant nutrition, lack or insufficient quantity of a required element.

defoliation: loss of leaves from a tree or other plant by biological, chemical, or mechanical means (as opposed to natural shedding).

degree day: measure of heat accumulation over time that allows comparison of the daily average temperature and a given temperature base. Used to plan, monitor, or schedule plant or pest management (see *growing degree days*).

desiccation: severe water loss or drying out of living tissue; dehydration.

determinate growth: development to a point characterized by a fixed number of leaves or other lateral organs (contrast with *indeterminate growth*).

diagnosis: 1) process of identifying a causal agent by analyzing signs, symptoms, site conditions, patterns, climate, cultural history, and other factors. 2) the conclusion reached from the process.

diameter: length of a straight line through the center of a circle and extending to the perimeter at both ends.

diameter at breast height (DBH): tree diameter measured at 4.5 ft (1.4 m) above the soil line in the United States and some other countries; at 1.3 m in the United Kingdom, Canada, and other countries (in forestry applications); and 1.5 m in U.K. arboriculture.

diameter tape: measuring tape scaled such that when it encircles a tree trunk, the diameter can be read directly.

dieback: condition in which the branches in a tree die from the tips toward the main stem.

differential diagnosis: a method used by practitioners to identify the cause of symptoms or condition. It involves distinguishing between a particular disease or condition from others presenting with similar characteristics.

differentiation: process in the development of cells in which they become specialized for various functions.

diffuse-porous: pattern of wood development in which the vessels and vessel sizes are distributed evenly throughout the growth ring (contrast with *ring-porous*).

dioecious: having male and female flowers borne on separate individuals of the same species (contrast with *monoecious*).

directional pruning: selective removal of branches to guide and/or discourage growth in a particular direction.

discoloration: wood response of a tree to microorganisms, including bacteria and non-decay-causing fungi, resulting in dead, darkened wood with little strength.

disease: result of interaction over time of a pathogen or abiotic agent with a susceptible host in a conducive environment to interfere with one or more physiological functions or structures.

disease complex: combination of factors involved in development of a disease.

disease triangle: conceptual model showing three factors required for plant disease: susceptible host, pathogen or abiotic agent, and a conducive environment.

disorder: abnormal condition that impairs the performance of one or more vital functions. Often associated with noninfectious agents or abiotic factors.

division phylum: taxonomic group below the kingdom level but above the class level.

dominant leader: stem of a tree that grows larger than all other stems and branches.

dormancy: period of reduced physiological activity in a plant, with the potential for reactivation of growth.

dormant: in a period of dormancy.

dormant bud: bud originally developed in a leaf axil and connected to the pith by a bud trace that has not been stimulated to mature and grow. Some buds remain dormant throughout the life of a woody plant.

dormant oils: see *horticultural oils*.

double reduction cut: A type of reduction cut whereby the retained lateral branch also receives a reduction cut.

drag: resistance to wind.

drill-hole fertilization: applying fertilizer by drilling holes in the soil within the root zone (contrast with *broadcast fertilization* and *liquid fertilization*).

drill resistance testing: use of a small-diameter drill bit on an electric drill, measuring the change of torque resistance and indicating wood density as the drill bit passes through a trunk or limb; in tree risk assessment, used to measure the extent and possible location of decay.

drip irrigation: method of minimizing evaporation and runoff by applying water slowly through small emitters.

drip line: imaginary line defined by the branch spread of a single plant or group of plants, or the outer edge of the crown, projected onto the ground.

drought: extended period of unusually low precipitation and less moisture in the soil.

duty of care: legal obligation that requires an individual to apply reasonable actions when performing tasks that may potentially harm others.

dynamic load: any load or force that changes over time. These types of loads exert forces onto a structure such as trees (contrast with *static load*).

E

earlywood: portion of an annual ring (growth ring) that forms after a period of dormancy, characterized by large-diameter cells and thin walls (in ring-porous species). Also called springwood (contrast with *latewood*).

easement: legal, nonpossessory interest in real property that conveys use or partial use, but not ownership, of all, or more typically a portion, of an owner's property.

ecology: study of the relationships between organisms and other living and nonliving elements of their environment.

economic threshold: pest population or plant damage level at which the cost of inaction is less than the cost of action (see *action threshold*).

ecosystem: complex system of living organisms and their environment.

ecosystem services: benefits people obtain from ecosystems such as trees providing shade, carbon sequestration, and improved air and water quality.

edaphic: related to soil and its properties including the soil itself and its characteristics.

embolism: air bubbles in the xylem that typically enter during periods of drought or injury, also call tyloses.

endogenous: having an internal cause or origin; growing or originating from within an organism.

engineered soil: a soil that has been altered or manufactured for a specific application with specific properties. It consists of specified ratios of sand, silt, clay, and organic or man-made materials. It can be custom blended to meet different needs including planting media.

epicormic shoot: shoot arising from a dormant bud or from newly formed adventitious tissue.

epidermis: outer tissue of leaves, stems, roots, flowers, and seeds.

epinasty: downward twisting, curling, or bending of a leaf blade or shoot that may result from disease or herbicide injury.

epiphyte: nonparasitic plant that uses another plant as a host for mechanical support but does not derive nutrients or water from it (contrast with *parasite* and *saprophyte*).

eradication: total removal of a species from a particular area. May refer to pathogens or insect pests or to unwanted plants.

erosion: wearing away of the land by the movement of water, wind, or ice.

establish: to grow, or cause a plant to grow, successfully in a new place. In forestry, may refer to initial planting and care; in landscaping, may refer to care after planting until the plant is growing successfully.

ethics: body of moral principles or values governing a group or individual's conduct.

ethylene: gaseous plant hormone that triggers fruit ripening and plant senescence.

evapotranspiration: loss of water by evaporation from the soil surface and transpiration by plants.

evergreen: tree or other plant that does not shed all of its foliage annually, but rather progressively over a period of years (contrast with *deciduous*).

excurrent: pattern of tree branching characterized by a dominant leader and an upright or pyramidal, cone-shaped crown (contrast with *decurrent*).

exogenous: developing from external factors; originating from outside an organism.

exotic species: species of plant or other organism not native or indigenous to a region. May or may not be invasive (contrast with *introduced species, native species,* and *naturalized species*).

extreme (risk rating): failure is *imminent* with a *high* likelihood of impacting the target, and the consequences of the failure are *severe.*

exudation: an oozing from pores or breaks in cell membranes.

F

failure (of tree or tree part): breakage of stem, branch, or roots, or loss of mechanical support in the root system.

failure potential: in tree risk assessment, the professional assessment of the likelihood of a tree or tree part to fail within a defined time frame.

fertilization: application of essential elements to a plant, usually through the soil.

fertilizer: material containing one or more of the elements essential for plant growth, development, and reproduction.

fertilizer analysis: composition of a fertilizer, expressed as a percentage by weight of total nitrogen (N), available phosphoric acid (P_2O_5), soluble potash (K_2O), and other nutrients.

fertilizer burn: injury to a plant resulting from excess fertilizer salts in the surrounding soil.

fertilizer ratio: ratio of total nitrogen (N), available phosphoric acid (P_2O_5), and soluble potash (K_2O); for example, the ratio of a 30:10:10 fertilizer is 3:1:1.

fiber: elongated, tapering, thick-walled cell that provides strength to wood.

field capacity: maximum soil moisture content following the drainage of water due to the force of gravity.

fissure: long, narrow opening or split.

flagging: symptom in which leaves on a branch wilt and may ultimately turn brown without falling from the shoot.

flexure wood: response growth triggered by the continued flexing of a tree stem or branch.

flower: organ of sexual reproduction in angiosperms.

flush cut: pruning cut made through and/or removing the branch collar, causing unnecessary injury to the trunk or parent stem.

foliage: leaves of a plant.

foliar analysis: laboratory analysis of the mineral content of foliage.

foliar application: applying a fertilizer, pesticide, or other substance on foliage.

foliar burning: symptom of phytotoxicity characterized by browning and shriveling of foliage, especially at the leaf margin. Leaf scorch.

force: any action or influence causing an object to accelerate/decelerate. Calculated as mass multiplied by acceleration.

forestry: management of forests, using silvicultural treatments, to provide a variety of forest benefits including timber, aesthetics, wildlife habitat, and/or recreational opportunities.

frass: fecal material and/or wood dust or shavings produced by insects.

freeze-thaw crack: see *frost crack.*

frequent occupancy: target zone is occupied for a large portion of the day or week.

friable: soil that can easily crumble or be pulverized into finer particles.

frost crack: vertical split in the wood of a tree, generally near the base of the bole, caused by internal stresses and low temperatures (see *southwest injury*).

fruiting body: reproductive structure of a fungus that usually develops in diseased tissues. The presence of certain species may indicate decay in a tree. Many fruiting bodies are small and can only be seen with a hand lens, and require microscopic examination to determine the species of fungus present (conks, brackets, mushrooms).

functional goals: in landscape design, the set of goals pertaining to the future needs and practical purpose of the site.

functional value: value of the environmental, aesthetic, or architectural functions of a plant.

fungal fruiting structures: reproductive structures of a fungus (conks, brackets, mushrooms).

fungus: from the kingdom Fungi, including yeasts, molds, those that form mushrooms or conks, smuts, and rusts. Usually multicellular and filamentous, fungi lack the chlorophyll and vascular tissue of plants and reproduce asexually or sexually to form spores often borne in various types of fruiting bodies.

fungi: plural of fungus (see *fungus*).

G

genera: plural of genus (see *genus*).

genus: taxonomic group, composed of species having similar fundamental traits. Botanical classification under the family level and above the species level.

geographic information system (GIS): computer application using longitude and latitude coordinates to place objects or specified locations on maps for reference to and/or analysis of other data describing the objects or locations; combines spatial and descriptive data, in different layers.

geotropism: plant growth produced as a response to the force of gravity, either positive as in the direction of gravity (roots) or negative as in opposite the direction of gravity (shoots).

gibberellins: group of plant hormones involved in cell elongation and other physiological processes.

girdling: restriction, compression, or destruction of the vascular system within a root, stem, or branch that inhibits movement of sap in the phloem.

girdling root: root that encircles all or part of the tree trunk or the tree's other roots, constricting the vascular tissue and inhibiting secondary growth and the movement of water and photosynthates.

grade: 1) *n.* surface level of the ground. 2) *n.* the slope, or percentage of change, in the surface level of the ground; important for drainage and for the safe operation of equipment. 3) *v.* to change or groom the surface level or contours of the ground. 4) *n.* and *v.* quality determination for nursery stock, timber, or other crops.

grade change: topographic alteration to the surface of the ground.

grading: 1) phase of construction, landscaping, or site development that provides a site with intended grades (contours, elevations, and slopes). Involves cutting and filling (see *mass grading*). 2) act or process of applying quality determinations for nursery stock, timber, or other crops.

graft: 1) to join together scion and stock (or root stock) from the same or different plants in order to combine desirable characteristics or to affect a repair (see *bridge graft*). 2) junction between root stock and scion wood. 3) the joined tissues formed by grafting or natural processes.

graft union: junction between root stock and scion wood. Often evident by a thickening or crook in the trunk at the union, or a change in bark pattern.

gravitational water: water that drains from soil macropores due to the force of gravity (compare to *available water, field capacity, permanent wilting point,* and *saturation point*).

gravitropism: directional growth of a plant in response to gravity.

gravity: force that attracts a body toward the center of the earth.

green infrastructure: living plants and natural materials within developed areas. Can be incorporated with gray infrastructure (utilities, paved surfaces, and buildings) to perform important ecological functions, such as stormwater management, erosion control, temperature buffering, and more.

ground: 1) *n.* the Earth's surface and everything beneath it. 2) *n.* the grade or elevation of the surface of the Earth.

groundcover: low-growing plants that require minimal maintenance; used in landscaping instead of turf or mulch.

ground-penetrating radar: nondestructive technique that uses radar technology (electromagnetic waves) to image the subsurface.

groundwater: water naturally stored underground in aquifers or that flows through and saturates soil and rock, supplying springs and wells (see *water table*; contrast with *surface water*).

growing degree days: a measure of heat accumulation used to predict plant and insect development rates, such as the date that a flower will bloom or when an insect will emerge from dormancy.

growing medium: one or more materials, such as soil, peat moss, perlite, ground bark, or sand, in which plants are grown, especially in nursery production.

growth rate: speed at which something grows per unit of time, typically determined by examining tree rings or internodal distances between terminal buds.

growth rings: rings of xylem that are visible in a cross section of the stem, branches, and roots of some trees. In temperate zones, the rings typically represent one year of growth and are sometimes referred to as annual rings.

guard cells: pair of specialized cells that regulate the opening and closing of a stomate (see *stomata*) due to a change in water pressure within cells.

gummosis: exudation of sap, gum, or resin often in response to disease or insect damage.

gymnosperm: plants with exposed seeds, usually within cones. The classes Ginkgopsida and Coniferopsida are members of the group (contrast with *angiosperm*).

gypsum: hydrated calcium sulfate. Sometimes recommended as a soil conditioner or a source of calcium to raise pH values.

H

habit: characteristic form or manner of growth.

habitat: environment suitable for sustaining a population of a given organism.

hand lens: small, handheld magnifying glass; a loupe.

hand pruning shears: tool used for pruning twigs and small branches less than 0.5 in. (1.3 cm) diameter (see *secateurs*).

hand pull test: load test that involves installing a line in a tree, and then pulling and releasing the line several times to move the tree or branch. Most commonly used in a pre-work inspection but can be used as part of an advanced tree risk assessment.

hanger: broken or cut branch that is hanging in a tree.

hardened off: 1) plant tissue that is acclimated to the cold or a new environment. 2) a process that acclimates balled and burlapped trees to water stress when dug with foliage.

hardiness: genetically determined ability of a plant to survive winter growing conditions; often referring specifically to low temperatures.

hardpan: compacted soil layer nearly impervious to water, air, and roots.

hardscape: constructed elements of a landscape, such as walls, pathways, and seats made of wood, stone, and/or other materials.

hardwood: 1) wood of an angiosperm tree, as distinguished from that of a conifer. 2) referring to an angiosperm or broadleaf tree.

harm: personal injury or death, property damage, or disruption of activities.

hazard: 1) situation or condition that is likely to lead to a loss, personal injury, property damage, or disruption of activities; a likely source of harm. 2) in trees, a tree or tree part identified as a likely source of harm. 3) in tree care or forestry operations, the presence of a condition or situation that may cause harm or injury to workers or others.

heading cut: pruning cut that removes a branch or stem between nodes (leaving a stub), to a bud, or to a live branch that is less than one-third the diameter of the stem being removed.

healing: physiological, regenerative process not known to occur in plants and animals.

heartwood: central wood in a branch or stem characterized by being composed of dead cells, more resistant to decay, generally darker, and harder than the outer sapwood. Trees may or may not have heartwood (contrast with *sapwood*).

heartwood rot: any of several types of fungal decay of tree heartwood, often beginning with infected wounds in the living portions of wood tissue; also called heart rot.

heat island: an urban area that is warmer than its surrounding rural areas due to urban activities and greater heat retention by urban structures and development.

heat tolerance: ability of a plant to endure high temperatures.

herbicide: a pesticide used to kill, slow, or suppress plant growth by interfering with botanical pathways.

high (likelihood of impact): failed tree or tree part is likely to impact the target. This is the case when there is a constant target, with no protection factors, and the direction of fall is toward the target.

high (risk rating): defined by its placement in the risk rating matrix; consequences are *significant* and likelihood is *very likely* or *likely*, or consequences are *severe* and likelihood is *likely*.

horizon: layer of soil within the soil profile.

hormone: see *plant hormone*.

horticultural oils: highly refined petroleum oils applied to plants to control certain insects and other pests by disrupting their respiration.

horticulture: the art and science of cultivating ornamental plants, flowers, fruits, and vegetables.

host: living organism from which a parasite obtains nutrition.

humus: dark-colored, stable form of organic matter that remains after most of plant or animal residues have decomposed.

hybrid: plant resulting from a cross between two different species or two highly inbred lines within a single species.

hydrology: study of the properties, distribution, and effects of water on the Earth's surface, underground, and in the atmosphere.

hydrophilic: the physical or chemical property of a molecule or substance that is attracted to water (and may be dissolved in water).

hydrophobic: the physical or chemical property of a molecule or substance that repels water.

hygroscopic: retained water that is held too tightly in soil for uptake by plants.

I

imminent likelihood of failure: failure has started or is most likely to occur in the near future, even if there is no significant wind or increased load.

impact: to strike a target or cause a disruption that affects activities.

impervious: surface that does not allow for infiltration and/or penetration of liquids.

improbable likelihood of failure: tree or tree part is not likely to fail during normal weather conditions and may not fail in extreme weather conditions within the specified time frame.

included bark: bark that becomes embedded in the union between branch and trunk or between codominant stems. Lacks wood connections, resulting in a weak structure.

increment borer: device used to take core samples from trees to determine age or detect problems such as decay (see *microdrill resistance testing*).

indeterminate growth: continued growth due to sustained activity of the apical and lateral meristems (contrast with *determinate growth*).

infectious: capable of being spread from plants to result in infection of other plants or organisms.

infestation: presence of a pest population on a host plant or group of plants.

infiltration: downward entry of water into the soil (contrast with *percolation*).

infiltration rate: speed at which water penetrates the soil.

injection: insertion of a liquid substance into a plant or soil through a port.

inoculants: in soil science, microorganisms (e.g., mycorrhizal fungi) that are introduced into the soil in order to improve growth or plant health.

inoculum: part of a pathogen that may enter the host or germinate to produce a structure that enters the host to cause disease.

inorganic: compound or substance not containing carbon. Not containing organic material (contrast with *organic*).

inorganic fertilizer: mineral fertilizer, not coming from plant or animal sources (contrast with *organic fertilizer*).

insecticidal soaps: soap-based pesticides approved for application to plants to kill insects and certain mites by disrupting the cell membranes or the insect's respiratory tracheas.

insecticides: substances toxic to insects.

isohydric: trees and plants that limit water loss through the leaves in order to maintain a constant amount of water in the leaves.

inspection: organized and systematic examination.

inspection frequency: the number of inspections per given unit of time (e.g., once every three years).

inspection interval: time between inspections.

integrated pest management (IPM): systematic method of managing pests that uses cultural, biological, physical, and chemical methods to manage pests to acceptable levels.

integrated vegetation management (IVM): system of managing plant communities based on the principles of integrated pest management in which managers identify compatible and incompatible vegetation, consider action thresholds, evaluate control methods, and select and implement controls to achieve specific objectives. The choice of control methods is based on the anticipated effectiveness, environmental impact, site characteristics, safety, security, economics, and other factors.

internodal: between the nodes on a stem.

internode: region of the stem between two successive nodes (contrast with *node*).

interveinal chlorosis: yellowing of interveinal tissue.

introduced species: organisms not native or endemic to a region; typically introduced by human activity in the modern era (contrast with *exotic species, native species,* and *naturalized species*).

invasive species: an alien species whose introduction causes or is likely to cause economic or environmental harm or harm to human health.

ion: atom or a group of atoms with a positive or negative charge.

ISA Certified Arborist: arborist who has passed an independent exam administered by the International Society of Arboriculture and who maintains the credential through continuing education.

i-Tree: suite of software products and management tools that allows the user to inventory the urban forest and analyze its costs, benefits, and management needs.

J

No terms included.

K

No terms included.

L

land disturbances: disruptions to a terrestrial site, community, or ecosystem that alters the physical environment.

landscape function: purpose or intended use of a plant in the landscape (e.g., shade, aesthetics, screening).

land-use history: actions, events, or changes that have taken place on a site.

larva: immature life stage of an insect between the egg and the pupa stage.

larvae: plural of larva (see *larva*).

latent bud: bud held in dormancy for more than one year by hormones originating from the terminal bud. Term sometimes used to refer to dormant buds or adventitious buds.

lateral: secondary or subordinate branch or root.

lateral bud: vegetative bud on the side of a stem (contrast with *terminal bud*).

lateral root: root that arises by cell division in the pericycle of the parent root and then penetrates the cortex and epidermis.

latewood: portion of an annual ring (growth ring) that forms during summer, characterized by small-diameter cells with thick walls; summer wood (contrast with *earlywood*).

leach(ing): 1) tendency for elements or compounds to wash down through and/or out of the soil. 2) tendency for elements or compounds to wash into the soil.

leader: primary terminal shoot or trunk of a tree. Large, usually upright stem. A stem that dominates a portion of the crown by suppressing lateral branches.

leaf: primary photosynthetic organ of a plant that is connected to a stem by a petiole.

leaf scar: scar left on the twig after a leaf is shed.

leaflet: separate part of a compound leaf blade.

lean: predominant angle of the trunk from vertical.

lenticel: small opening in the bark that permits the exchange of gases.

level(s) of assessment: categorization of the breadth and depth of analysis used in an assessment.

lever arm: distance between the applied force (or center of force) and the point where the object will bend or rotate. The lever arm is the distance from the tree trunk to the load application point or a point on the branch.

liability: something for which one is responsible. Legal responsibility.

LiDAR: acronym for light detection and ranging (see *light detention and ranging*).

lifting: (crown) raising or objectively pruning to elevate the crown

lignification: process in which secondary cell walls are formed, making cell walls thicker and stronger by deposition of lignin.

lignin: organic substance that impregnates secondary cell walls to thicken and strengthen the cell and, at times, to reduce susceptibility to decay and pest damage.

likelihood: chance of an event occurring. In the context of tree failures, the term may be used to specify (1) the chance of a tree failure occurring, (2) the chance of impacting a specified target, or (3) the combination of the likelihood of a tree failing and the likelihood of impacting a specified target.

likelihood of failure: chance of a tree or tree part failure occurring within the specified time frame.

likelihood of failure and impact: chance of a tree or tree part failure occurring and impacting a target within the specified time frame.

likelihood of impact: chance of a tree or tree part failure impacting a target within the specified time frame.

likely (likelihood of failure and impact): defined by its placement in the likelihood matrix; *imminent* likelihood of failure and *medium* likelihood of impact, or *probable* likelihood of failure and *high* likelihood of impact.

limb: large, primary branch of a tree.

limitations: restraints or factors that restrict the precision, applicability, or extent of something.

limited visual inspection/assessment (Level 1): visual assessment from a specified perspective such as foot, vehicle, or aerial (airborne) patrol of an individual tree or a population of trees near specified targets to identify specified conditions or obvious defects of concern.

limiting factor: environmental factor that when absent or in low quantities limits the growth, abundance, or distribution of a plant.

line clearance: pruning, trimming, repairing, maintaining, removing, treating, or clearing of trees or the cutting of brush (vegetation management) that is within the minimum approach distance of electric supply lines and equipment; vegetation management work performed by qualified line-clearance arborists or qualified line-clearance arborist trainees for the construction or maintenance of electric supply lines and/or the electric utility right-of-way corridor.

lion's tailing: the improper practice of removing all or most secondary and smaller branches from the interior portion of the crown leaving most live foliage at the edge of the canopy.

Results in poor branch taper, poor wind load distribution, and a higher risk of branch failure.

liquid fertilization: applying liquid formulations of fertilizer by injection into the root zone of a tree or by application to soil surface or to foliage.

live crown ratio (LCR): ratio of the length of the crown containing live foliage to the overall height of the tree.

load: a general term used to indicate the magnitude of a force, bending moment, torque, pressure, or such, applied to a structure.

loam: soil texture classification based on a certain ratio of sand, silt, and clay. Considered ideal for plant growth.

lobe: leaf segments that project outward, creating voids (sinuses) between the segments.

low (likelihood of impact): there is a slight chance that the failed tree or tree part will impact the target.

low (risk rating): defined by its placement in the risk rating matrix; consequences are *negligible* and likelihood is *unlikely*, or consequences are *minor* and likelihood is *somewhat likely*.

M

macroinfusion: technique to introduce substances directly into the xylem of a tree to treat or prevent diseases, disorders, or pest problems through a relatively large-diameter hole in the trunk or a cut root (contrast with *implant, macroinjection,* and *microinjection*).

macroinjection: trunk injection technique that introduces a relatively large volume of substance directly into the xylem of a tree under pressure (contrast with *implant, macroinfusion,* and *microinjection*).

macronutrient: essential element that is required by plants in relatively large quantities, such as nitrogen (N), phosphorus (P), potassium (K), and sulfur (S) (contrast with *micronutrient*).

macropore: relatively large space between soil particles that is usually air filled and allows for water movement and root penetration (contrast with *micropore*).

mallet: broad-headed hammer made of wood, plastic, or resin used in the context of tree risk assessment for sounding a tree to detect internal decay; also called sounding hammer.

management plan: document that describes how tree preservation and/or maintenance goals are to be accomplished within a defined time frame; includes tasks, priorities, best management practices, standards, specifications, budgets, and staffing analyses within the context of the requirements of the appropriate agency or municipality.

mass damping: process by which the amplitude of oscillations is reduced; in trees, motion created by the forces of wind may be reduced through branch movement.

mature height: maximum height that a plant is likely to reach under favorable growing conditions.

mechanical stress: measure of the internal forces acting within a deformable body; force per unit area.

mechanics: study of forces and their effects on bodies at rest or in motion.

medium (likelihood of impact): failed tree or tree part could impact the target but is not expected to do so.

meristem (meristematic tissue): undifferentiated tissue in which active cell division takes place; found in the root tips, buds, cambium, cork cambium, and latent buds.

metamorphosis: biological transformation in the life cycle of an organism, going from an immature to an adult stage.

microenvironment: small area with physical and ecological characteristics that distinguish it from its immediate surrounding area.

microinfusion: technique to introduce a relatively small volume of substance directly into the xylem of a tree to treat or prevent disorders or pest problems by passive or low-pressure means (contrast with *implant, macroinjection,* and *microinjection*).

micronutrient: essential element that is required by plants in relatively small quantities, such as iron (Fe), manganese (Mn), zinc (Zn), copper (Cu), and boron (B) (contrast with *macronutrient*).

mineral: naturally occurring, inorganic solid that has a definite chemical composition and possesses characteristic physical properties; certain minerals are considered essential elements for tree growth and development.

minor (consequences): minor personal injury, low- to moderate-value property damage, or small disruption of activities.

mitigate: to lessen or make less severe a hazardous condition or dangerous situation.

mitigation: the act of making a condition or consequence less severe; in tree risk management, the process for reducing risk.

mitigation options: in tree risk management, alternatives for reducing risk.

mobile target: target that is in motion or intermittently moving.

moderate (risk rating): defined by its placement in the risk rating matrix; consequences are *minor* and likelihood is *very likely* or *likely*, or likelihood is *somewhat likely* and consequences are *significant* or *severe*.

molecule: one or more atoms. The smallest amount of a substance that keeps its chemical properties.

moment: turning, bending, or twisting force exerted by a lever, defined as the force (acting perpendicular to the lever) multiplied by the length of the lever.

monitoring: keeping a close watch. Performing regular checks or inspections.

monocotyledon: plant with an embryo that has one single seed leaf (cotyledon). Examples are grasses and palms (contrast with *dicotyledon*).

monoculture: cultivation or planting of a single species on agricultural land, in a forest setting, or within an urban landscape.

monoecious: species with male and female flowers borne on the same plant (contrast with *dioecious*).

morphology: study of the form and structure of plants and other living organisms (contrast with *anatomy, physiology,* and *taxonomy*).

mortality spiral: combination of events or conditions that accelerate decline and, if left untreated, may cause the eventual death of a tree.

movable target: target that can be relocated.

mulch: material that is spread on the soil surface to reduce weed growth, retain soil moisture, moderate temperature extremes, reduce compaction from pedestrian or vehicle traffic, and/or improve aesthetic appearance of the landscape. Mulch can be composed of chipped, ground, or shredded organic material such as bark, wood, or recycled paper (see *green mulch*).

municipal arborist: individual specializing in the fields of arboriculture and urban forestry and having responsibility for the management of all or part of planted and naturally occurring greenspaces on public land in communities (see *city forester, tree officer, tree warden,* and *urban forester*).

mushroom: fleshy, spore-bearing, often stemmed, fruiting body of fungi in the Basidiomycota subdivision.

mycelium: network or mass of filamentous cells (hyphae) that make up the vegetative body and fruiting bodies of a fungus.

mycorrhizae: a fungus that grows in association with the roots of a plant, creating a symbiotic association between the fungi and absorbing roots of plants. The word "mycorrhiza" means fungal root.

N

native species: plants endemic (indigenous) to a region within the modern era. Naturally occurring and not introduced by humans (contrast with *exotic species, introduced species,* and *naturalized species*).

natural enemy: predator, parasite, or pathogen that targets an organism.

necrosis: localized or general death of cells or parts of a living organism.

necrotic: displaying necrosis.

needle: slender leaf of a conifer.

negligence: failure to use reasonable care, resulting in damage or injury to another.

negligible (consequences): no personal injury, low-value property damage, or disruptions that can be replaced or repaired.

neutral plane: in biomechanics, a conceptual plane within a trunk, branch, or root, which, when loaded by a bending force, bends so that one side is in compression, the other is in tension, and the plane is neither.

node: point on a stem from which leaves, branches, and aerial roots are attached (contrast with *internode*).

noninfectious: disorders that are not caused by a pathogen and cannot be passed from one host to another.

non-target organism: organism that is not intended to be damaged or killed by a pesticide but that may still be susceptible to its effects.

noxious: harmful, injurious, detrimental, or undesirable.

nursery stock: plant obtained from or grown in a nursery and intended to be planted out in the landscape.

nutrient: substance that provides nourishment essential for the growth and maintenance of life; in plants, often used synonymously with essential element.

nutrient cycling: movement of mineral elements (sometimes called nutrients) within an ecosystem as organic matter decomposes, releasing bound nutrients back to plants.

nutrient deficiency: condition in which the supply or availability of an essential element causes cessation of critical plant processes, resulting in visible, physical symptoms such as chlorotic leaves or necrotic margins, and which, if left untreated, will reduce tree health and may eventually lead to premature death.

O

occasional occupancy: target zone is occupied by people or other targets infrequently or irregularly.

occupancy rate: amount of time targets are within a target zone.

Occupational Safety and Health Administration (OSHA): in the United States, the federal agency responsible for establishing and enforcing safety work rules according to the Occupational Safety and Health Act.

oozing: seeping or exudation from a tree cavity or other opening.

open-grown: tree that has grown with exposure to wind and other elements from all directions.

orientation: the upward, downward, or lateral angle of departure.

organic: containing carbon and of animal or vegetable origin, especially when referring to a fertilizer or pesticide.

organic fertilizer: fertilizer derived from plants or animals (contrast with *inorganic fertilizer*).

organic layer: layer of organic matter at the soil surface.

organic matter: material derived from the growth (and death) of living organisms. The organic components of soil (see *compost, green mulch, humus,* and *mulch*).

osmosis: diffusion of water through a semipermeable membrane from a region of higher water potential (lower salt concentration) to a region of lower water potential (higher salt concentration).

osmotic pressure: pressure generated by the movement of water across the cell membrane caused by the difference between solute concentrations on each side.

overextended branch: branch that extends outside the normal crown area.

owner/manager: person or entity responsible for tree management, or the controlling authority regulating tree management.

P

palm: monocotyledonous plant, usually with an unbranched trunk; its leaves (fronds) are arranged around the terminal bud, and the vascular bundles are scattered.

parasite: organism living in or on another living organism (host) from which it derives nourishment to the detriment of the host (see *parasitoid;* contrast with *epiphyte* and *saprophyte*).

parasitoid: organism living in or on another living organism during its immature stage, killing, sterilizing, and/or consuming the host.

parenchyma cells: thin-walled, living cells capable of dividing and essential in photosynthesis, radial transport, energy storage, and production of defense compounds.

parent branch: larger branch or stem from which a smaller, lateral branch arises.

parent material: soil bedrock or base material from which a soil profile develops.

pathogen: causal agent of disease. Usually refers to microorganisms..

peeling: 1) the removal of dead frond bases without damaging living trunk tissue. 2) result of a poor pruning cut that allows the bark to tear below the branch attachment.

penetrometer: device for assessing soil compaction; can be dynamic or static.

perched water table: accumulation of water in an upper soil layer above the actual water table, resulting when drainage in an area is impeded by an impermeable layer, such as rock or heavy clay, or when the interface between soil layers has become stratified (i.e., with sharp contrast in soil pore sizes) (see *water table*).

percolation: movement of water through the soil (contrast with *infiltration*).

percolation rate: the measure of a soil's capacity to absorb water applied to the surface during the period of application.

perennial: plants that live for more than two growing seasons (contrast with *annual* and *biennial*).

periderm: the outer layers of tissue of woody roots and stems, consisting of the cork cambium and the tissues produced by it, such as bark.

peri-urban: populated area immediately around the core area of a city.

permanent branches: in pruning, branches left in place for the life of the tree, often forming the initial scaffold framework of a tree (contrast with *temporary branches*).

permanent wilting point: point at which a plant cannot pull any more water from the soil and suffers permanent damage (compare to *available water, field capacity, gravitational water,* and *saturation point*).

permeable: allowing liquids or gases to pass through, as with a membrane or material.

permit: written order granting permission to conduct a specified task or action.

personal protective equipment (PPE): personal safety gear such as helmet, safety glasses, hearing protection, gloves, and leg protection, including chaps.

pervious: penetrable, allowing passage through; often used to describe surfaces.

pH: unit of measure describing the alkalinity or acidity of a solution. Negative log of the hydrogen ion concentration. Measured on a scale from 0 to 14. Greater than 7 is alkaline, less than 7 is acid, and 7 is neutral (pure water) (see *acidity* and *alkalinity*).

phellogen: meristematic tissue located in the bark that generates the phelloderm on the inside and the phellem on the outside. Cork cambium.

phenology: relationship between the climate and biological events, such as flowering or leafing out in plants.

pheromone: chemical substance produced by an animal (usually in reference to insects) that serves as a method of communication with other individuals of the same species.

phloem: plant vascular tissue that transports photosynthates and growth regulators. Situated on the inside of the bark, just outside the cambium. Is bidirectional (transports up and down) (contrast with *xylem*).

photoperiod: length of daylight and/or darkness required for certain developmental processes and growth in plants.

photosynthate: general term for the sugars and other carbohydrates produced during photosynthesis.

photosynthesis: process in green plants (and in algae and some bacteria) by which light energy is used to form glucose (chemical energy) from water and carbon dioxide.

photosynthesize: to form sugar (glucose) from water and carbon dioxide using the energy of light.

phototropism: influence of light on the direction of plant growth. Tendency of plants to grow toward light.

physiology: in arboriculture, the study of the life function of a tree.

phytochemicals: chemicals produced by plants with biological activity playing a role in plant growth and defense against pathogens and pests.

phytotoxic: term to describe a compound that is poisonous to plants.

phytotoxicity: level of toxicity of a compound to plants.

pigment: substance that appears colored as a result of the reflection of certain light wavelengths.

pinching: a heading cut performed on a very young tree or plant (see *heading cut*).

pith: central core of a stem. Often a lighter color than surrounding tissue.

plant growth regulator: compound effective in small quantities that affects the growth and/or development of plants. May be naturally produced (hormone) or synthetic (see *plant hormone*).

Plant Health Care (PHC): comprehensive program to manage the health, structure, and appearance of plants in the landscape.

plant hormone: substance produced by a plant that, in low concentrations, affects physiological processes such as growth and development, often at a distance from the substance's point of origin (see *plant growth regulator*).

planting: installing a plant in the landscape.

planting hole/pit: hole dug in the soil to plant a tree or other plant.

planting site: location chosen to plant a tree, including its surroundings.

pole pruner: long-handled, scissors-like tool used to make pruning cuts.

pole saw: long-handled tool with a pruning saw on the end, used to cut branches from a distance.

pollarding: semiformal pruning system that maintains crown size by initial heading of branches on young trees or young portions of older trees, followed by removal of sprouts to their point of origin at appropriate intervals, without disturbing the resulting pollard heads.

pore space: air- or water-filled spaces between soil particles.

porous: having spaces or holes through which liquid or air may pass.

positive indicator: a sign that decay is definitely present.

possible (likelihood of failure): failure may be expected in extreme weather conditions, but it is unlikely during normal weather conditions within the specified time frame.

pot bound: see *root bound*.

potential indicator: indicator that decay might be present.

precipitation: any form of water that falls to the Earth's surface such as rain, snow, or sleet.

predator: any organism that preys on another organism.

prescription fertilization: basing fertilization recommendations on plant needs as determined by conducting soil and/or foliar nutrient analysis, setting plant health goals, and selecting a fertilizer to achieve the goals.

primary branch: a branch attached directly to the trunk extending to the tip following the largest stem at each fork.

primary elements: nitrogen (N), phosphorus (P), and potassium (K). Included in various ratios in fertilizer (see also *essential elements, macronutrient,* and *micronutrient*).

primary growth: root and stem growth in length. Occurs in apical and lateral meristems.

prioritizing targets: process for ranking targets according to importance or value.

probability: measure of the chance of occurrence expressed as a number between 0 and 1, where 0 is impossibility and 1 is absolute certainty. Often expressed as a percentage (0 to 100%).

probable (likelihood of failure): failure may be expected under normal weather conditions within the specified time frame.

probe: stiff, small-diameter rod, stick, or wire that is inserted into a cavity or crack to estimate its size or depth.

Proctor test: test used to determine the relationship between soil moisture content and soil density.

progressive reduction: phased reduction in crown height and/or spread over a period of time.

prop: rigid brace, acting in compression, to support a tree, tree branch, or utility pole.

propagation: process of increasing plant numbers, both sexually and asexually.

protection factors: structures, trees, branches, or other factors that would prevent or reduce harm to targets in the event of a tree failure.

protection zone: see *tree protection zone* and *branch protection zone*.

provenance: geographic origin of a plant, differing from the range of a species or where a species can be expected to be hardy.

proximate cause: legal term that describes an act or omission that can be attributed as cause of loss and/or injury.

pruning: removing branches (or occasionally roots) from a tree or other plant, using approved practices, to achieve a specified objective.

pruning cycle: time schedule for pruning events over time. It is the time between pruning sessions established to provide appropriate removal of live tissue without compromising tree health.

pruning dose: the amount of live tissue removed at one pruning event relative to the total amount on the tree prior to pruning expressed as a percentage of the foliage (in summer) or buds (in winter) removed; can be referenced on a whole-tree basis or on one stem or branch only.

pruning saw: hand saw designed for pruning trees, generally with a bowed blade and teeth.

Q

qualification: education or training program consisting of a learning event designed to train participants to achieve specified learning outcomes, including an assessment of achievement of those outcomes; assessment-based certificate program.

quick-release fertilizer: fertilizer with less than 50% water-insoluble nitrogen (contrast with *slow-release fertilizer*).

R

radial transport: lateral movement of substances, perpendicular to the longitudinal axis of the tree or stem.

radial trenching: technique for relieving soil compaction and improving rooting conditions around a tree by digging or excavating trenches in a spoke-like pattern radially from the trunk. Air excavation minimizes root damage.

radius: distance from the center to the perimeter of a circle. One half of diameter.

raising: selective pruning of lower branches to provide vertical clearance (see *crown raising*).

ram's horns: inward curling formation of wound wood resembling the horns of a ram.

rare occupancy: target zone is not commonly used by people or other mobile/movable targets.

ray: parenchyma tissues that extend radially across the xylem and phloem of a tree and function in transport, storage, structural strength, and defense.

reaction force: force of equal magnitude exerted in the opposite direction of a force acting on a body.

reaction wood: wood formed in leaning or crooked stems or on lower or upper sides of branches as a means of counteracting the effects of gravity (see *compression wood* and *tension wood*).

reaction zone: natural boundary formed chemically within a tree to separate damaged wood from existing healthy wood. Important in the process of compartmentalization (contrast with *barrier zone*).

reactive force: force generated in response and opposite to another force. Often demonstrated when operating a chain saw.

recommendations: one or many alternatives that are promoted to achieve a desired outcome, based on professional judgment.

reducing a primary branch: start primary branch reduction by first pruning (using the specified reduction cut diameter) the end of the branch after following the largest stem at each fork. The second step is to reduce the next largest diameter branch using the specified cut diameter. Continue this process cutting the third largest branch, et cetera, until the specified amount is removed.

reduction pruning: pruning to decrease height and/or spread of a crown or reduce the length of a branch.

reduction cut: shortens a branch by removing the terminal portion back to a lateral branch of equal or smaller diameter. The cut should be made just beyond the lateral branch and the remaining branch should be one-third or larger than the size of the branch removed.

removal cut: eliminates the target branch back to the trunk or a larger branch, just outside the branch collar and branch bark ridge.

remove a primary branch: cutting a branch off at the trunk.

remove a secondary branch: removing a branch growing from a primary branch.

replacement cost: cost to replace an improvement with a similar item having equivalent functional utility. Previously referred to as *cost of cure*.

residual risk: risk remaining after mitigation.

resistance: 1) in plants, the tendency to withstand certain pests, diseases, or fungal problems; pest resistance. 2) in insects, the ability to withstand certain insecticides; survival of just a few genetically resistant insects that reproduce can lead to populations that are resistant. 3) in other pests, the ability to withstand certain pesticides. 4) measure of the difficulty in cutting or penetrating a material when using a tool or instrument.

resistance-recording drill: device consisting of a specialized micro-drill bit that drills into trees and graphs resistance to penetration; used to detect internal differences in the wood, such as decay.

resistant varieties: plant varieties that are tolerant of, or not susceptible to, certain diseases or pest problems or abiotic disorders.

resource allocation: in plant physiology, distribution and use of photosynthates for various plant functions and processes.

respiration: in plants, process by which carbohydrates are converted into energy by using oxygen.

response growth: new wood produced in response to loads to compensate for higher strain experienced (adaptive growth); includes reaction wood (compression and tension), flexure wood, and wound wood.

restoration: 1) pruning to improve the structure, form, and appearance of trees that have been vandalized, damaged, or improperly trimmed. 2) management and planting to restore altered or damaged ecosystems or landscapes. 3) replacement or reproduction of damaged or destroyed plants.

retain and monitor: recommendation to keep a tree and conduct follow-up assessments after a stated inspection interval.

retrenchment: natural process during which an overly mature tree reduces its crown and increases its girth to consolidate resources and increase longevity; the deliberate process of reducing tree height to mimic natural processes.

rhizosphere: soil area immediately adjacent to, and affected by, plant roots. Typically has a high level of microbial activity.

rib: longitudinal bulge of response wood growth.

right-of-way: defined area of land, usually a linear strip, reserved for the passage of traffic (e.g., paths and roadways) or the construction, maintenance, and operation of various aboveground or underground utilities. Rights-of-way may be granted by easement rights and may cross a single property or many properties (highways, railroads, and utility corridors are common examples).

rights-of-way: plural of right-of-way.

ring-porous: pattern of wood development in which the large-diameter vessels are concentrated in the earlywood.

riparian: relating to or situated on the banks of a river or "relating to wetlands adjacent to rivers and streams. Also called bottomland.

risk: the combination of the likelihood of an event and the severity of the potential consequences. In the context of trees, risk is the likelihood of a conflict or tree failure occurring and affecting a target, and the severity of the associated consequences.

risk analysis: systematic use of information to identify sources of risk and to estimate the risk.

risk assessment: the process of risk identification, analysis, and evaluation.

risk categorization: the process of assigning risk and risk factors to categories based on severity or hierarchy.

risk evaluation: the process of comparing the assessed risk against given risk criteria to determine the significance of the risk.

risk management: application of policies, procedures, and practices to identify, evaluate, mitigate, monitor, and communicate risk.

risk perception: the subjective perceived level of risk from a situation or object, often differing from the actual level of risk.

risk rating: the level of risk combining the likelihood of a tree failing and impacting a specified target, and the severity of the associated consequences.

risk rating matrix: tool for ranking and displaying risks by assigning ratings for consequences and likelihood.

risk tolerance: degree of risk that is acceptable to the owner, manager, or controlling authority.

root: plant organ, usually underground, that serves as anchorage and also absorbs and conducts water and minerals.

root collar: base of the trunk where roots and trunk merge that becomes swollen as most trees grow; also referred to as root crown, root flare, and trunk flare.

root ball: soil containing all or a portion of the roots that are moved with a plant when it is planted or transplanted.

root barrier: membranes or sheets installed vertically in the soil to limit or direct the growth of tree roots.

root bound: condition in which plant roots are overcrowded in a container or site and root growth is restricted. Often accompanied by circling and girdling roots.

root cap: group of cells protecting the apical meristem at the root tip.

root collar: flared area at the tree trunk base where the roots and trunk come together (see *root crown, root flare,* and *trunk flare*).

root collar excavation: process of removing soil to expose and assess the root collar (root crown) of a tree.

root crown: area where the main roots join the plant stem, usually at or near ground level. Also known as root collar.

root flare: see *trunk flare*.

root graft: natural union of two roots, either from the same plant or two different plants. Can result in disease transmission.

root hairs: modified epidermal cells of a root that absorb the majority of water and minerals.

root mat: dense network of roots. In palms, near the base of the stem.

root plate: combination of large structural and smaller roots and soil near the base tree's trunk, largely responsible for holding the tree erect.

root protection zone: surface area of tree root concentration to be protected from construction damage, usually soil compaction damage. Rigid fencing is most effective for restricting access to this area. Also known as *tree protection zone*.

root pruning: severing roots selectively. 1) in transplanting, the process of cutting roots to increase the density of root development within the root ball. 2) in tree conservation and preservation, the process of cutting roots to prevent tearing and splintering of remaining roots. 3) in tree disease management, severing tree roots to prevent disease transmission through root grafts.

root rot: decay located in the roots; root decay is usually developed from the bottom up, and crown symptoms may or may not be visible.

root zone: 1) horizon or layer within the soil profile where roots exist. 2) volume of soil containing tree roots. 3) horizontal spread of tree roots from the trunk. Typically the root zone of a tree extends well beyond the drip line.

ruderal: a plant that is a pioneer species and first to colonize disturbed ground or land.

S

safety: process of preventing injury, damage, or loss.

saline: water solution containing dissolved salt.

saline soils: soils with a high concentration of soluble salts. Can cause poor plant growth.

salinity: amount or percentage of salt in the soil.

salt index: osmotic potential ratio of a fertilizer compared to sodium nitrate, based on the relative value of 100. The higher the salt index, the more likely plant damage will occur.

sand: soil particles with a size between 0.06 mm and 2.0 mm in diameter (contrast with *clay* and *silt*).

sanitation: cultural practice of removing dead, infested, or diseased plant parts to reduce the spread of insects or disease.

sap: term used for plant fluids found in cells or transported through the vascular system.

sap leakage: dripping of sap from a pruning cut, rubbing, or other wound (see *bleeding*).

saprophyte: organism that lives on and may act to decay dead organic matter (contrast with *epiphyte* and *parasite*).

sapwood: outer wood (xylem) that has living cells that are active in longitudinal transport of water and solutes (contrast with *heartwood*).

sapwood rot: decay located in the sapwood. Bark and/or cambium may be damaged or dead. Signs of this classification of rot are usually numerous but small, fruiting bodies along the bark's surface are common.

saturation point: point at which a soil or an aquifer will no longer absorb any amount of water without losing an equal amount.

secondary branch: a branch growing from a primary branch.

scaffold: pertaining to tree architecture or form, a strong and properly spaced arrangement, framework, or system of branches throughout the crown.

scaffold branches: permanent branches that form the scaffold or structure of a tree.

scion: top part of a graft, with leaves and buds, that is grafted to the rootstock.

sclerophylly: where plants form hard leaves by building parenchyma tissues and increasing protective structures like the upper cuticle and both the upper and lower epidermis.

scope of work: defined project objectives and requirements.

scorch: browning and shriveling of foliage, especially at the leaf margin.

seam: lines formed where two edges of bark meet at a crack or wound.

secateurs: hook and blade or scissors-type hand pruning tool intended for cutting single, small-diameter stems. Also called pruning shears, bypass pruners, or hand pruners.

secondary growth: increase in root and stem girth or diameter. Occurs at lateral or secondary meristems in some vascular plants such as dicots.

secondary nutrients: mineral elements, such as calcium (Ca) and magnesium (Mg), required in moderate amounts by plants (see *essential elements, macronutrient, micronutrient, and primary elements*).

secondary pest: insect or other pest that frequently develops on a plant stressed and weakened by another factor.

secondary pest outbreak: increase in a secondary pest population following a reduction in the population of natural predators or parasites due to broad-spectrum pesticide application.

selective management: methods used to target undesirable vegetation while retaining desirable vegetation.

semi-ring-porous: pattern of wood development displaying characteristics of both ring-porous and diffuse-porous woody plants.

senescence: process of aging. Process preceding leaf drop in deciduous plants.

severe (consequences): serious personal injury or death, high-value property damage, or major disruption of important activities.

shake: separation of wood at the growth rings or rays, usually along the barrier zone that forms in the compartmentalization process (CODIT). Also called ring shake and radial shake.

shall: as used in the ANSI standards (United States), denotes mandatory requirement.

shear: 1) *n.* in mechanics, a force, stress, or failure mode characterized by the sliding of parts of a material, especially laminar material such as wood, side by side. 2) *n.* tool used to cut small-diameter plant material, including secateurs and snips, as well as long-bladed hand tools and power tools used to cut hedges. 3) *v.* to cut; often used to describe cutting foliage or stems to a single plane, as in a hedge.

shear plane crack: crack at the neutral plane between tension and compression stresses.

shearing: cutting leaves, shoots, and branches to a desired plane, shape, or form, using tools designed for that purpose, as with topiary (hedge).

shell wall: remaining solid wood around a cavity or internal wood decay.

ship auger: type of drill bit with an open spiral form. Used to drill holes in trees for cable or bracing installation.

shoot: new stem or branch growth on a plant.

should: as used in the ANSI standards (United States), denotes an advisory recommendation that ought to be applied unless there is a compelling reason not to.

shrub: woody plant, generally smaller than a tree, typically with multiple stems branching from or near the ground.

sieve cells: long, slender phloem cells in gymnosperms.

sieve tube elements: specialized phloem cells involved in photosynthate transport. Exist only in angiosperms.

sign: physical evidence of a causal agent (e.g., insect eggs, borer hole, frass, mycelium, fruiting body) (contrast with *symptom*).

significant (consequences): substantial personal injury, moderate- to high-value property damage, or considerable disruption of activities.

silt: soil particles with a grain size between 0.004 mm and 0.062 mm (coarser than clay particles but finer than sand) (contrast with *clay* and *sand*).

silviculture: a branch of forestry concentrating on establishing, caring for and reproducing stands of trees for a variety of forest uses including wildlife habitat, timber production, and outdoor recreation.

sink: plant part that uses or stores more energy than it produces (contrast with *source*).

sinker roots: downward-growing roots that provide anchorage and take up water and minerals. Especially useful during periods of drought.

site analysis: 1) consideration or evaluation of the conditions, restrictions, and environment of a planting site. 2) consideration or evaluation of a construction or development site requiring a tree conservation or preservation plan. 3) consideration during a tree risk assessment.

skirt: dead or dying palm fronds gathered down the stem of a palm, which may be removed for aesthetic or safety reasons.

soft rot: decay of plant tissues characterized by the breakdown of tissues within the cell walls (contrast with *brown rot* and *white rot*).

soil: layers of sand, clay, silt, and organic material on the surface of the earth that support plants.

soil aeration: see *aeration*.

soil amendment: material added to soil to improve its physical, chemical, and/or biological properties.

soil analysis: analysis of soil to determine pH, mineral composition, texture, structure, salinity, organic matter, and other characteristics.

soil auger: 1) device for removing cores of soil for inspecting or testing. 2) device (drill bit) used to drill holes in the soil, often used for vertical mulching or for the drill-hole method of fertilizing.

soil bulk density: dry mass of soil per unit volume used as a measure of compaction.

soil compaction: compression of the soil, as a result of pedestrian, vehicle, or heavy-equipment traffic, that breaks down soil aggregates and reduces soil volume and total pore space, especially macropore space.

soil depth: the vertical extent of soil present belowground.

soil horizon: layer or zone of the soil profile with physical, chemical, and biological characteristics that differ from adjacent layers.

soil inoculant: soil amendments that use beneficial microbes to promote plant health.

soil organic matter: amount of organic material in soil, measured as a percentage by dry weight.

soil particle: smallest grains of soil categorized by size into clay, silt, or sand.

soil probe: any one of a number of instruments used to take soil cores or samples. Usually some variation of a metal tube with a sharpened or serrated point and a T-shaped handle.

soil profile: vertical section through the soil and all of the soil horizons.

soil reaction: pH (acidity or alkalinity) of a soil.

soil salinity: measure of the soluble salts in a soil.

soil structure: arrangement of soil particles into aggregates.

soil texture: relative fineness or coarseness of a soil due to particle size (sand, silt, and clay).

soil-drench application: method of applying chemicals to trees through the soil.

somewhat likely (likelihood of failure and impact): defined by its placement in the likelihood matrix; *imminent* likelihood of failure and *low* likelihood of impact, or *probable* likelihood of failure and *medium* likelihood of impact, or *possible* likelihood of failure and *high* likelihood of impact.

sonic assessment: process of measuring wood density or other mechanical properties using an instrument that transmits, receives, and records the velocity of stress (sound) waves through wood.

sonic tomography: see *tomography*.

sooty mold: fungus that appears as a black coating on the surface of leaves, branches, fruits, and other surfaces (including those of buildings) resulting from deposits of sugary excrement from aphids and scale insects.

sounding: process of striking a tree with a mallet or other appropriate tool and listening for tones that indicate dead bark, a thin layer of wood outside a cavity, or cracks in wood.

source: in physiology, plant part that produces carbohydrates. Most green parts are sources because the presence of chlorophyll is indicative of photosynthesis, including mature leaves and green bark (contrast with *sink*).

spatial reality is recognition of the plant sizes in a three-dimensional perspective. This includes plant shape and how it fits within the confines of the intended space of the design.

species: taxonomic group of organisms composed of individuals of the same genus that can reproduce among themselves and have similar offspring.

species diversity: measure of the number and variety of different species found in a given area.

specifications: detailed plans, requirements, and statements of particular procedures and/or standards used to define and guide work.

stakeholder: 1) person or group that has an interest in, or is affected by, an activity or decision. 2) person that either creates policy, is affected by those policies, or perceives to be affected by those policies.

staking: supporting a tree with stakes and ties. Usually used in reference to newly planted trees.

standard: established or widely recognized authority of acceptable performance.

standard of care: degree of care that a reasonable person should exercise in performing duty of care; a measurement used to assess whether an individual acted in a reasonable manner.

starch: chain of sugar molecules that serves as a form of energy storage in plants (see *carbohydrate*).

static cable system: cabling system that utilizes rigid materials, such as rods and steel cable, to limit movement and provide constant support of limbs.

static load: constant load exerted by a mass due to its weight (contrast with *dynamic load*).

static pull test: in advanced tree risk assessment, a load test that measures outermost fiber strain in the stem or branches, and/or inclination at the root flair, in response to a controlled pull.

static target: target than cannot easily be relocated.

stem: a woody structure bearing foliage and buds that gives rise to other stems and branches.

stem-girdling root: root that grows tangentially across a stem, causing vascular constriction (see *girdling root*).

stomata: small apertures, between two guard cells on the undersides of leaves and other green plant parts, through which gases are exchanged and water loss is regulated.

stomates: plural of stomata (see *stomata*).

stormwater runoff: water originating from precipitation (rain or melting snow and ice) that flows above ground rather than infiltrating into the soil. May occur if the soils is frozen or saturated or if the rate at which precipitation falls is greater than the infiltration rate of the soil.

strength loss: degradation of the ability to withstand mechanical stress.

stress: 1) in Plant Health Care, a factor that negatively affects the health of a plant; a factor that stimulates a response. 2) in mechanics, a force per unit area.

structural cells: modular system consisting of integrated support structures filled with soil that serves as both a foundation for paved surfaces and a hospitable environment for tree root growth.

structural defect: feature, condition, or deformity of a tree that indicates a weak structure or instability that could contribute to tree failure.

structural pruning: pruning to develop a strong trunk and branch framework to support the tree crown which is less likely to fail. This pruning strategy influences the orientation, spacing, growth rate, strength of attachment, and ultimate size of branches and stems.

structural roots: large, woody tree roots that anchor and support the trunk and crown. Roots characterized by secondary thickening and relatively large diameter, giving form to the root system and functioning in anchorage and support.

structural soil: custom blended soil mix that meets certain specifications for pH, texture, infiltration, nutrient value, metals, or other parameters.

structural support system: device or mechanism providing supplemental support to individual branches and/or entire trees.

stub: portion of a branch or stem near the trunk and remaining after a stub cut, branch breakage, or branch death.

stub cuts: improper pruning cuts made too far outside the branch bark ridge or branch collar that leave branch tissue attached to the stem.

stunting: growth reduction of plants or plant parts.

suberin: waxy substance present in some cell walls.

subordinate: 1) *v.* prune to reduce the size and ensuing growth of a branch in relation to other branches or leaders. 2) *adj.* dominated by other trees, branches, or parts; suppressed.

subordinating: slowing the growth rate on a stem or branch by removing its terminal or end portion, or by removing lateral branches from the end (or a combination of both), so other portions of the tree grow faster.

subsurface application: placement of fertilizer or other material below the soil surface.

sucker: vigorous, upright epicormic shoot that arises from latent buds below soil level or the graft union (contrast with *water sprout*).

sudden branch drop: sudden, unanticipated failure of a tree branch with little or no discernable defect; often associated with long, horizontal branches and warm temperatures.

sunburn: injury to bark and cambium caused by a combination of radiant energy from light and insufficient soil moisture.

sunscald: injury to bark tissues on the trunk and branches caused by rapid changes in temperature, especially on warm days and cool nights in winter (contrast with *scorch*).

suppression: 1) in pest management, management practices intended to reduce the pest population and associated plant injury to a tolerable level.

susceptibility: potential for infection and/or injury from exposure to a biotic or an abiotic factor.

suspended pavement: paving option that transfers the load to the subsoil rather than to the top of the soil surface.

sustainability: ability to maintain ecological, social, and economic benefits over time.

southwest (SW) injury: type of bark cracking that occurs on the lower part of the trunk on the southwest side of the tree when there is a sudden temperature drop (e.g., the sun going behind a cloud during the winter). The rapid freeze-thaw cycle causes the bark to expand and contract, resulting in a fissure in the trunk.

sweep: corrected tree lean characterized by a leaning lower trunk and a top that has grown back toward vertical.

symbiosis: association of two different types of living organisms that is not detrimental to either organism and is beneficial to one, and usually both, of the organisms. Also called *symbiotic.*

symplast: entire mass of protoplasm of all the cells in a plant, interconnected by plasmodesmata (contrast with *apoplast*).

symptom: plant reaction to a disease or disorder (e.g., spot, wilt, dieback) (contrast with *sign*).

systemic: 1) when referring to a substance: moving throughout an organism after absorption. 2) when referring to a condition, disease, disorder, or pest that affects the entire organism.

systemic pesticide: pesticide that moves throughout a tree after it has been injected or absorbed (often by roots or foliage, or injected cambium).

T

taper: change in diameter over the length of trunks, branches, and roots.

taproot: central, vertical root growing directly below the main stem or trunk that may or may not persist into plant maturity.

target: 1) people, property, or activities that could be injured, damaged, or disrupted by a tree failure. 2) intended application site or pest.

target management: acting to control the exposure of targets to risk.

target pruning: process of branch removal to achieve a specified goal in which the pruning cuts are made at nodes and in relation to the positions of the branch collar and branch bark ridge. Also called *natural pruning.*

target value: monetary worth of something; the importance or preciousness of something.

target zone: area where a tree or tree part is likely to land if it were to fail.

target-based actions: risk mitigation actions aimed at reducing the likelihood of impact in the event of tree failure.

temperate: region lying between the tropics and permafrost zones where temperature ranges and conditions permit plant growth.

temporary branches: branches (generally lower branches) that are left in place or subordinated but will be removed later (contrast with *permanent branches*).

tensiometer: instrument used to measure soil moisture.

tension: in mechanics, the action of forces to stretch or pull apart any material or substance (contrast with *compression*).

tension stress: stretching or pulling force on any material or substance.

tension wood: form of reaction wood in angiosperms that forms on the upper side of branches or the trunks of leaning trees (contrast with *compression wood*).

terminal bud: bud at the tip of a twig or shoot. Also called apical bud (contrast with *lateral bud*).

thinning: selective removal of live branches to reduce density of branches and foliage. Thinning is not an acceptable practice.

threshold: 1) in integrated pest management, pest population levels requiring action. 2) in risk assessment and management, level of risk requiring action.

time frame: in tree risk assessment, the time period for which an assessment is defined.

tissue: group of cells with similar structure that have a special function.

tomograph: device used to build a detailed image of tissue at a particular plane and depth (see *tomogram*).

tomography: use of multiple sensors placed around a trunk or limb to record sound or magnetic waves traveling through the wood, with measurements resulting in a picture of internal density characteristics. Typically used to measure the extent of decay.

topiary: a formal pruning system that uses a combination of pruning, supporting, and training branches to orient a plant into a desired shape. Hedging is a subset of topiary.

topping: reduction of tree size by cutting live branches and leaders to stubs, without regard to long-term tree health or structural integrity.

topping cut: hinge cut used when removing the top from a tree in the tree removal process.

topography: land and water features of an area, including changes in elevation.

topsoil: surface layer of soil that may be rich in nutrients from decaying plants and bacterial material, often removed when lots are graded in preparation for construction or development.

torsion: action of twisting or being twisted.

tracheid: elongated, tapering xylem cell that is dead at maturity and is adapted for the support and transport of water and elements.

translocation: movement of sugars in the phloem.

transpiration: water vapor loss, primarily through the stomata of leaves.

transplant shock: plant stress following transplant; characterized by reduced growth, wilting, dropping foliage, or death.

transplanting: process of moving a plant to a new location. This applies to tree spade moving trees and the act of installing balled and burlapped plants.

tree: woody plant possessing a woody trunk that perennially renews its growth, typically with one erect perennial stem (trunk) at least three inches in diameter at a point 4-1/2 feet above the ground, a formed crown of foliage, and a mature height of at least 13 feet. Some trees may have multiple stems or trunks.

tree architecture: the overall form and shape of a tree. Excurrent and decurrent forms are common.

tree-based actions: risk mitigation actions aimed at reducing the likelihood of tree failure.

tree biology: study of the life processes of trees and other woody plants.

tree biomechanics: the study of the action of mechanical and physical forces on living trees and their reaction to those forces.

tree failure: uprooting or mechanical breakage of a tree, its parts, and/or surrounding soil.

tree growth regulator: chemical that can be applied to trees that slows terminal growth by reducing cell elongation.

tree inventory: record of trees within a designated area that provides specified identification and condition information to be used for management decisions and actions.

tree ordinance: tree-specific law that defines a public agency's authority, describes required conditions or actions, establishes penalties for nonconformance, and identifies who is responsible for enforcement and oversight. In Canada, tree bylaw.

tree population: a defined set, group, or collection of trees.

tree preservation: process of protecting trees from damage related to construction activity.

tree protection zone (TPZ): defined area within which certain activities are prohibited or restricted to prevent or minimize potential injury to designated trees, especially during construction or development.

tree risk assessment: a systematic process to identify, analyze, and evaluate tree risk.

tree risk evaluation: the process of comparing the assessed risk against given risk criteria to determine the significance of the risk.

tree risk management: application of policies, procedures, and practices to identify, evaluate, mitigate, monitor, and communicate tree risk.

tree guard: tube placed around a tree seedling for protection and growth enhancement.

tree spade: mechanical equipment to dig, transport, and replant trees with a sufficiently large volume of roots and soil.

tree stress: factor that affects the health or condition of a tree.

tree support system: mechanical system used to limit movement or provide supplemental support to leaders, branches, or entire trees.

tree value: 1) appraised monetary value placed on a tree. 2) nonmonetary benefit(s) or significance of a tree.

tree well: wall constructed around a tree to maintain the original grade between the trunk and the wall, with a raised grade beyond the wall. Used to maintain adequate oxygen levels in the root zone and to protect the trunk from being buried too deep.

tree wrap: material used to wrap the trunks of newly planted or transplanted trees or to protect thin-barked mature trees when they are newly exposed to the sun (see *sunburn* and *sunscald*).

trenching: linear or curvilinear open excavation, often used to install utilities or structural footings. Can cause tree root damage (contrast with *horizontal boring* and *tunneling*).

triangular cable system: tree cabling system that joins three branches, with three cables forming a triangle (contrast with *box cable system* and *direct cable system*).

tropism: tendency of growth or variation of a plant in response to an external stimulus, such as gravity (geotropism) or light (phototropism).

trunk: the stem that arises from the ground following the largest stem at forks.

trunk flare: transition zone from trunk to roots where the trunk expands into the buttress or structural roots. Also called root flare.

trunk formula technique: technique for developing a cost basis that involves extrapolating the purchase cost of a nursery-grown tree up to the size of the subject tree being valued.

trunk injection: technique to introduce substances directly into the xylem of a tree to treat or prevent diseases, disorders, or pest problems.

trunk taper: natural, gradual reduction in diameter of the trunk of a tree from bottom to top (see *taper*).

turgor: swelling in a plant cell caused by its fluid contents.

twig: small, woody branch, stem, or shoot.

tyloses: protrusions of portions of the adjacent parenchyma cells that enter and block xylem vascular cells when those cells become inactive or injured.

U

unacceptable risk: degree of risk that exceeds the tolerance of the owner, manager, or controlling authority.

unavailable water: soil water that is held so tightly to the soil particles that it cannot be absorbed by plant roots (contrast with *available water*).

unlikely (likelihood of failure and impact): defined by its placement in the likelihood matrix; *possible* or *probable* likelihood of failure and *low* likelihood of impact, or *possible* likelihood of failure and *medium* likelihood of impact, or *improbable* likelihood of failure with any likelihood of impact rating, or any likelihood of failure rating with *very low* likelihood of impact.

urban forest: the sum of all woody and associated vegetation in and around dense human settlements, from small communities to metropolitan regions, including street, residential, and park trees, greenbelt vegetation, trees on unused public and private land, trees in transportation and utility corridors, and forests on watershed lands.

urban forest management plan: document that describes how urban forestry goals are to be accomplished within a defined time frame; includes tasks, priorities, best management practices, standards, specifications, budgets, and staffing analyses.

urban forester: individual trained in or practicing urban forestry.

urban forestry: overall management of trees and associated plants in the built environment.

utility arboriculture: sector of the arboriculture industry that provides services to utilities. Often requires specialized equipment and training of workers. Includes utility line-clearance pruning, right-of-way clearing and maintenance, and integrated vegetation management.

utility pruning: pruning around or near utility facilities with the object of maintaining safe and reliable utility service.

utility vegetation manager: professional with the proper experience, education, and training to successfully establish or supervise an integrated vegetation management program.

V

variety: naturally occurring subdivision of a species having a distinct difference and breeding true to that difference (contrast with *cultivar*).

vascular cambium: lateral meristem from which secondary xylem and secondary phloem originate (see *cambium*).

vascular discoloration: darkening of the xylem or phloem of woody plants in response to disease, insect boring, or injury.

vascular system: phloem and xylem, the parts of a tree that conduct water and minerals or organic compounds.

vascular tissue: tissue that conducts water or nutrients.

vascular wilt: disease caused by xylem-colonizing pathogens that interfere with water transport and induce loss of turgor and rigidity of leaves or young stems.

vector: in pathology, biotic or abiotic agent that transmits a pathogen.

verbal report: information delivered orally.

vertical mulching: an aeration or fertilization technique. Drilling vertical holes in the soil and filling them with materials to improve aeration.

very likely (likelihood of failure and impact): defined by its placement in the likelihood matrix; *imminent* likelihood of failure and *high* likelihood of impact.

very low (likelihood of impact): the chance of the failed tree or tree part impacting the specified target is remote.

vessel: end-to-end, tube-like, water-conducting cells in the xylem of angiosperms.

veteran tree: tree which, because of its great age, size, or condition, is of exceptional cultural, landscape, or nature conservation value (see *ancient tree* and *champion tree*).

vigor: the genetic capacity of a tree to grow and resist stress.

virus: ultramicroscopic, nonliving infectious agent (consisting of nucleic acid and protein) that can replicate only in living cells of other organisms. Numerous viruses cause diverse symptoms of different tree parts.

vista pruning: selective pruning to enable a view from a predetermined point.

visual tree assessment (VTA): method of assessing the structural integrity of trees using external symptoms of mechanical stress (such as bulges, reactive growth, etc.) and defects (cracks, cavities, etc.).

vitality: state of overall plant health referring to the ability of a plant to deal effectively with stress.

volatilization: conversion of a solid or liquid into a gas or vapor.

W

water-holding capacity: ability of a soil to hold moisture. Varies with the amount of sand, silt, and clay present.

water sprout: upright, epicormic shoot arising from the trunk or branches of a plant above the root graft or soil line. Incorrectly called a sucker (contrast with *sucker*).

weight: force created by the action of the earth (gravity) on an object.

white rot: fungal decay of wood in which both cellulose and lignin are broken down (contrast with *brown rot* and *soft rot*).

wilt: 1) *n.* loss of turgor and subsequent drooping of leaves and young stems; a symptom. 2) *n.* infectious disease caused by a particular agent on a particular host or range of hosts. 3) *v.* to lose turgor or cell fluid.

wind exposure: exposure to the forces of wind.

wind load: the force on a tree or structure resulting from the impact of wind.

windthrow: uprooting and overthrowing of a tree caused by wind.

winter burn: specific to conifers and broad-leaved evergreens. When soil freezes and moisture cannot be absorbed through the roots and/or transported through the trunk, the foliage dies and becomes brown or reddish brown.

winter kill: injury from cold winter temperatures.

wire basket: type of metal basket used to support the root ball of a balled and burlapped tree or a tree dug with a tree spade.

witch's broom: plant disorder characterized by a shortening of the internodes and a proliferation of terminal shoots forming a dense, brush-like mass of twigs.

wood: hard, fibrous inner part of tree trunks, branches, and stems. The secondary xylem of seed plants.

wood decay: the process of wood degradation by microorganisms.

woody plant: plant whose stem (and branches) have a wood-like consistency (e.g., trees, shrubs, and vines as well as palms, bamboos, and other woody monocots).

wound dressing: compound applied to tree wounds or pruning cuts consisting of various ingredients.

wound wood: lignified, differentiated tissues produced on woody plants as a response to wounding (contrast with *callus*).

written report: document with text, images, and/or references, delivered in print or electronic form, to communicate assessment results or other information.

X

xylem: main water- and mineral-conducting (unidirectional, up only) tissue in trees and other plants. Provides structural support. Arises (inward) from the cambium and becomes wood after lignification.

Y

No terms included here.

Z

No terms included here.

REFERENCES AND CITATIONS

Appraisal Foundation. (2018). *Uniform standards of professional appraisal practice.*

Appraisal Foundation. (2020). *Uniform standards of professional appraisal practice.*

Appraisal Institute. (2015). *The dictionary of real estate appraisal* (6th ed.). Appraisal Institute.

Basu, S., Ramegowda, V., Kumar, A., & Pereira, A. (2016). Plant adaptation to drought stress. *F1000Research, 5*, F1000 Faculty Rev-1554. https://doi.org/10.12688/f1000research.7678.1

Barnes, E., Usborne, R., Stone, A., & Sadof, C. (2021). Lessons from a 10-yr invasive species webinar program: Emerald Ash Borer University. *Environmental Entomology, 50.* https://doi.org/10.1093/ee/nvab002.

Beckerman, J., Bessin, R., Welty, C., Athey, K., Wahle, E., Lewis, D., Long, E., Joshi, N., Guedot, C., Meyer, S., Strang, J., & Gauthier, N. (2023). *Midwest fruit pest management guide.* https://ag.purdue.edu/department/hla/extension/_docs/id-465.pdf

Beckerman, J. & Creswell, T. (2022). *Symptoms and signs for plant problem diagnosis—An illustrated glossary.* Purdue University Extension.

Beezley, Kenneth & Dahle, Gregory & Miesbauer, Jason & Walker, Matthew. (2020). Importance of the Root-Stem Transition Zone in the Stability of Trees.

Bethge, K., Mattheck, C., & Hunger, E. (1996). Equipment for detection and evaluation of incipient decay in trees. *Journal of Arboriculture, 20,* 13–37.

Bratkovich, S. (2010). *Urban forestry: An evolving discipline.*

Brienen, R. J. W. et al. (2020). Forest carbon sink neutralized by pervasive growth-lifespan trade-offs. *Nature Communications, 11,* 4241.

Bond, J. (2013). *Tree inventories* (2nd ed.). International Society of Arboriculture.

Bond, Jerry. 2024. Inside the TPZ. *Arborist News 33*(3):32–41.

Bond, Jerry, and E. Buchanan. 2006. Tree inventories. Arborist News 15(3):12–17

Clark, J. R. & Matheny, N. (2010). The research foundation to tree pruning: A review of the literature. *Arboriculture & Urban Forestry, 36,* 110–120.

Coder, Kim. (1997). *Crown pruning effects on roots.* University of Georgia.

Coder, Kim. (2023). *Control of shoot/root balance in trees.* University of Georgia.

Costello, L. R. & Jones, K. S. (2003). *Reducing infrastructure damage by tree roots: A compendium of strategies.* International Society of Arboriculture.

Costello, L. R., Perry, E. J., Matheny, N. P., Henry, J. M., & Geisel, P. M. (2003). *Abiotic disorders of landscape plants: A diagnostic guide.* University of California Agriculture and Natural Resources.

Council of Tree and Landscape Appraisers. (2018). *Guide for plant appraisal* (10th ed). International Society of Arboriculture.

David Henry (Agronomist, A&L Great Lakes Laboratories) and Dr. Cliff Johnston (Professor of Soil Chemistry, Purdue University) expert review and technical assistance with illustration 3.11.

Doccola, J. J., Smitley, D. R., Davis, T. W., Aiken, J. J., & Wild, P. M. (2011). Tree wound response following systemic insecticide trunk injection treatments in green ash (*Fraxinus pennsylvanica* marsh.). *Arboriculture & Urban Forestry, 37, 6–12.*

Dujesiefken, D. & Stobbe, H. (2002). The Hamburg Tree Pruning System—A framework for pruning of individual trees. *Urban Forestry & Urban Greening, 1*(2), 75–82. https://doi.org/10.1078/1618-8667-00008

Dujesiefken, D., Drénou, C., Oven, P., & Stobbe, H. (2005). Arboricultural practices. In Konijnendijk, C., Nilsson, K., Randrup, T., & Schipperijn, J. (Eds.), *Urban forests and trees.* Springer. https://doi.org/10.1007/3-540-27684-X_16

Dujesiefken, D. & Liese, W. (2015). *The CODIT principle: Implications for best practices.* International Society of Arboriculture.

Dunster, J. A., Smiley, E. T., Matheny, N., Lilly, S. (2017). *Tree risk assessment manual* (2nd ed.). International Society of Arboriculture.

Fei, S., Desprez, J., Potter, K., Jo, I., Knott, J., & Oswalt, C. (2017). Divergence of species responses to climate change. *Science Advances, 3.* https://doi.org/10.1126/sciadv.1603055

Fite, K. & Smiley, E. T. (2016). *Best management practices: Managing trees during construction* (2nd ed.). International Society of Arboriculture.

Foth, H. D. (1978). *Fundamentals of soil science* (6th ed.). Wiley and Sons.

Franzmeier, D., McFee, W., Graveel, J., & Kohnke, H. (2016). *Soil science simplified* (5th ed.). Waveland Press.

Gilman, E. (2002). *An illustrated guide to pruning.* Delmar Publishers.

Gilman, E. F. & Grabosky, J. C. (2006). Branch union morphology affects decay following pruning. *Arboriculture and Urban Forestry, 32*(2), 74–79.

Gilman, E. F. (2011). *An illustrated guide to pruning* (3rd ed.). Delmar Cengage Learning.

Gilman, E. F., Kempf, B., Matheny, N., & Clark, J. (2013). *Structural pruning: A guide for the green industry.* Urban Tree Foundation.

Grabosky, J. C. & E. F Gilman. (2007). Response of two oak species to reduction pruning cuts. *Urban Forestry & Urban Greening, 6,* 159–167.

Harris, R. W. (1992). *Arboriculture: Integrated management of landscape trees, shrubs, and vines.* Prentice Hall, Inc.

Harris, R. W., Clark, J. R., & Matheny, N. P. (2004). *Arboriculture: Integrated management of landscape trees, shrubs, and vines* (4th ed.). Prentice Hall.

Hauer, Richard & Koeser, Andrew & Miesbauer, Jason & Kleinhuizen, David. (2021). Impacts of Wire Basket Retention and Removal on Whole Tree Stability and Long-term Growth. 39. 41–46. 10.24266/0738-2898-39.1.41.

Hauer, R., Koeser, A., Miesbauer, J., & Kleinhuizen, D. (2021). Impacts of wire basket retention and removal on whole tree stability and long-term growth. *Journal of Environmental Horticulture,* 39(1), 41–46. https://doi.org/10.24266/0738-2898-39.1.41

Havlin, J. L., Beaton, J. D., Tisdale, S. L., & Nelson, W. L. (Eds.). (1999). *Soil fertility and fertilizers: An introduction to nutrient management* (6th ed.). Prentice Hall.

Hayes, E. 2002. *Evaluating trees for defects* (2nd ed.). Safetrees.

Hilbert, D. R., North, E. A., Hauer, R. J., Koeser, A. K., McLean, D. C., Northrop, R. J., Andreu, M., & Parbs, S. (2020). Predicting trunk flare diameter to prevent tree damage to infrastructure. *Urban Forestry & Urban Greening, 49*(2020), 126645.

Hirons, A. D. & Sjöman, H. (2018). *Tree biology.* Wiley-Blackwell.

James, Ken & Moore, John & Slater, Duncan & Dahle, Gregory. (2017). Tree biomechanics. *CAB Reviews Perspectives in Agriculture Veterinary Science Nutrition and Natural Resources.* 12. 10.1079/PAVSNNR201712038.

Jim, C. Y. (2019). Soil volume restrictions and urban soil design for trees in confined planting sites. *Journal of Landscape Architecture, 14*(1), 84–91.

Johnson, G. & Hauer, R. (2000). *A practitioner's guide to stem girdling roots of trees.* University of Minnesota Extension Service.

Kane, B. & Koeser, A. (2022). Introduction for special issue on trees, risk, and storms: Implications for urban forest management. *Urban Forestry & Urban Greening, 69,* 127496. https://doi.org/10.1016/j.ufug.2022.127496

Kelby, F. & Smiley, T. (2016). *Best management practices: Managing trees during construction* (2nd ed.). International Society of Arboriculture.

Kelby, F., Smiley, T., McIntyre, J., & Wells, C. E. (2011). Evaluation of a soil decompaction and amendment process. *Arboriculture & Urban Forestry, 37*(6), 293–300.

Klein, R., Koeser, A., Hauer, R., Miesbauer, J., Hansen, G., Warner, L., Dale, A., Watt, J. (2021). Assessing the consequences of tree failure. *Urban Forestry & Urban Greening, 65,* 127307. https://doi.org/10.1016/j.ufug.2021.127307

Klein, Ryan & Koeser, Andrew & Hauer, Richard & Miesbauer, Jason & Hansen, Gail & Warner, Laura & Dale, Adam & Watt, John. (2021). Assessing the Consequences of Tree Failure. *Urban Forestry & Urban Greening.* 65. 127307. 10.1016/j.ufug.2021.127307.

Koeser, A. K., Gilman, E. F., Paz, M., & Harchick, C. (2014). Factors influencing urban tree planting program growth and survival in Florida, United States. *Urban Forestry & Urban Greening, 13,* 655–661.

Koeser, A. K., Hauer, R., Klein, R., & Miesbauer, J. (2017). Assessment of likelihood of failure using limited visual, basic, and advanced assessment techniques. *Urban Forestry & Urban Greening, 24,* 71–79. https://doi.org/10.1016/j.ufug.2017.03.024

Koeser, A., Vogt, J., Hauer, R., Northrop, R., & Peterson, W. (2016). The cost of not maintaining trees: Findings and recommendations from an international symposium and summit. *Arboriculture & Urban Forestry,* 42. 377–388. https://doi.org/10.48044/jauf.2016.032

Kramer, P. J. & Kozlowski, T. T. (1979). *Physiology of woody plants.* Academic Press.

Kurzeja, P. S. (2015). *Determining the cause of abiotic leaf scorch: Getting to the root of the problem* [Master of Science thesis, Michigan State University.]

Lilly, S. J. (2010). *Arborists' certification study guide* (3rd ed.). International Society of Arboriculture.

Lilly, S. J., Gilman, E. F., & Smiley, E. T. (2019). *Best management practices: Pruning* (3rd ed.). International Society of Arboriculture.

Lindsey, P. & Bassuk, N. (1991). Specifying soil volumes to meet the water needs of mature urban street trees and trees in containers. *Journal of Arboriculture, 17.* https://doi.org/10.48044/jauf.1991.040

Luley, C. J. (2005). *Wood decay fungi common to urban living trees in the Northeast and Central United States.* Urban Forestry.

Manion, P. D. (1991). *Tree disease concepts.* Prentice-Hall.

Martinson, H., Raupp, M., & Frank, S. (2020). How urban forest composition shapes the structure and function of arthropod communities. In Barbosa, P. (Ed.), *Urban ecology: Its nature and consequences.* CABI. https://doi.org/10.1079/9781789242607.0015

Matheny, N. P. & Clark, J. R. (1998). *Trees and development: a technical guide to preservation of trees during land development.* International Society of Arboriculture.

Matheny, N. & Ecklund, D. (2020). Soil volume requirements for trees in dry climates. *Western Arborist.*

Mattheck, K. & Breloer, H. (1994). *The body language of trees, a handbook for failure analysis.* Her Majesty's Stationary Office.

Mattheck C., Bethge, K. A., & Albrecht, W. (1997). How to read the results of Resistograph M. *Journal of Arboriculture, 21,* 331–336.

Miller, R. W., Hauer R. J., & Werner, L. P. (2015). *Urban forestry: Planning and managing urban greenspaces* (3rd ed.). Waveland Press.

Moore, G. M. (1998). Tree growth regulators: Issues of control, matters of management. *Journal of Arboriculture, 24,* 10–18.

Morris, H., Brodersen, C., Schwarze, F., & Jansen, S. (2016). The parenchyma of secondary xylem and its critical role in tree defense against fungal decay in relation to the CODIT model. *Frontiers in Plant Science, 7*(1665), 1–18.

Niklas, K. J. (1992). *Plant biomechanics: An engineering approach to plant form and function.* University of Chicago Press

Nowak, D., Walton, J., Dwyer, J., Kaya, L., & Myeong, S. (2005). The increasing influence of urban environments on U.S. forest management. *Journal of Forestry, 103*(8), 377–382.

Nowak, D., & Greenfield, E. (2018). Declining urban and community tree cover in the United States. *Urban Forestry & Urban Greening, 32*: 32–55. https://doi.org/10.1016/j.ufug.2018.03.006

Ostry, M. E., Venette, R. C., & Juzwik, J. (2011). Decline as a disease category: Is it helpful? *Phytopathology, 101,* 404–409.

Pallardy, S. G. (2008). *Physiology of woody plants* (3rd ed.). Academic Press.

Peper, P. J., McPherson, E. G., Simpson, J. R.; Vargas, K. E., & Xiao, Q. (2009). *Lower Midwest*
community tree guide: benefits, costs, and strategic planting. Gen. Tech. Rep. PSW- GTR-219. Albany, CA: U.S. Department of Agriculture, Forest Service, Pacific Southwest Research Station.

Percival, G. C., Graham, S., & Franklin, E. (2023). The influence of soil decompaction and amendments on soil quality. *Arboriculture & Urban Forestry,* 49(4), 179–189. https://doi.org/10.48044/jauf.2023.012

Pérez-de-Lis, G. (2016). Effects of climate on wood formation of Quercus robur L. and Quercus pyrenaica Willd.

Piana, M. R., Pregitzer, C. C., & Hallett, R. A. (2021). Advancing management of urban forested natural areas: toward an urban silviculture? *Frontiers of Ecology and the Environment.* https://doi.org/10.1002/fee.2389

Purcell, L. (2017). Tree pruning: What do trees think? In *Forestry and Natural Resources.* Purdue Extension.

Randrup, T. B., & Dralle, K. (1997). Influence of planning and design on soil compaction in construction sites. *Landscape and Urban Planning, 38*(1–2), 87–92.

Raupp, M. & Gonthier, P. (2017). Biotic factors: Pests and diseases. In Ferrini, F., Konijnendijk van den Bosch, C. C., & Fini, A. (Eds.), *Routledge Handbook of Urban Forestry.* Routledge. https://doi.org/10.4324/9781315627106

Reducing Infrastructure Damage by Tree Roots: A Compendium of Strategies. 2003. Costello, L. R.; Jones. Western Chapter, International Society of Arboriculture, Cohasset, CA.

Rinn, F. (2011). Basic aspects of mechanical stability of tree cross-sections. *Arborist News, 20*(1), 52–54.

Roman, L. A., Battles, J. J., & McBride, J. R. (2014). Determinants of establishment survival for residential trees in Sacramento County, CA. *Landscape and Urban Planning, 129,* 22–31.

Roppolo Jr., D. & Miller, R. W. (2001). Factors predisposing urban trees to sunscald. *Journal of Arboriculture, 7,* 246–254.

Sade, N., Gebremedhin, A., & Moshelion, M. (2012). Risk-taking plants: Anisohydric behavior as a stress-resistance trait. *Plant Signaling & Behavior, 7*(7):767–70.

Sadof, Clifford & Gula, Scott & Purcell, Lindsey & Ginzel, Matthew. (2022). Factors Affecting Efficacy of Trunk-Injected Emamectin Benzoate to Manage Emerald Ash Borer. *Arboriculture & Urban Forestry.* 48. 165–175. 10.48044/jauf.2022.013.

Sadof, C., Gula, S., Purcell, L., & Ginzel, M. (2022). Factors affecting efficacy of trunk-injected emamectin benzoate to manage emerald ash borer. *Arboriculture & Urban Forestry, 48,* 165–175. https://doi.org/10.48044/jauf.2022.013

Sadof, C., Mockus, L., & Ginzel, M. (2021). Factors influencing efficacy of an area-wide pest management program in three urban forests. *Urban Forestry & Urban Greening, 58,* 126965. https://doi.org/10.1016/j.ufug.2020.126965.

Scharenbroch, B. C., Smiley, E. T., & Kocher, W. (2014). *Best management practices: Soil management for urban trees.* International Society of Arboriculture.

Scheyer, J. M. & Hipple, K. W. (2005). Urban Soil Primer. United States Department of Agriculture, Natural Resources Conservation Service, National Soil Survey Center, Lincoln, Nebraska.

Schwarze, F. (2008). Diagnosis and prognosis of the development of wood decay in urban trees. ENSPEC Australia.

Sharon, J., Lilly, C. G., Bassett, J. K., & Purcell, L. (2022). *Arborists' certification study guide* (4th ed.). International Society of Arboriculture.

Shigo, A. L. & Marx, H. (1977). *Compartmentalization of decay in trees.* Agric. Info. Bull. No. 405. Washington, DC: U.S. Department of Agriculture, Forest Service. 73 p.

Shigo, A. (1991). *Modern arboriculture.* Shigo and Trees, Associates.

Shigo, A. L. (1989). *A worldwide photo guide to pruning.* Shigo and Trees, Associated.

Shigo, A. L. (1986). *A new tree biology dictionary.* Shigo and Trees, Associates.

Shortle, W. C. & Dudzik, K. R. (2012). *Wood decay in living and dead trees: A pictorial overview. Gen. Tech Rep. NRS-97.* Newtown Square, PA: U.S. Department of Agriculture, Forest Service, Northern Research Station.

Sinclair, W. A. & Hudler, G. W. (1988). Tree declines: Four concepts of causality. *Journal of Arboriculture, 14*(2), 29–35.

Sinclair, W. A. & Lyon, H. H. (2005). *Diseases of trees and shrubs* (2nd ed.). Cornell University Press.

Sjöman, H., Hirons, A. D., & Bassuk, N. L. (2015). *Tree selection for green infrastructure: A guide for specifiers.* Trees & Design Action Group.

Sjöman, H. (2015). Urban forest resilience through tree Selection—Variation in drought tolerance in Acer. *Urban Forestry and Urban Greening, 14,* 858–865.

Slater, D. (2016). *Assessment of tree forks.* Arboricultural Association.

Slater, D., Bradley, R. S., Withers, P. J., & Ennos, A. R. (2014). The anatomy and grain pattern in forks of hazel (Corylus avellana L.) and other tree species. *Trees, 28,* 1437–1448.

Smiley, T. (2001). Air excavation to improve tree health. *TreeCare Industry,* 45–47.

Smiley, E. T., Matheny, N., & Lilly, S. (2011). *Best management practices: Tree risk assessment.* International Society of Arboriculture.

Smiley, E. T., Lilly, S. J., & Werner, L. (2013). *Best management practices: Tree and shrub fertilization* (3rd ed.). International Society of Arboriculture.

Smiley, E. T. & Lilly, S. (2014). *Best management practices: Tree support systems—Cabling, bracing, guying, and propping* (3rd ed.). International Society of Arboriculture.

Smiley, E. T., Graham Jr., A. W., & Cullen, S. (2015). *Best management practices: Tree lightning protection systems* (3rd ed.). International Society of Arboriculture.

Smiley, E. T., Matheny, N., & Lilly, S. (2017). *Best management practices: Tree risk assessment* (2nd ed.). International Society of Arboriculture.

Smith, K. T. (2006). Compartmentalization today. *Arboricultural Journal, 29,* 173–184.

Smith, Kevin. (2007). Pruning wounds and occlusion [2]. Journal of Forestry -Washington-. 105. 229–230. 10.1093/jof/105.5.229a.

Smith, Kevin. (2014). Origin of buds, branches, and sprouts. Tree Care Industry. 25. 22–27.

Smith, Kevin. (2018). Big trees, old trees, and growth-factor tables. Tree Care Industry. 29. 36–38.

Truog, E. (1946). Soil reaction influence on availability of plant nutrients. *Soil Science Society of America Proceedings, 11,* 305–308.

Trowbridge, P. J. & Bassuk, N. L. (2004). *Trees in the urban landscape: Site assessment design and installation.* John Wiley & Sons.

Urban, J. (1992). Bringing order to the technical dysfunction within the urban forest. *Journal of Arboriculture, 18*(2), 85–90.

Urban, J. (2008). *Up by roots: Healthy soils and trees in the built environment.* International Society of Arboriculture.

Vogt, J. M., Watkins, S. L., Mincey, S. K., Patterson, M. S., & Fischer, B. C. (2015). Explaining planted-tree survival and growth in urban neighborhoods: A social-ecological approach to studying recently-planted trees in Indianapolis. *Landscape & Urban Planning, 136,* 130– 143.

Vogt, J., Hauer, R., Fischer, B., & Widney, S. (2018). *The cost of not maintaining trees.* https://doi.org/10.13140/RG.2.2.15779.96804.

von Gadow, K. (2002). Adapting silvicultural management systems to urban forests. *Urban Forestry & Urban Greening, 1,* 107–13.

Watson, G. W. & Himelick, E. B. (1997). *Principles and practice of planting trees and shrubs.* International Society of Arboriculture.

Watson, G. (2014). *Best management practices: Tree planting* (2nd ed.). International Society of Arboriculture.

Watson, G. W. & Himelick, E. B. (2013). *The practical science of planting trees.* International Society of Arboriculture.

Wiseman, P. E. & Raupp, M. J. (2016). *Best management practices: Integrated pest management* (2nd ed.). International Society of Arboriculture.

Wolf K. L., Lam S. T., McKeen J. K., Richardson G. R. A., van den Bosch, M., & Bardekjian, A. C. (2020). Urban trees and human health: A scoping review. *International Journal of Environmental Research and Public Health, 17*(12), 4371. https://doi.org/10.3390/ijerph17124371

INDUSTRY STANDARDS AND BEST MANAGEMENT PRACTICES

American National Standards Institute. (2017). *ANSI Z133 Safety Standard.* American National Standards Institute.

American National Standards Institute. (2014). *American standard for nursery stock.* American Hort.

American National Standards Institute. (2013). *Tree, shrub, and other woody plant management—Standard Practices (Supplemental Support Systems) (ANSI A300, Part 3).* Tree Care Industry Association.

American National Standards Institute. (2014). *American National Standard— Tree, Shrub, and Other Woody Plant Management—Standard Practices (Lightning Protection Systems) (ANSI A300, Part 4).* Tree Care Industry Association.

American National Standards Institute. (2016). *American National Standard—Tree, Shrub, and Other Woody Plant Management—Standard Practices (IPM) (ANSI A300, Part 10).* Tree Care Industry Association.

American National Standards Institute. (2017). *American National Standard for Arboricultural Operations—Safety Requirements (ANSI Z133-2017).* International Society of Arboriculture.

American National Standards Institute. (2017). *American National Standard—Tree, Shrub, and Other Woody Plant Management—Standard Practices (Pruning) (ANSI A300, Part 1).* Tree Care Industry Association.

American National Standards Institute. (2017). *American National Standard—Tree, Shrub, and Other Woody Plant Management—Standard Practices (Tree Risk Assessment a. Tree Failure) (ANSI A300, Part 9).* Tree Care Industry Association.

American National Standards Institute. (2018). *American National Standard— Tree, Shrub, and Other Woody Plant Management—Standard Practices (Planting and Transplanting) (ANSI A300, Part 6).* Tree Care Industry Association.

American National Standards Institute. (2018). *American National Standard— Tree, Shrub, and Other Woody Plant Management—Standard Practices (ANSI A300, Part 2).* Tree Care Industry Association.

Florida Department of Agriculture and Consumer Services. (2022). *Florida Grades and standards for nursery plants.* https://sfyl.ifas.ufl.edu/media/sfylifasufledu/hillsborough/docs/pdf/environmentalhort/grades-and-standards-for-nursery-plants-2022.pdf

Scharenbroch, B. C. & Smiley, E. T. (2021). *Best management practices: Soil management for urban trees* (2nd ed.). International Society of Arboriculture.

Smiley, E. T., Graham Jr., A. W., Cullen, S. (2015). *Best management practices: Tree lightning protection systems* (3rd ed.). International Society of Arboriculture.

LP PUBLICATIONS

BCMA Guide for Exam Preparation. S. Lilly. 2020. ISA Board Certified Master Arborist (BCMA) certification. (2021, softcover, 356 pp.) Contributor and technical reviewer.

Lilly, S., Bassett, C., Komen, J., & Purcell, L. (2022). *Arborist Certification Study Guide, 4th* International Society of Arboriculture.

Purcell, L. A. (2009). Putting power behind urban forestry. *Arborist News, 18*(6), 26–27.

Purcell, L. A. (2015). Ethical arboriculture. *Arborist News, 24*(3), 46–48.

Purcell, L. A. (2017). Tree pruning. *Arborist News, 26*(2), 30–33.

Purcell, L. A. (2019). Pruning for performance and failure prevention. *Tree Care Industry Association Magazine, XXX*(2).

Purcell, L. A. (2020). Tree pruning essentials: Part 1. *Tree Care Industry Association Magazine, XXXI*(4).

Purcell, L. A. (2020). Tree pruning essentials: Part 2. *Tree Care Industry Association Magazine, XXXI*(6).

Purcell, L. A. (2020). Introduction to tree appraisal. *Tree Care Industry Association Magazine, XXXI*(11).

Purcell, L. A. (2020). Tree defect identification. *Arborist News, 29*(6).

Purcell, L. A. (2021). Plant health care: The science—Host, species, and the things that affect them. *Tree Care Industry Association Magazine, XXXII*(8).

Purcell, L. A. (2022). Elemental deficiencies. *Tree Care Industry Association Magazine, XXXIII*(2).

Purcell, L. A. (2022). Business ethics in arboriculture. *TCIA Magazine, 33*(8).

Sadof, C. S., Purcell, L. A., Quesada, C., Bishop, F. J., & Zhang, Z. W. (2011). Evaluating restoration capacity and costs of managing the emerald ash borer with a web-based cost calculator in urban forests. *Arboriculture & Urban Forestry, 37*(2), 74–83.

Sadof, C., Gula, S., Purcell, L., & Ginzel, M. (2022). Factors affecting the efficacy of trunk injected Emamectin Benzoate to manage emerald ash borer. *Arboriculture & Urban Forestry, 48*(3), 165–175.

INDEX

NOTE: Page numbers followed by italicized letters *f* or *t* indicate figures and tables respectively.